STP 1017

Computerization and Networking of Materials Data Bases

Jerry S. Glazman and John R. Rumble, Jr., editors

ASTM
1916 Race Street
Philadelphia, PA 19103

Library of Congress Cataloging-in-Publication Data

Computerization and networking of materials data bases/Jerry S.
Glazman and John R. Rumble, Jr., editors.
 Papers from the 1st International Symposium on Computerization and
Networking of Materials Property Data Bases, sponsored by ASTM
Committee E-49 on Computerization of Material Property Data, held at
Philadelphia, Pa., Nov. 2-4, 1987.
 "ASTM publication code number (PCN) 04-010170-63"—T.p. verso.
 Includes bibliographies and index.
 ISBN 0-8031-1191-6
 1. Materials—Data bases—Congresses. I. Glazman, Jerry S.
II. Rumble, John R. III. International Symposium on Computerization
and Networking of Materials Property Data Bases (1st: 1987:
Philadelphia, Pa.) IV. ASTM Committee E-49 on Computerization of
Material Property Data. V. Series: ASTM special technical
publication; 1017.
TA404.25.C66 1989
025'.0662—dc19   88-35987
                                                                   CIP

Copyright © by AMERICAN SOCIETY FOR TESTING AND MATERIALS 1989

NOTE

The Society is not responsible, as a body,
for the statements and opinions
advanced in this publication.

Peer Review Policy

Each paper published in this volume was evaluated by three peer reviewers. The authors addressed all of the reviewers' comments to the satisfaction of both the technical editor(s) and the ASTM Committee on Publications.

The quality of the papers in this publication reflects not only the obvious efforts of the authors and the technical editor(s), but also the work of these peer reviewers. The ASTM Committee on Publications acknowledges with appreciation their dedication and contribution of time and effort on behalf of ASTM.

Foreword

The 1st International Symposium on Computerization and Networking of Materials Property Data Bases was held 2-4 Nov. 1987 at Philadelphia, PA. The symposium was sponsored by ASTM Committee E-49 on Computerization of Material Property Data. John R. Rumble, Jr., National Institute of Standards and Technology, and Jerry S. Glazman, Combustion Engineering, served as chairmen of the symposium. John R. Rumble, Jr. and Jerry S. Glazman are editors of the resulting publication.

Contents

Overview   1

Standards for Materials Data Bases

Standards for Computerized Material Property Data—ASTM Committee E-49—
J. GILBERT KAUFMAN   7

Designation, Identification, and Characterization of Metals and Alloys—
J. H. WESTBROOK   23

VAMAS Activities on Materials Data Banks—KEITH W. REYNARD   43

National and International Data Base Activities

The National Materials Property Data Network, Inc.—A Cooperative National Approach to Reliable Performance Data—J. GILBERT KAUFMAN   55

European Activities Towards the Integration and Harmonization of Materials Data Systems—HERMANN KRÖCKEL AND GÜNTER STEVEN   63

Materials Data Activities in China—YUNWEN LU AND SHOUSAN FAN   75

Japanese Progress in Materials Data Bases—SATOSHI NISHIJIMA, YOSHIO MONMA, AND MASAO KANAO   80

Use of Materials Data Bases in France—CLAUDE BATHIAS AND BENARD MARX   92

CODATA Activities on Materials Data—ANTHONY J. BARRETT   99

Emerging Issues

Uniform Treatment of Integrated CAD/CAM Data and Metadata—STANLEY Y. W. SU AND ABDULLAH ALASHQUR   109

Distributed Data Bases on the Factory Floor—CITA M. FURLANI, DON LIBES, EDWARD J. BARKMEYER, AND MARY J. MITCHELL   126

Information Systems Design for Material Properties Data—JOHN L. McCARTHY   135

Capture of Published Materials Data—WALTER GRATTIDGE 151

Expert Systems Interfaces for Materials Data Bases—SHUICHI IWATA 175

An Interactive Inquiry System for Materials Data Bases Using Natural Clustering—
ROBERT A. PILGRIM, PHIL M. JOHNSON, AND PATRICK M. FALCO, JR. 185

IMPACT OF MATERIALS DATA BASES

Data Base R&D for Unified Life Cycle Engineering—HARRIS M. BURTE AND
CLAYTON L. HARMSWORTH 197

Computerized Materials Data in Aerospace Applications—C. DALE LITTLE AND
THOMAS E. COYLE 200

The Business of Materials Data Banks—JANE E. MARTINI-VVEDENSKY 211

Socioeconomic Barriers in Computerizing Materials Data—JOHN R. RUMBLE, JR. 216

MATERIALS DATA BASE PROJECTS

Engineering Plastics via the Dow MEC Data Base—LIGAYA S. PETRISKO 229

RUST: A Coupon Corrosion Test Data Base for Metals and Nonmetals—
B. J. MONIZ AND T. C. WOOL 239

Generation and Use of Composite Material Data Bases—KENNETH RANGER 253

Designation and Characterization of Composite Materials—JOSEPH K. LEES,
BEVERLY K. ROBERTS, AND ROBERT J. MICHAUD 265

Building Blocks for an On-Line Materials Data Base—HUI H. LI AND CHO-YEN HO 272

Consideration of a Preliminary Data Base for MIL-HDBK-17B—CRYSTAL H. NEWTON 280

PC-Access to Ceramic Phase Diagrams—PETER K. SCHENCK AND JENNIFER R. DENNIS 292

Creating a Materials Data Base Builder and Producing Publications for Ceramic Phase
Diagrams—HELEN M. ONDIK AND CARLA G. MESSINA 304

COOPERATIVE MATERIALS DATA BASE PROGRAMS

Corrosion Data for Materials Performance Characterization—DAVID B. ANDERSON
AND GLENN J. LAVERTY 317

Development of Data Bases for the ASM/NBS Data Program for Alloy Phase
Diagrams—
WILLIAM W. SCOTT, JR., HUGH BAKER, AND LINDA KACPRZAK 322

Welding Information Systems—JERALD D. JONES AND H. H. VANDERVELDT 329

ACTIS: Towards a Comprehensive Tribology Data Base—SAID JAHANMIR,
STEPHEN M. HSU, AND RONALD G. MUNRO 340

Index 349

Overview

Using computers to deliver materials information is now a reality. Many groups that previously published are now using data bases to make available to the general public property data for metals, alloys, composites, polymers, and plastics. Many more data bases are being developed for individual companies.

In spite of all this activity, building, distributing, and using materials data bases is not yet routine or necessarily easy. Of all the materials data bases that now exist, no two of them are compatible in any significant way. The user interfaces are different, the nomenclature varies, and different data are collected, even for the same test methods. Unfortunately, all this incompatibility lies on top of the chaotic hardware and software situation.

ASTM has established Committee E49 on Computerization of Material Property Data to develop standards and guidelines for materials data bases. The goal is not to impose rigidity or hinder innovation, but rather to develop a common basis for handling materials data on computers. The key questions being addressed are as follows:

1. How can materials be described and identified in data bases?
2. What data items must be reported with test results to make them meaningful?
3. How should these data be reported? In what format?
4. What information should the user interface contain?
5. How can the developed guidelines be used in data base building?
6. How can data from two different data bases be accessed and combined?
7. How do you indicate the quality of data in data bases?

The standards will allow builders of materials data bases to draw on the experience of the community in answering these and similar questions. Users will be able to access different data bases more uniformly. Distributors of data bases will be able to maintain compatibility from one data base to another.

Enough progress has been made in the two years the Committee E49 has operated to make it worthwhile to involve the materials data base community in an open forum to exchange ideas on what has been done and what needs to be done next. That forum became the First International Symposium on Computerization and Networking of Materials Property Data Bases, held in Philadelphia, PA, on 2-4 Nov. 1987, and attended by 128 experts from 11 countries. The presentations covered many aspects of materials data bases:

- Standards activities
- National and international activities
- Emerging issues
- Impact
- Data base projects
- Cooperative data base programs

The papers in this volume are grouped according to these categories. Before describing them, some general conclusions can be drawn.

The state of the art of materials data bases had advanced rapidly with respect to the coverage of materials. Of the major classes of structural engineering materials, polymers, metals, and composites all have substantial data bases. These are being distributed both publicly and within private organizations. The general situation seems to be the same in the United States, the European Community, Japan, and China.

Aside from the coverage, the situation for materials data bases can best be described as lots of ideas and many goals but few successes. Some important points can be made:

- Standards for materials data bases have considerable support, and there is a strong desire for international compatibility of these standards. The standards, however, are not yet in place.
- Most industrial countries have made a commitment to develop an integrated network of individual data bases. Implementation varies considerably and most efforts have found that the work is going more slowly and costs more than originally anticipated.
- The impact of materials data bases is recognized to be indirect, thus making its calculation very difficult. This is further reflected by the difficulty some data base providers have had in getting started.
- Personal computer data bases of materials data are much further advanced than networks. Although individual PC data bases have attractive manipulation and display features, they do not now contain very much data.
- Very few data bases have a full range of test data and most do not include property data as a function of temperature or other conditions.
- Expert systems linked to user interfaces and integration of materials data bases into other computerized engineering tools are high-priority goals, but more research must be done.
- Significant problems still exist with respect to capturing the richness and complexity of materials data.
- In the United States, cooperative data base programs have been very successful in drawing upon industrial and government resources to improve the quality and accessibility of materials data.

Standards for Materials Data Bases

In his paper, Kaufman describes the activities of ASTM Committee E49 on Computerization of Materials Property Data. A full discussion is given of the types of standards being developed. Westbrook gives a detailed example of one E49 standards area, the identification of metals and alloys. Reynard discusses international aspects of materials data base standards, focusing on the recent activities of the Versailles Project on Advanced Materials and Standards (VAMAS). Great concern has been expressed by many people in this field that standards developed by individual countries must be compatible, and VAMAS, in conjunction with E49, has been defining the issues involved.

National and International Materials Data Base Activities

Almost every industrialized country has a major effort to make materials data bases available by on-line computer networks. The papers in this section give an overview of notable examples of these efforts. Kaufman discusses the National Materials Property Data Network, Inc., a cooperative effort between industry and technical societies in the United States to build such a network. Kröckel and Steven present the efforts of the EEC on a similar network that involves data bases from many European countries. While the goals of the two groups are about the same, the approaches are quite different as are the sources of support.

Three papers describe aspects of other national efforts for materials data bases. Lu and Fan cover activities in China, which include an impressive list of areas where work has started. Nishijima, Momma, and Kanao discuss work in Japan, especially at the National Research

Institute of Metals. Finally, Bathias and Marx give the results of a recent survey done in France on the need and acceptance of materials data bases.

The paper by Barrett covers the activities of the Committee on Data for Science and Technology (known as CODATA) of the International Council of Scientific Unions. CODATA provides a forum for cooperative data work on an international scale, and its task group on materials data bases will be important in future years.

Emerging Issues

Materials data are rich and complicated. The data themselves are an integral part of engineering and manufacturing. The interpretation of these data, especially in the context of various applications, comes close to being an art.

The set of papers in this section deals with these issues from the perspective of computerization: How to capture the complexity? How to integrate materials data bases with other software? How to incorporate expert systems?

Most of the ideas presented here are new and have not been incorporated in working systems. Some are drawn from other fields. Some are embryonic. They are important ideas and, even though some of the papers require serious reading and rereading, this will be well worthwhile. Su and Furlani have been leaders in the integration of engineering data bases, especially in computer-aided design (CAD) and computer-aided manufacturing (CAM). It is the dream of many to link materials data bases into CAD/CAM systems, and these papers give important background.

McCarthy and Grattidge have played key roles in the development of prototype Materials Information for Science and Technology system in the United States and discuss important aspects of handling materials metadata and data capture. Many of the issues raised are still problems that must be solved before large-scale materials data networks will exist.

In his paper, Iwata presents some thought-provoking ideas on the need for expert systems as part of the interface to materials data bases. The final paper of this section is by Pilgrim et al. covers an important approach to querying materials data bases.

Impact of Material Data Bases

Little has been written as to the impact that materials data bases will have on engineers in their work, or the related issue, the lack of perceived impact. Burte and Harmsworth describe how data bases affect the unified life cycle engineering concepts being developed by the U.S. Air Force. Little and Coyle cover how materials data bases relate to other work in the aerospace industry.

The other two papers look at the impact issue from the point of view of overcoming the lack of perceived impact. Martini-Vvedensky discusses the problems of starting a business based on materials data bases, while Rumble looks at various socio-economic issues that provide barriers to progress.

Materials Data Base Projects

This section of papers contains descriptions of a variety of materials data bases, built for many reasons. Petrisko, Moniz, Ranger, and Lees et al. all describe materials data bases built for their companies. Li and Ho and Newton and Gall discuss data bases built for the U.S. government. Schenck and Dennis, and Ondik and Messina cover work of the American Ceramics Society and NBS on phase diagram data bases.

Cooperative Materials Data Base Programs

Over the last decade, several large data programs in the United States have addressed industrial needs for high-quality, easily accessible materials data. These have been cooperative programs between technical societies, industry, and government.

Anderson and Laverty discuss the corrosion data program of the National Association of Corrosion Engineers (NACE) and the National Bureau of Standards (NBS). Scott et al. describe the Alloy Phase Diagram Program of ASM International and NBS, which has had many contributions from other countries. Jones and Vanderveldt cover the welding data program of the American Welding Institute. To end this section, Jahanmir, Hsu and Munro discuss the new tribology data program involving NIST, the Department of Energy, the American Society of Mechanical Engineers, and American Society of Lubrication Engineers.

Summary

The editors hope that the readers of this volume will come away with an understanding of the present-day status of materials data bases, both in the United States and internationally. There are many ideas, old and new, that should be useful. In future years, we look forward to seeing these ideas become reality.

Jerry S. Glazman
Combustion Engineering, Inc. Windsor, CT 06095; symposium chairman and editor.

John R. Rumble, Jr.
National Institute of Standards and Technology, Gaithersburg, MD 20899; symposium chairman and editor.

Standards for Materials Data Bases

J. Gilbert Kaufman[1]

Standards for Computerized Material Property Data—ASTM Committee E-49

REFERENCE: Kaufman, J. G., "**Standards for Computerized Material Property Data—ASTM Committee E-49,**" *Computerization and Networking of Materials Data Bases, ASTM STP 1017,* J. S. Glazman and J. R. Rumble, Jr., Eds., American Society for Testing and Materials, Philadelphia, 1989, pp. 7-22.

ABSTRACT: ASTM Committee E-49 on the Computerization of Materials Property Data was instituted in 1986 to provide guidance to individuals and organizations building machine-readable data bases and to aid in the development of compatible and consistent sources capable of a reasonable exchange of information. In the relatively short interval since then, progress has been made in several areas, most notably in the development of formats for the characterization of materials and reporting of test data. Progress has been most rapid in the more mature and stable material classes like metals and polymers. The relative newness and lack of stability of terminology and test procedures for advanced materials will require more time and effort before major strides are apparent. ASTM Committee E-49 is also taking on the significant challenges of data exchange formats and guidelines for the evaluation of data.

Continued rapid progress in all of these areas is essential if we are to maximize the opportunity to improve the quality of materials related decisions through the use of easily accessible and exchangeable numeric performance data.

KEY WORDS: standards, materials, data base, ASTM, computer

ASTM Committee E-49 on the Computerization of Materials Property Data was formed in 1986 [1,2] in response to an increased recognition of the great resource value of well-documented materials property data and of their importance in high-quality decision making in materials selection and design [3-5]. It was one of two major developments in recent years in response to the critical need for easier direct access to more reliable numeric performance data, the other being the formation of the National Materials Property Data Network, Inc. [6,7]. The latter, known now as the MPD Network or sometimes simply as MPD, is a not-for-profit corporation providing U.S. engineers and scientists with easy on-line access to high-quality, machine-readable, numeric data. It was established to fill the critical need for materials selection and design decisions based upon more accurate and reliable performance information. This need was identified in several expert studies, including one by the National Materials Advisory Board [8-11].

The second development was the formation of ASTM Committee E 49, to develop standard classifications, guides, practices, and terminology for building and accessing materials property data bases. This committee is working with other ASTM technical committees and organizations that develop standards related to materials and their properties to bring some consensus to this explosive areas of activity. Committee E-49 will establish guidelines to aid those individuals and organizations building data bases or intelligent knowledge systems to ensure that high lev-

---

[1] President, National Materials Property Data Network, Inc., 2540 Olentangy River Rd., P.O. Box 02224, Columbus, OH 43202.

els of quality and reliability are maintained and that compatibility with other sources is assured. The latter is essential if we are to be able to readily and cost-effectively share and compare data across organization lines.

The motivation for a cohesive national resource optimizing our ability to share data is particularly compelling now. Many of our traditional sources of data, such as materials producers and government laboratories, have either cut back their activities in this area or eliminated them completely. In some cases the organizations themselves no longer exist. Pressures to restructure and cut costs in the traditional materials, engineering, and heavy industry segments plus the pressure on reducing the national deficits will keep the United States in this situation for some time. Yet to meet foreign competition, we must maintain our materials know-how at a high level. The development of easily accessible electronic sources of materials property data is one significant way to do so.

The second compelling reason for a major effort in this important area is the proliferation of new, high-performance structural materials, notably composites of metals, polymers, or ceramics, or both, composing the group commonly referred to as "advanced materials." The large numbers of basic materials, plus the seemingly endless combinations of procedures by which they are assembled, make assimilation and comparison of data for these new products difficult if not impossible. Yet to shorten the time needed to commercialize these new products and establish competitive production positions, it is essential to develop guidelines and standards for handling data for these materials in high-quality, numeric, machine-readable data bases.

Finally, the effort to standardize procedures for dealing with numeric materials property data in machine-readable form is vital. This is because the development of systems for organizing and storing numeric data for retrieval greatly increases the value of our materials research dollars. In computerized systems, the individual test results are kept "alive" and accessible in well-documented form for ready comparison with newly developing materials research. A mechanism will also exist to resurrect and archive "lost" data for earlier programs, the results of which are often resident in untraceable company or laboratory files or in the drawers of the original researchers, never to be looked upon again. A concerted effort to incorporate the results of the more valuable research programs in searchable electronic sources can substantially reduce the new testing required to establish performance comparisons for competitive materials for new applications.

Background of ASTM Committee E-49

Several other groups, including American National Standards Organization (ANSI), International Standards Organization (ISO), and Institute of Electrical and Electronics Engineers (IEEE), as well as ASTM Committee E-31 on Computerized Systems [12] have had a significant role in establishing standards for computerized information systems. But ASTM Committee E-49 is the only one to specifically focus on the problems that arise in dealing with numeric/factual materials property data. It is worth noting the breadth of activity of these other groups, however, as indicated by Table 1 [12].

Committee E-49 held its formation meeting in March, 1986, primarily as a result of the concerted efforts of the National Materials Property Data Network and the National Bureau of Standards to move toward consensus (rather than arbitrary) standards for developing and accessing/transferring data. The scope of Committee E-49 is as follows:

> The promotion of knowledge and development of standard classifications, guides, practices, and terminology for building and accessing computerized material property data bases.
> This committee will work in concert with ASTM technical committees and other organizations that develop standards related to materials and their properties.

TABLE 1—*Standards for computerized systems and network development and operation* [12].

These standards for network development are recommended for adoption and use:
Terminology
    ANSI X3/TR-1—Dictionary for Information Systems
    ASTM E 1013—Terminology Relating to Computerized Systems
Overview
    ASTM E 622—Generic Guide for Computerized Systems
Project Definition
    Draft E3103-5—Guide for Project Definition for Computerized Systems
Functional Requirements
    ASTM E 623—Guide for Developing Functional Requirements for Computerized Systems
Functional Design
    ASTM E 730—Guide for Developing Functional Designs for Computerized Systems
    ISO 7498—Basic Reference Model for Open Systems Interconnection
Implementation Design
    ASTM E 624—Guide for Developing Implementation Designs for Computerized Systems
    ASTM E 625—Guide for Training Users of Computerized Systems
System Assembly, Installation, and Testing
    ASTM 3 731—Guide for Procurement of Commercially Available Computerized Systems
    IEEE 730—Standard for Software Quality Assurance Plans
    IEEE 828—Standard for Software Configuration Management Plans
    IEEE 829—Standard for Software Test Documentation
    IEEE 830—Guide to Software Requirements Specifications
    IEEE Draft P983—Guide for Software Quality Assurance
    IEEE Draft P1012—Standard for Software Verification Plans
    IEEE Draft P1042—Guide for Software Configuration Management
    IEEE Draft P1059—Guide for Software Verification and Validation
    EIA, ANSC X3, UL, NFPA Standards as required
System Evaluation
    ASTM E 626—Guide for Evaluating Computerized Systems
System Documentation
    ASTM E 627—Guide for Documenting Computerized Systems
    ASTM E 919—Specification for Software Documentation for a Computerized System
    ANSI X3.88—Computer Program Abstracts
These standards for network operation are recommended for study and imitation:
    ANSI X9.2—Interchange Message Specification for Debit and Credit Card Message Exchange among Financial Institutions
    IATA—Standards for data messages

The second paragraph of this scope statement emphasizes that the activities of the group do not infringe on those of other committees and organizations with expertise on materials, test methods, properties, and applications. Rather, E-49 will cooperate closely with those groups, providing the uniformity and standardization required to ensure the generation of cohesive and compatible computerized sources, and an outlet for the communication of those standards to the technical community.

Examples of current cooperative efforts with other organizations include the effort to address the characterization and presentation of data for ceramics with the American Ceramic Society. Guidelines for dealing with data for welded joints are also being established through collaboration with the American Welding Society and American Welding Institute. These functions are accomplished by the active participation of representatives of those organizations on Committee E-49. The role of the committee in such instances will be to influence the standards development work of these other groups, endorse their efforts, and incorporate them into the appropriate ASTM standards.

The activities of ASTM Committee E-49 also have the strong support and participation of the

National Bureau of Standards, notably the Standard Reference Data Program and the Institute of Material Science and Engineering, as well as of other government laboratories as well, including Lawrence Berkeley and Sandia Laboratories.

Close association and cooperation with international groups is being sought. Strong links have already been established with the Commission of the Economic Community data base network program [13]; the European Aluminum Association; VAMAS [14]; and CODATA [15], the Committee on Data for Science and Technology of the International Council of Scientific Unions (ICSU). ASTM Committee E 49 is filling a key role defined in the recent VAMAS report [14], in which the need for strong leadership in the field was cited. A cooperative position with ISO is also being sought, a goal that has been hampered by some recent personnel changes in that organization.

Four Subcommittees of E-49 have been established (Table 2). High priorities have been put on the development of data recording and computer storage formats, and therefore on the options for presenting the data. Preliminary guidelines for the working subcommittees and task groups have been written and distributed. Several individual task group activities are already underway to address specific material descriptors and test data reports. Each of the subcommittee activities will be discussed in turn below.

Standards for Material Characterization and Test Reporting

Materials Descriptors

Preliminary guidelines developed by Committee E 49 suggest that the following categories of information will probably be required to characterize all materials for computerized data systems:

- material class (metal, polymer, composite, ceramic, etc.),
- specific material within class (stainless steel, SiN, etc.),
- material designation (industry standard or experimental),
- material condition (industry standard or broad class),
- material specification (ASTM or another or both),
- material producer or source of material,
- producer lot number or number assigned by source or data base builder or both lot and assigned number,
- product form (casting, forging, laminate, compact, and so forth)
- material composition (individual lot for which data is shown), and
- fabrication history (major elements of fabrication procedure, especially those most likely to have a specific impact on material properties or performance).

TABLE 2—*E 49 subcommittees.*

E49.01—identification of materials: identifying and selecting those descriptors and coding systems necessary to adequately describe and differentiate materials for a computerized materials data base.
E49.02—reporting of material property data: development of guidelines for presentation of material properties data to be included in computerized materials properties data bases.
E49.03—terminology: oversight of the development of a standard terminology for use in building and accessing a computerized materials property data base, including definitions, nomenclature, abbreviations, and units.
E49.04—data base interfaces and functionalities: researching and developing guidelines for the transfer, processing, and presentation of materials data among distributed data bases and users, including areas such as data exchange formats, data security, search and presentation capabilities, and data quality and reliability.

The next part of the activity becomes "personalizing" the fields or records to make them appropriate for characterizing first a material class, that is, metals or polymers, and then the individual material subgroups, that is, aluminum alloy products. The first will be "generic" to a particular material class, recognizing the unique aspects of their descriptions that must be considered to maximize the value of the data source. The latter format (for the aluminum alloy product example, see Ref 2) becomes the version that might appear in an aluminum alloy standard or specification as a guideline to builders of data bases containing the properties of aluminum alloys/products.

Not all of the items called for above will be available for every lot of material for which data is shown in each data base. Such listings should not be misconstrued to mean that items such as a detailed fabrication history will be required for each lot included. Rather, these are listings of those items that would be desirable to include if available, as they are likely to be of value to at least some users of the data base.

There are, of course, certain minimum information requirements to assure adequate quality and reliability in data bases, and those minimum sets of fields will be identified as well. For aluminum alloys, for example, it is essential to know the alloy and temper designations and the product forms for the associated properties/test results to be of any value at all.

Examples of two "generic" guidelines for material identification can be presented as a result of the early work of the committee. Tables 3 and 4 show the latest draft standards for metals and polymers, respectively. Note that these are not yet standards, but they have been through ballots of the originating task groups and subcommittee and so their general content has been pretty well established.

Work on similar guidelines for ceramics and composites has progressed, but is not quite as far along, primarily because of the newer technologies involved and the resultant lack of maturity in materials designations and test methods. Task groups are active in both areas, however, and preliminary guidelines will be available within a year.

Reporting of Test Data

The initial E-49 guidelines suggest that the following types of information will be valuable for inclusion in data bases containing individual test results:

- general type of test data (tension, emmissivity, etc.),
- specific test method (ASTM standard test method designation or the equivalent),
- specimen type, dimensions, location and orientation within original lot/sample,
- test independent variables (temperature, time, loading rate, environment, prior exposures, and so forth),
- property descriptors, if variable (for example, percent offset for tensile yield strength),
- specific test results, including all raw data (loads, extensions, and load-deformation curves, as well as calculated strengths, and so forth), and
- all validity requirements and any individual data or assessments needed to establish validity.

As in the case for material descriptors, all of these individual items are not required for every data base. However, all are desirable if available and will be useful to some users of any specific data base. Likewise, there will always be a certain few records that are essential if the data are to be of any value at all, and these must be identified as well.

Using the guidelines noted above, experts from the committees with direct responsibility for specific test methods (ASTM Committee E-28 for Method on Tension Testing of Metallic Material [E 8]) are working with representatives of Subcommittee E49.02 to develop specific formats for each type of test data. These formats will eventually be published in E-49 summary guidelines and perhaps as appendixes to the appropriate methods.

TABLE 3—*Generic data format identification of metals and metal alloys.*

Field Number	Field Name/Description	Units
1	accession number	string
2	broad material class	string
3	subdivision of class	string
4	finer subdivision of class	string
5	common name	string
6[a]	material designation/division of standard	string
6[a]	primary standard identification	string
6[a]	organization	string
6[a]	key to supplementary material info	string
7[a]	primary form	string
7[a]	secondary form	string
7[a]	part specific identification/number	string
8[a]	thickness/short transverse/diameter	mm
9	width/long transverse	mm
10	length/longest	mm
11	key to supplementary size info	string
12[a]	process designation	string
12[a]	process standard	string
12[a]	organization	string
12[a]	key to supplementary process info	string
13	country of origin	string
14[a]	manufacturer	string
14[a]	manufacturer's plant	string
14[a]	manufacturer's designation	string
14[a]	lot identification/number	string
15	date of manufacture	string
16	key to supplementary source info	string
17[a]	service/fabrication history (if any)	string
17[a]	key to supplementary service info	string
18	supplementary identification notes	string

[a] Indicates essential information.
[b] Indicates multiple entries may be necessary.
NOTE: not a standard, presented for information only.

An example of what such a format might look like is illustrated in Table 5, where the draft standard data format for tension testing per ASTM Method E 8 is presented (note this not yet an accepted standard, but the second draft of a format now under study). Formats for several other mechanical properties are also under development; a summary is shown in Table 6.

The corrosion data format standardization is an outgrowth of a NACE/NBS Corrosion Data Workshop. It is being carried forth by an ASTM GO1/NBS committee [20], with which E 49 is also cooperating.

Format for Standard Data Records

The examples in Tables 3, 4 and Fig. 1 provide a style guideline for the standard formats. This particular form, which may change with further study, has been conceived of to aid in structuring a data base containing those records. It is not necessarily the style likely to be found in a testing lab. The basic elements of such a presentation are the field number (not an essential

TABLE 4—*Generic data format for identification of polymers.*

Field Number	Description
1	Material Class (Polymer)
2	Subdivision of Class or Polymer Class (Thermoplastic)
3	Finer Subdivision of Class or Generic Family Classification of Polymers (Acrylonitrile-Butadiene-Styrene)
4	Family Abbreviation (ABS)
5	Monomer or Blend Composition (20% Acrylonitrile, 10% Butadiene, 70% Styrene)
6	Manufacturer's Designation (Trade Name, Grade)
7	Manufacturer
8	Production Process (Mass, Suspension, or Emulsion)
9	Production Plant (Where Produced)
10	Data of Manufacture
11	Condition of the Product as Tested (V or X)
12	Modifier (Glass)
13	Modifier Composition (10% Short Glass)
14	Commercial Specification (ASTM D1788 ABS 3-2-2)
15	Sample Form (Sheet)
16	Fabrication Type (Extrusion)
17	Fabrication Conditions (Process Temp = 500 F, Press = 1000 PSI, Mold Temp = 120 F

[a] Essential Information.
NOTE: not a standard, presented for information only.

element, but a tool for keeping track of size and location in the format), a short (five-letter) name constituting an abbreviation or "mnemonic" for the field, the full field title, and either a set of terms that would be the logical entries in the field (category sets) or the units of the numbers that might be placed in the field.

Discussion continues on the degree of universality and meaningfulness that can be expected of the mnemonics. The situation is complicated by the desire to make these applicable on an international level, as it is impossible to have a single, five-letter abbreviation which has some mnemonic aid in more than one or two languages. The tendencies to date have been to make these shortened English terms, but it remains to be seen whether or not this is a useful approach on an international level, and with so many terms to cover without duplication and confusion, if it is even useful to try.

Formats for the Testing Laboratory

The most reliable way to assure the availability of data in machine-readable form is to collect them electronically from the experiment when it is conducted. This is being done increasingly. The next best thing is to collect it in a form that reflects all of the desired information from the standard formats being defined in E 49 activity. Still, the use of forms like that in Table 5 are not very helpful in the testing laboratory. Versions such as that in Table 7 are much more likely to be accepted in that environment. Because such forms may vary with the testing system used, the ASTM Committee E-49 does not intend to define such laboratory data sheets, but to provide the background documents in the form of Table 5.

Terminology

Subcommittee E 49.03 on Terminology has begun to come to grips with several aspects of the problems associated with the lack of consistency in terms, definitions, abbreviations, and symbols for different material classes and different groups of properties. Its primary objectives in

TABLE 5—*Recommended standard data format for computerization of plane-strain fracture toughness test data per ASTM Test Method E 399.*

Field No.[a]	Field Name and Description	Category Sets, Values or Units[b]
TEST AND MATERIALS IDENTIFICATION		
1.*	Material identification	(This information will be supplemented by material description guideline.)
2.*	Lot identification	
3.*	Data source identification	
4.*	Type of test	Plane-Strain Fracture Toughness
5.*	ASTM, ISO or other applicable standard method number	ASTM E 399
6.*	Material yield strength	MPa (psi)
7.*	Material elastic modulus	GPa (psix106)
SPECIMEN INFORMATION		
8.*	Specimen identification	alpha-numeric string
9.*	Specimen type	see Table X1.1
10.*	Specimen orientation	see Table X1.2
11.	Specimen location	see Table X1.3
12.*	Specimen thickness B	mm (in.)
13.*	Specimen width (depth) W	mm (in.)
14.	Specimen span length S	mm (in.)
15.*	Loading hole offset (arc-shaped)	mm (in.)
16.*	Inner radius (arc-shaped)	mm (in.)
17.*	Outer radius (arc-shaped)	mm (in.)
FATIGUE PRECRACKING		
18.	Fatigue cracking maximum load PF_{max}	N (lbf)
19.	Fatigue maximum stress intensity Kf_{max}	MPa (sqrt m)
20.	Fatigue cracking load ratio R	
21.	Cycles to complete fatigue cracking	$\times 10^3$
22.	Fatigue crack length, edge, fce1	mm (in.)
23.	Fatigue crack length, center, fcc	mm (in.)
24.	Fatigue crack length, edge, fce2	mm (in.)
KW DETERMINATION		
25.	KQ loading rate	N/min (lbf/min)
26.	KQ test chart slope	%
27.	KQ candidate load, PQ	N (lbf)
28.*	Candidate plane-strain intensity factor, KQ	MPa (sqrt m) (psi (sqrt in.))
29.	Maximum load, P_{max}	N (lbf)
30.	Maximum stress intensity factor, K_{max}	MPa (sqrt m) (psi (sqrt in.))
TEST RESULTS AND ANALYSIS		
31.	Total crack length, edge, a1	mm (in.)
32.	Total crack length, quarter, a2	mm (in.)
33.	Total crack length, center, a3	mm (in.)
34.	Total crack length, quarter, a4	mm (in.)
35.	Total crack length, edge, a5	mm (in.)
36.*	Average crack length	mm (in.)
37.	Fracture appearance	% oblique
38.	Fatigue crack plane angle to crack plane	%
39.	$2.5 \times (KQ/TYS)^2$	mm (in.)
40.	P_{max}/PQ	
41.	a/W	

TABLE 5—Continued.

Field No.[a]	Field Name and Description	Category Sets, Values or Units[b]
42.	Kf_{max}/KQ	
43.	Kf_{max}/E	sqrt m (sqrt in.)
44.	Minimum fatigue precrack length	mm (in.)
45.	Maximum difference between $a2$, $a3$, $a4$	mm (in.)
46.	Difference between $a1$ and $a5$	mm (in.)
47.	KQ stressing rate	MPa (sqrt m) (psi (sqrt in.))

TEST PARAMETERS AND PROCEDURES

48.*	Test temperature	degree C (degree F)
49.	Test environment	alpha-numeric string
50.	Test humidity	%
51.	Test date	MMDDYY

TEST VALIDATION

52.	Is $B > 2.5(KQ/TYS)2$?	yes or no
53.	Is $a > 2.5(KQ/TYS)2$?	yes or no
54.	Is $a/W = 0.45 - 0.55$?	yes or no
55.	Is $P_{max}/PQ < 1.10$?	yes or no
56.	Is $Kf_{max}/KQ < 0.6$?	yes or no
57.	Is $Kf_{max}/E < 0.02$?	yes or no
58.	Is max difference between $a2$, $a3$, $a4 < 0.10a$?	yes or no
59.	Is difference between $a1$, and $a5 < 0.10a$?	yes or no
60.	Is min fatigue precrack > 0.050 in?	yes or no
61.	Is fatigue crack plane angle < 10 deg?	yes or no
62.	Is loading rate = 30000-150000 psi (sqrt in)?	yes or no
63.	Is KQ test chart slope = 0.7-1.5?	yes or no
64.*	Is KQ valid measure of K_{Ic}? (all criteria met?)	yes or no
65.	Specimen strength ratio	

NOTE: *denotes essential information for computerization of test results.
[a]Field numbers are for reference only. They do not imply a necessity to include all these fields in any specific database nor imply a requirement that fields used be in this particular order.
[b]Units listed first are for SI; those in parentheses are English.

TABLE 6—Mechanical properties formats.

ASTM Test Method	Cooperating ASTM Committee	Reference
Tension (E 8)	E-28	(16)
Compression (E 9)	E-28	—
Notched Bar Impact (E 23)	E-28	—
Plane Strain Fracture Toughness (E 399)	E-24	(16, 17)
Fatigue Crack Growth (E 647)	E-24	(18)
Creep and Stress Rupture (E 139)	E-28	(19)
Various Corrosion Properties	G-1	(20)

TABLE 7—*Input layout for computerization of plane strain fracture toughness (ASTM E 399).*

```
INPUT LAYOUT FOR COMPUTERIZATION OF PLANE STRAIN FRACTURE TOUGHNESS
ASTM METHOD E399

1.Database identification_____
2.Material identification_____
3.Lot identification
4.Source identification
5.Material tensile yield strength_____    psi
6.Material elastic modulus         _____    10**6psi

SPECIMEN MEASUREMENTS
7.Specimen identification_____    8.Specimen location_____
9.Specimen orientation    _____   10.Specimen type    _____
11.Specimen width,W             _____in.
12.Specimen thickness,B         _____in.
13.Specimen span length,S       _____in.

FATIGUE PRECRACKING
14 Maximum cyclic load, Pfmax                       _____lbs
15.Maximum cyclic stress intensity factor, Kfmax, psi(sqrt in)____#
16.Cyclic load ratio R ___
17.Cycles to complete precracking, *10**3, ____
Fatigue crack length, in. 18.Edge____ 19.Center____ 20.Edge____

KQ DETERMINATION
21.KQ loading rate, Prate, lbs/min ____ 22.KQ test chart slope ___
23.KQ candidate load, PQ, lbs ____
24.Candidate plane strain fracture toughness, KQ,psi(sqrt in)____#
25.Maximum load, Pmax, lbs ____
26.Maximum stress intensity factor, Kmax, psi(sqrt in) _____#

POST-TEST SPECIMEN MEASUREMENTS
Total crack length, in. 27.Edge ____ 28.Quarter ____ 29 Center ____
                       30.Quarter ____ 31.Edge ____
32.Average crack length, a, in. ____
33.Fracture appearance, % oblique ___
34.Fatigue crack plane angle, deg ___

OTHER VALIDITY CRITERIA INDICES  #
35.Specimen thickness, crack length: 2.5(KQ/TYS)**2, in. ____
36.Pmax/PQ ___   37. a/W ___   38. Kfmax/KQ ___ 39.Kfmax/E ___
40.Minimum fatigue precrack length, in. ___
41.Maximum difference between a2,a3,a4, in. ___
42.Difference between a1 and a5, in. ___
43.KQ stressing rate, Krate, psi(sqrt in)/min ___

VALIDITY CRITERIA MET ( Y-YES, N-NO ) #
44.B=/>2.5(KQ/TYS)**2 ___            45.a=/>2.5(KQ/TYS)**2 ___
46.a/W=0.45-0.55 ___    47.Pmax/PQ</=1.1 ___   48.Kfmax/KQ</=0.6 ___
49.Kfmax/E</=0.002 ___              50.Max diff a2,a3,a4</=0.10a ___
51.Diff between a1,a5</=0.10a
52.Minimum fatigue precrack length=/>0.050 in     ___
53 Fatigue crack plane angle</=10deg              ___
54.Loading rate = 30000-150000 psi(sqrt in)       ___
55.KQ test chart slope = 0.7-1.5                  ___
56.KQ valid measure of plane strain fracture toughness,
                         KIC, psi(sqrt in)  Y-YES, N-NO ___

TEST CONDITIONS
57.Test temperature, degC_____    58.Test environment_____
59.Test humidity,% _____        60.Test date        _____
61.Special notes_____

#-NOTE:Option exists to have calculation and reporting of flagged
items (#) handled as part of computer analysis of raw data.
```

the short term are to assemble complete glossaries of these elements and begin to identify those areas in most critical need of harmonization. In addition, the task of assembling and standardizing terms related to computerized data bases is itself vital to developing a consistent approach to the problem. It is thus also receiving high priority in E 49.03 efforts.

A preliminary glossary of terms important to the computerization of data has been assembled to aid the subcommittee in moving forward. However, no progress has been made in standardizing these terms nor in dealing with the technical and political complexities of eliminating inconsistencies in nomenclature and terminology across materials producing and testing industries.

To illustrate the breadth of this problem, even within rather specific and mature industries, the nomenclature issue is a complex one. In the U.S. steel industry, for example, there are a number of technically acceptable alternative designations for specific compositions, and no one organization has taken on the responsibilities of correlating/collating these into a single set. The Unified Numbering System [21] provides the broadest coverage, but it is not used in general commerce and even it has many voids for important grades. Furthermore, there is no standard system of designating the condition or temper or any consistent characterization of basic fabrication practice comparable, to the temper designation system employed in the Aluminum Industry for example [22]. Even in ASTM specifications, terms like "grade," "class," and "type" are used in different ways for different alloys.

As indicated earlier, the problem is exaggerated for materials such as composites and advanced ceramics because of the relative newness and lack of stability in procedures and terminologies. Nevertheless, order can be brought to these areas before they get out of hand. E49.03 will be making every effort to do so.

Interfaces

Among the most complex and yet most critical elements of E-49 activities are those of Subcommittee E 49.04 on interfaces and related functionalities, as these deal with the heart of the issues of data exchange and comparison. These issues are key to the long-range goal of having a cohesive system of data sources nationally and internationally. Several existing standard data exchange formats as well as a new, possibly simpler format are already under study by this group. Further discussion of interface issues is beyond the scope of this paper.

Evaluation and Analysis of Data for Quality and Reliability

One of the other major challenges taken up by E 49.04 is that of creating definitions and procedures for the evaluation and analysis of materials property data. While many organizations and individuals consistently cite the need for "evaluated" data, the term has had many different interpretations. There has been no consistent procedure used by individuals or groups carrying out such evaluations.

When the implementation of the National Materials Property Data Network was begun, one of the first issues taken up by the MPD Technical Committee [23] was that of guidelines for the evaluation and analysis of data. One of the most significant contributions made by that Technical Committee was to begin to deal with this issue. While no formal guidelines have yet been adopted, some overall perspective was brought to the subject, as outlined below. These will be the starting point for the E 49.04 effort. It remains to be seen whether or not all of these areas are within the scope of the ASTM Committees mission.

First, there are several basic terminology ground-rules to cover, as there remains no consensus on the use of terms like "evaluation," "validation," and the like. One of the most lucid discussions of this subject is by Barrett [24], though unfortunately a few aspects of his conclusions are inconsistent with long-standing ASTM usage. For purposes of this paper, this terminology is included here.

Evaluation

This broad term is used here to encompass all aspects of the qualitative (subjective) and quantitative assessment of the quality and reliability of an individual datum or groups of data. It also encompasses the statistical or parametric analyses (or both) of data to refine or generate more information from groups of data. The four specific elements of data evaluation as defined here are

(1) the subjective assessment of data quality,
(2) validation,

(3) analysis, and
(4) certification.

The first is simply included in the definition of evaluation, and the other three are as defined below. (By Barrett's usage, "evaluation" is a broad term, but it includes only the subjective judgement and not "validation" and "analysis" of data as defined below.)

Validation—This is the specific process of quantitatively measuring the degree to which individual test results themselves or test conditions meet the validity criteria spelled out in standard test methods. (Barrett uses "validation" in place of "certification" below.)

Analysis—This is the specific process of statistically or parametrically manipulating groups of data to refine them or to derive more general performance values (properties or property values as opposed to individual test results). This would also include graphic analysis and comparison with theoretical expectations in a quantitative fashion.

Certification—This involves a group of experts or at least authorities reviewing and analyzing data by whatever means appropriate to develop and approve (certify or ratify) performance properties for use in design or quality control. (This is "validation" in Barrett's discussion, and not encompassed in the broader "evaluation" term.)

The only significant conflict between current ASTM usage and Barrett's is in the validation/certification tandem. ASTM has long used and published the validation term for the assessment of the degree to which data are generated according to standard procedures. That is the usage referred to above. Barrett's choice reflects long-standing usage by Engineering Sciences Data Unit, Ltd. (ESDU) on behalf of the British Aerospace Industry, so the conflict will not be easily resolved. However, it seems that there are two separate functions to be described. Thus the definitions above at least serve the purpose of distinguishing among them.

Also, there are several types of numeric/factual information covered by the general term "data":

- specific results of individual tests; "raw test data,"
- data sets; groups of individual test results that as a group are descriptive of some measured characteristic,
- average or typical properties/property values; the result of analysis of one or more data sets to develop performance indicators of normal or average behavior of the material, and
- design values; the result of some type of statistical analysis of a group or groups of data/data sets, certified for use in design by an appropriate group.

The evaluation process deals with these various types of data. In fact, it deals with the procedures involved in the development of several of those types, as well as with the ongoing maintenance of the accuracy and integrity of the values.

There are at least four major aspects to the evaluation processes. Two deal with the data source and two with various types of numeric data:

I. Evaluation of the data source owner/maintainer,
II. Appraisal and validation of individual raw test data, and
III. Analytical treatment of data sets to develop properties,
IV. Characterization of the data source; creation of data base descriptors,

I. Evaluation of Owner/Maintainer of Data Sources

The following factors are considered in evaluating the builders, owners, and operators of data sources, both from the viewpoints of their capability and of their service:

A. Organization, Supervision and Management

(1) experience of the organization as a whole,
(2) experience of specific personnel managing data source,

(3) specific assigned responsibilities
  (4) clarity of definition and documentation of procedures
  (5) quality control procedures, and
  (6) degree of auditability

B. *Data Base-Related Factors*
  (1) type of computer,
  (2) quality of data base management system,
  (3) compatibility of language and terminology with standards,
  (4) accessibility for multiple users,
  (5) frequency of user access,
  (6) procedures for evaluation and verification, and
  (7) hotlines and related support services for users.

II. Appraisal of Individual Raw Data Records and Datasets

Five areas of evaluation are important in judging the reliability and utility of individual data and of the complete data sets from specific test programs:

A. *Assessment of Data Source (Testing Organization)*
  (1) experience,
  (2) qualifications of personnel,
  (3) integrity,
  (4) bias,
  (5) equipment and calibration practices,
  (6) use of standards, and
  (7) general reputation.

B. *Completeness of Material Description*
  (1) producer,
  (2) heat/lot identification,
  (3) commercial/experimental status,
  (4) name/UNS number,
  (5) specifications,
  (6) condition/temper/type/grade/class,
  (7) product form,
  (8) chemical analysis,
  (9) melting practice/source material (if special, that is, vacuum cast or powder metallurgy),
  (10) process/thermal history,
  (11) microstructure, and
  (12) background properties (that is, tensile properties).

C. *Completeness of Test Method Description*
  (1) test standard,
  (2) specimen type, size, shape, finish,
  (3) specimen location, orientation,
  (4) specimen relation to end-use situation,
  (5) loading rate,
  (6) temperature; temperature measurement,
  (7) environment; monitoring method, and
  (8) precision and accuracy.

D. *Completeness of Test Data*
   (1) report format,
   (2) completeness of coverage of logical variables,
   (3) tests run to completion/according to test plan,
   (4) replications,
   (5) units/conversions documented,
   (6) consistency of results/anomalies/outliers, and
   (7) fracture type/description.

E. *Validation*
   (1) validity criteria identified,
   (2) data to satisfy criteria provided,
   (3) in the absence of validity measures, other indices of reliability, and
   (4) consistency with previous data/theoretical expectations.

III. Analysis/Derivation of Material Properties

The final set of evaluation steps cover the types of analyses that may be considered for the development of material property values, that is, average or statistically derived values indicative of material performance, from groups of individual test results:

A. *Graphical Procedures*
   (1) Trend analysis and
   (2) Interpolation/extrapolation.

B. *Statistical Procedures*
   (1) average,
   (2) minimum (state basis, for example, 95% confidence of 99% exceedance),
   (3) standard deviation, and
   (4) distribution analysis.

C. *Parametric Modeling*
   (1) time-temperature parameters,
   (2) stress-strain life parameters, and
   (3) comparison with theoretical models.

D. *Relationship to Other Data Sets/Sources*
   (1) comparison with like data sets and
   (2) combination with other data sets as appropriate.

Given the several evaluation procedures and practices above, users of individual data sources are entitled to summaries of information about those sources, including a minimum of the following information:

IV. Characterization of Data Source

A. *Type of Data*

   (1) original test data ("raw test data"),
   (2) derived data (that is, calculated), with basis,
   (3) properties derived from data sets, with basis,

(4) handbook typical value,
(5) handbook minimum values, with basis,
(6) producer data, or
(7) data extractions from the literature.

B. *Test Methods*

(1) well documented in data source,
(2) traceable via references, or
(3) unknown.

C. *Traceability of Original Data*

(1) available in data source,
(2) traceable via references, or
(3) not traceable.

D. *Degree of Evaluation*

(1) by expert group (that is, MPC, ASME),
(2) by individual expert,
(3) published in reviewed journal, or
(4) no evaluation.

E. *Periodicity and Nature of Updating*

F. *Support Index*

(1) fully supported; hot-line available,
(2) limited support, or
(3) no support available.

Obviously, these guidelines do not represent the final word on the subject. It is hoped that the broader representation that results from the ASTM consensus approach to the work of all ASTM committees will result in an improved definition of the broad categories above and the specific criteria. Several of the areas above, such as the evaluation of data bases as a whole and of data base builders and suppliers, may be beyond the scope of E 49 and of ASTM in general. However, both are quite compatible with ASTM's focuses on consumer issues and laboratory accreditation, and so may well receive some attention.

Summary

ASTM Committee E-49 on the Computerization of Materials Property Data was instituted in 1986 to provide guidance to individuals and organizations building machine-readable data bases and to aid in the development of compatible and consistent sources capable of reasonable exchange of information. In the relatively short interval since then, progress has been made in several areas, most notably in the development of formats for the characterization of materials and reporting of test data. Progress has been most rapid for more mature and stable material classes like metals and polymers. The relative newness and lack of stability of terminology and test procedures for advanced materials requires more time and effort before much progress is apparent.

ASTM Committee E-49 is also taking on the significant challenges of data exchange formats and guidelines for the evaluation of data, in the latter case building upon the work of the Technical Committee of the National Material Property Data Network.

References

[1] Kaufman, J. G., "Standardization for Materials Property Data Bases and Networking," *Standardization News*, Feb. 1986, pp. 28-33.
[2] Kaufman, J. G., "Towards Standards for Computerized Material Property Data and Intelligent Knowledge Systems," *Standardization News*, March 1987, pp. 38-43.
[3] Ambler, E., "Engineering Property Data—A National Priority," *Standardization News*, Aug. 1985, pp. 46-50.
[4] U.S. Department of Commerce, "The Economic Effects of Fracture in the United States," National Bureau of Standards Special Publications 647-1 and -2, Washington, DC, 1983.
[5] U.S. Department of Commerce, The Economic Effects of Metallic Corrosion in the United States," National Bureau of Standards Special Publications 511-1 and -2, Washington, DC, 1978.
[6] Kaufman, J. G., "The National Materials Property Data Network: Response to a Critical Need," *Chemical Engineering Data Sources*, D. A. Jankowski and T. B. Selover, Eds., AIChE Symposium Series No. 247, Vol. 82, 1986.
[7] Kaufman, J. G., "The National Materials Property Data Network, Inc.—The Technical Challenges and The Plan," *Materials Property Data—Applications and Access*, MPD-Vol. 1, PVP-Vol. 111, ASME, New York, 1986, pp. 159-163.
[8] "Materials Data Management—Approaches to a Critical National Need," National Materials Advisory Board (NMAB) Report 405, National Research Council, National Academy Press, Washington, DC, Sept. 1983.
[9] "Computerized Materials Data Systems," *Proceedings of the Workshop at Fairfield Glade, 7-11 November 1982*, J. H. Westbrook and J. Rumble, Jr., Eds., National Bureau of Standards, Washington, DC, 1983.
[10] "Material Property Data for Metals and Alloys: Status of Data Reporting, Collecting, Appraising, and Disseminating," Numeric Data Advisory Board, National Academy Press, Washington, DC, 1980.
[11] "Computerized Aerospace Materials Data," J. H. Westbrook and L. R. McCreight, Eds., *Proceedings of a Workshop on Computerized Materials Property and Design Data for the Aerospace Industry, June 1986*, AIAA, New York, Jan. 1987.
[12] Shilling, P. E., "Standards for the Materials Property Data Network," *Standards Engineering*, Vol. 38, No. 3, June 1986, pp. 55-57, 68.
[13] Krockel, H. and Steven, G., "EC Action Towards a European Pilot Network for Materials Data Information," *Materials Property Data—Applications and Access*, MPD-Vol. 1, PVP-Vol. 111, ASME, New York, 1986, pp. 175-182.
[14] "Factual Materials Databanks," J. Rumble, Ed., VAMAS Report, in press, 1988.
[15] "Materials Data Systems for Engineering," J. H. Westbrook, Ed., *Proceedings of a CODATA Workshop, Schluchsee, FRG, Sept 1985*, Fachinformationszentrum, Karlsruhe, 1986.
[16] Kaufman, J. G., "Data Validation for User Reliability," *Materials Property Data: Applications and Access*, MPD-1, PVP-Vol. 111, ASME, New York, 1986, pp. 149-158.
[17] Kaufman J. G. and Collis, S. F., "A Fracture Toughness Data Bank," *Journal of Testing and Evaluation*, March 1981, pp. 121-126.
[18] Mindlin, H., "Establishment of a Database for Environmentally Assisted Cracking," *Materials Property Data: Applications and Access*, MPD-1, PVP-Vol. 111, ASME, New York, 1986, pp. 113-120.
[19] Leyda, W. E. and Prager, M., "Data Requirements and Procedures for the Evaluation and Analysis of Stress Rupture Data," *Materials Property Data: Applications and Access*, MPD-1, PVP-Vol. 111, ASME, New York, 1986, pp. 97-104.
[20] Proceedings of a NACE/NBS Corrosion Data Workshop, Columbia, MD, Aug. 1985, in press.
[21] "Metals and Alloys in the Unified Numbering System," ASTM DS-56, SAE HSJ 1086, (published periodically), SAE, Warrendale, PA.
[22] "Aluminum Standards and Data," (published periodically), The Aluminum Association, Washington, DC.
[23] Brister, P. M., "Evaluation and Analysis of Computerized Material Property Data Bases," *Materials Property Data: Applications and Access*, MPD-1, PVP-111, ASME, New York, 1986, pp. 141-149.
[24] Barrett, A. J., "On the Evaluation and Validation of Engineering Data," Computer Handling and Dissemination of Data, P. S. Glaser, Ed., CODATA, Elsevier Science Publishers, North Holland, Amsterdam, 1987.

J. H. Westbrook[1]

Designation, Identification, and Characterization of Metals and Alloys

REFERENCE: Westbrook, J. H., "**Designation, Identification, and Characterization of Metals and Alloys,**" *Computerization and Networking of Materials Data Bases, ASTM STP 1017*, J. S. Glazman and J. R. Rumble, Jr., Eds., American Society for Testing and Materials, Philadelphia, 1989, pp. 23-42.

ABSTRACT: In contrast to the situation with polymeric and ceramic materials, alpha-numeric designation systems are relatively well established and well known for metals and alloys. Unfortunately, since such designations are mostly based on composition alone and differ between alloy groups, major problems still occur in building computer files for an information system of broad scope in materials and properties. This paper reviews existing systems of designation for metals and alloys and outlines the approach taken for characterization of materials in the MIST demonstration program being built for the National Materials Property Data Network.

Designation systems for metallic materials have been formulated by the trade associations (for example, Aluminum Association [AA], Copper Development Association [CDA], and the American Iron and Steel Institute [AISI]), by professional societies (for example, ASTM), by user groups (for example, Military and Federal), and by private companies (both users and producers). Naturally this has led to a chaotic situation with many ambiguities and uncertainties. A big step forward was taken by ASTM-Society of Automotive Engineers (SAE) who established a Unified Numbering System (UNS), which provided a logical and unambiguous code for thousands of commercially important metallic materials. Unfortunately, not all significant alloys are included; the coding relates to composition alone; and the scheme has not been endorsed outside the United States by such bodies as the International Standards Organization (ISO), or the European Standards Association.

Since in general properties are not defined by composition alone further characterization must be provided. Among the features that must be noted are form, size, method of synthesis, post-processing treatments, applications, and so forth. Illustrations of the incorporation of all these descriptors are taken from the metadata structure built for the Materials Information for Science and Technology (MIST) system.

KEY WORDS: designation, identification, characterization, unified numbering system, specification, international numbering system for metals, descriptor, validation file, metals, alloys

Acronyms

AA	Aluminum Association
AISI	American Iron and Steel Institute
AMS	Aerospace Materials Specification
ASD	Aluminum Standards and Data (publication of AA)
ASM	American Society for Metals
ASME	American Society of Mechanical Engineers
ASMH	Aerospace Structural Materials Handbook

---

[1] Principal consultant, Sci-Tech Knowledge Systems, Inc., 133 Saratoga Rd., Scotia, NY 12302.

ASTM   American Society for Testing and Materials
CDA   Copper Development Association
CEM   Consumable Electrode Melt
CODATA Committee on Data of the International Council of Scientific Unions
EB   Electron Beam (melt)
EF   Electric Furnace (melt)
ESR   Electro Slag Remelt
FED   Federal (Specification)
F_{tu} Allowable Tensile Stress (per Mil Handbook 5)
INSM   International Numbering System for Metals
ISO   International Standards Organization
JRC   Joint Research Center (of the European Economic Community)
MIL   Military (Specification)
MIST   Materials Information for Science and Technology (a project of NBS and Dept. of Energy)
NBS   National Bureau of Standards
NMPDN   National Materials Property Data Network, Inc.
OQ   Oil Quenched
Q + T   Quenched and Tempered
SAE   Society of Automotive Engineers
SPIRES A data base management system developed by Stanford University
UNS   Unified Numbering System
VAMAS   Versailles Advanced Materials and Standards
VAR   Vacuum Arc Remelt
VIM   Vacuum Induction Melt

Introduction

Broadly speaking, one consults a computerized materials data base or system of data bases for one of three purposes:

(1) to retrieve the properties data for a particular material,
(2) to find a short list of candidate materials with a specified set of properties, or
(3) to sort particular materials into groups serving some functional purpose or application area.

For all these purposes a designation or identification scheme is required, which will accomplish two things: (1) uniquely define and differentiate materials with significantly different properties and (2) allow for determination of material equivalency, thus permitting integration of independently generated test results.

The designation or identification of materials poses some particular problems in a computerized information system of broad scope and applicability. These problems derive partly from the fact that the computer's search and retrieval programs mindlessly look only for matches with character strings and partly from the fact that the user of the information, his application, and the context in which he operates cannot always be anticipated. To dig more deeply into the matter, we should examine some dictionary definitions.

- A *designation* is a distinguishing mark, title or appellation.
- An *identification* is something that identifies, that is, establishes the sameness in all that constitutes the objective reality of a thing described, claimed, or asserted.

Note the increased rigor and scope of the second definition. To put it in human terms, the bus driver in our community may be known to us only as "Joe," which suffices in that context, but to differentiate him from all the other Joe's in town, the postman must know him as "Joseph Q. Zwilsky, Jr.," and the Federal government has him in its multitudinous national records as "SS 099-12-3456," a truly unique identification.

Turning to the third term of the title, we find the following definition:

- A *characterization* is a description that makes a thing recognizable by peculiar marks, traits, or distinctive features.

Thus, in contrast to *identification,* which implies a singular code or feature, *characterization* connotes a conglomerate of features. As we shall see, both identification and characterization will be needed in a materials information system, but isolated designations outside a local context can be a source of trouble and confusion.

Present Naming Systems

To address the problem in the materials field, while alphanumeric designation systems are relatively well established and well known for metals and alloys (see appendix) (in contrast to the situation with polymeric and ceramic materials), such designations are mostly based on composition alone and differ in format and significance between alloy groups and between designating bodies. Thus confusions and ambiguities can readily arise. Consider the identical or similarly appearing designations:

356	CDA—leaded brass
356	Aluminum Association casting alloy
A356	ASTM specification for steel castings for steam turbines
A356	Aluminum Association casting alloy
A330	Sheffield Hollow Drill Steel Co., Ltd., Ni-Cr-Mo steel
3033	ASMH Code for a wrought magnesium alloy
384	AISI stainless steel grade
384	Aluminum Association casting alloy
A286	proprietary Fe-base superalloy
A285	Armco Steel Co. steel plate

This short list not only illustrates the ambiguities arising from different designating bodies (trade associations, professional societies, private companies, and government agencies), but also suggests that composition alone is not sufficient for identification; form, process, and application may also have to be considered. A not so obvious, but very common, occurrence is found in ASTM A356, which is not a single composition, but a family of several compositions commonly used for the stated purpose. To distinguish these materials, further descriptors must be added to the ASTM number.

Unified Numbering System (UNS)

To circumvent these difficulties and to achieve a single compositionally based system for all metals and alloys, ASTM and SAE jointly established a Unified Numbering System (UNS). Their first report on this system was issued in 1974, and the current 4th edition [1] lists over 3000 individual designations in 18 series. Each UNS designation consists of a single-letter prefix followed by five digits. In most cases, the letter is suggestive of the family of metals identified; for example, A for aluminum, P for precious metals, S for stainless steels. Although some of the digits in certain UNS designation groups have special assigned meanings, each series is independent of the others in regard to significance of digits, thus permitting greater flexibility and avoiding complicated and lengthy UNS designations (see Tables 1 and 2). Wherever feasible, and for user convenience, identification "numbers" from existing systems are incorporated into

TABLE 1—*Primary series of numbers for the UNS system.*

UNS Series	Metal
Nonferrous metals and alloys	
A00001–A99999	aluminum and aluminum alloys
C00001–C99999	copper and copper alloys
E00001–E99999	rare earth and rare earth-like metals and alloys (18 Items, see Table 2)
L00001–L99999	low melting metals and alloys (14 Items, see Table 2)
M00001–M99999	miscellaneous nonferrous metals and alloys (12 Items, see Table 2)
N00001–N99999	nickel and nickel alloys
P00001–P99999	precious metals and alloys (8 Items, see Table 2)
R00001–R99999	reactive and refractory metals and alloys (14 Items, see Table 2)
Z00001–Z99999	zinc and zinc alloys
Ferrous metals and alloys	
D00001–D99999	specified mechanical properties steels
F00001–F99999	cast irons
G00001–G99999	AISI and SAE carbon and alloy steels (except tool steels)
H00001–H99999	AISI H-steels
J00001–J99999	cast steels (except tool steels)
K00001–K99999	miscellaneous steels and ferrous alloys
S00001–S99999	heat and corrosion resistant (stainless) steels
T00001–T99999	tool steels
Welding filler metals	
W00001–W99999	welding filler metals, covered and tubular electrodes, classified by weld deposit composition (see Table 2)

the UNS designations. For example, carbon steel presently identified by The American Iron and Steel Institute as "AISI 1020" is covered by "UNS G10200."

It must be emphasized that the UNS system is based on composition *alone*. Thus a UNS designation is not in itself a unique identifier, since it establishes no requirements for form, condition, properties, or quality. It is a unified designation for a metal or an alloy for which controlling limits have been established in specifications published elsewhere.

Particular problems attach to the fact that one-to-one relationships do not in general exist between UNS numbers and those of other designation systems. The UNS reference work, which incorporates eleven different cross-referencing indices, illustrates this difficulty whether we go from a UNS Number to other designations or vice versa. Table 3 shows a multiplicity of referenced specifications for a single UNS number.

Conversely, a single code from one designating body, for example, ASTM may have several UNS numbers (Table 4). Finally, while the UNS system is large and growing, it by no means covers all alloys of commercial interest. For example, from a list of 172 relatively common and important alloys chosen for inclusion in the MIST demonstration system [2], 5% were found to have no assigned UNS number.

Specification Numbers

If the designations of manufacturers and trade associations are ambiguous and duplicative and the UNS system inadequately specific, why not use specification numbers in a computer-

TABLE 2—*Secondary division of some series of numbers for the UNS system.*

UNS Series	Metal
E00001–E99999 rare earth and rare earthlike metals and alloys	
E00000–E00999	actinium
E01000–E20999	cerium
E21000–E45999	mixed rare earths[a]
E46000–E47999	dysprosium
E48000–E49999	erbium
E50000–E51999	europium
E52000–E55999	gadolinium
E56000–E57999	holmium
E58000–E67999	lanthanum
E68000–E68999	lutetium
E69000–E73999	neodymium
E74000–E77999	praseodymium
E78000–E78999	promethium
E79000–E82999	samarium
E83000–E84999	scandium
E85000–E86999	terbium
E87000–E87999	thulium
E88000–E89999	ytterbium
E90000–E99999	yttrium
F00001–F99999 cast irons	gray, malleable, pearlitic malleable, and ductile (nodular) cast irons
K00001–K99999 miscellaneous steels and ferrous alloys	
L00001–L99999 low-melting metals and alloys	
L00001–L00999	bismuth
L01001–L01999	cadmium
L02001–L02999	cesium
L03001–L03999	gallium
L04001–L04999	indium
L50001–L59999	lead
L06001–L06999	lithium
L07001–L07999	mercury
L08001–L08999	potassium
L09001–L09999	rubidium
L10001–L10999	selenium
L11001–L11999	sodium
L12001–L12999	thallium
L13001–L13999	tin
M00001–M99999 miscellaneous nonferrous metals and alloys	
M00001–M00999	antimony
M01001–M01999	arsenic
M02001–M02999	barium
M03001–M03999	calcium
M04001–M04999	germanium
M05001–M05999	plutonium
M06001–M06999	strontium
M07001–M07999	tellurium
M08001–M08999	uranium
M10001–M19999	magnesium
M20001–M29999	manganese
M30001–M39999	silicon
P00001–P99999 precious metals and alloys	
P00001–P00999	gold

TABLE 2—Continued.

UNS Series	Metal
P01001-P01999	indium
P02001-P02999	osmium
P03001-P03999	palladium
P04001-P04999	platinum
P05001-P05999	rhodium
P06001-P06999	ruthenium
P07001-P07999	silver
R00001-R99999 reactive and refractory metals and alloys	
R01001-R01999	boron
R02001-R02999	hafnium
R03001-R03999	molybdenum
R04001-R04999	niobium (columbium)
R05001-R05999	tantalum
R06001-R06999	thorium
R07001-R07999	tungsten
R08001-R08999	vanadium
R10001-R19999	beryllium
R20001-R29999	chromium
R30001-R39999	cobalt
R40001-R49999	rhenium
R50001-R59999	titanium
R60001-R69999	zirconium
W00001-W99999 welding filler metals, classified by weld deposit composition	
W00001-W09999	carbon steel with no significant alloying elements
W10000-W19999	manganese-molybdenum low alloy steels
W20000-W29999	nickel low alloy steels
W30000-W39999	austenitic stainless steels
W40000-W49999	ferritic stainless steels
W50000-W59999	chromium low alloy steels
W60000-W69999	copper base alloys
W70000-W79999	surfacing alloys
W80000-W89999	nickel base alloys
Z00001-Z99999 zinc and zinc alloys	zinc

[a] Alloys in which the rare earths are used in the ratio of their natural occurrence (that is, unseparated rare earths). In this mixture, cerium is the most abundant of the rare earth elements.

TABLE 3—*Examples of multiplicity of referenced specifications for one UNS number. Material = UNS N06600, the Ni-Cr-base alloy, Inconel 600.*

Organization	Specification Numbers
AMS	5540, 5580, 5665, 5687, 7232
ASME	SB163, SB166, SB167, SB168, SB564
ASTM	B163, B166, B167, B168, B366, B516, B517, B564
FED	QQ-W-390
MIL	MIL-R-5031 (Cl 8), MIL-T-23227, MIL-N-23228, MIL-N-23229

TABLE 4—*Examples of single ASTM code having several UNS numbers. Material = ferritic stainless steel tubing.*

ASTM Specification	UNS Numbers
A268	S40800
A268	S41500
A268	S44735
A268	S44400
A268 (25-4-4)	S44635
A268 (26-3-3)	S44660
A268 (29-4)	S44700
A268 (29-4-2)	S44800
A268 (329)	S32900
A268 (405)	S40500
A268 (409)	S40900
A268 (410)	S41000
A268 (429)	S42900
A268 (430)	S43000
A268 (430 Ti)	S43036
A268 (439)	S43035
A268 (443)	S44300
A268 (446)	S44600
A268 (XM-27)	S44625 ob
A268 (XM-27)	S44627
A268 (XM-33)	S44626

ized materials information system? To consider this possibility, review the definition of the term, *specification* [3]:

> a document intended primarily for use in procurement which clearly and accurately describes the essential technical requirements for items, materials, or services, including the procedures by which it will be determined that the requirements have been met.

Here the operative or key word is "essential," for the most effective specification is that which accomplishes the desired result with the fewest requirements. Thus some specifications will be found limited to composition alone, while others will couple composition with requirements for form, melting practice, application, property levels, and so forth in various combinations. Also, as was noted above, some specifications embrace a number of different explicit compositions. Table 5 provides examples of this practice by considering just a few of the many specification numbers for AISI 4340, a common alloy steel.

Many specification numbers are made more explicit by coupling a grade or class to the basic specification designation; but others, without such an extension, simply connote a broad class of material meeting some particular property requirements thereby leaving the producer freedom to adjust composition, heat treatment, and so forth, to meet those requirements as he will. Thus, while specification designations are a useful adjunct or correlative reference for materials procurement or application in a computerized materials information system, they are not an adequate identification.

TABLE 5—Some specifications for 4340 Ni-Cr-Mo steel.

Issuing Body	Specification Designation	Form	Forming Process	Melting Practice	Heat Treatment	Property Requirement	Quality	Application
SAE	AMS 6414	bar, forging, tubing	...	CEM	...	...	...	...
	AMS 6415	bar, forging, tubing	...	air	...	...	...	...
	AMS 5330	casting	investment	air	...	...	...	...
	AMS 5331	casting	sand	air	...	...	...	...
ASTM	A290	forging, billet	...	air	...	defined by specification Class G thru L	...	reduction gear ring
	A547	wire	...	...	...	...	cold heading	...
	A646 Cl-I	forging, billet	...	VIM, CEM, EB, or ESR	...	...	premium aircraft quality	...
	Cl-II	forging, billet	...	EF vacuum degas	...	...	aircraft quality	...
	A711	forging, billet	...	air	...	...	...	bolts
	A732(10Q)	casting	investment	air	Q+T	180/145/5	...	...
Military	MIL-S-8844 Cl 1	bar, forging, tubing	...	CEM	...	...	...	...
	MIL-S-5000	bar, forging, tubing	...	air	...	...	...	...
	MIL-S-22141	casting	investment	air	...	...	...	...

International Numbering System for Metals

The international standards community has also been concerned with devising an International Numbering System for Metals (INSM). Discussions and proposals over a period of nearly ten years have evolved into a draft technical report [4], which is currently under review by the appropriate Technical Committees of the International Standards Organization (ISO). Briefly, this proposal envisions a system that can accommodate existing or future designations standardized at international, regional, or national levels. It further provides that the procedure for allocation of particular designations shall be under the jurisdiction of the appropriate ISO Technical Committee (for example, TC26 for copper based materials, TC17 for steels, and so forth). An eight position alpha-numeric code is proposed as shown in Fig. 1. The status block of Positions 1 and 2 indicates by code whether the standard is international, regional, or national and by what organization. The metal class of Position 3 indicates a broad metal group, for example, aluminum and its alloys, or gold and other precious metals. The remaining five positions constitute a designation block to be used for a code assigned by the appropriate technical committee. It is to be the responsibility of the technical committee to decide whether any significance is to be attributed to any of the characters in these positions and whether any further classification is required. The purpose of the "floating" alpha character in this block is to distinguish such parameters as form, process, special properties, and so forth. Typical examples (fictitious) of the application of this schema would be as shown in Table 6.

While an international metal numbering system is surely a desideratum to be striven toward, the present INSM proposal is defective in several respects:

- It does not accommodate the UNS system.
- The same material can have different INSM numbers.
- The number assignments are by independently acting, autonomous Technical Committees.
- The schema may not fully identify the material in certain instances, that is, materials with the same INSM number might have quite different combinations of properties.
- Most of the coding is nonsignificant and therefore not readily recognizable.

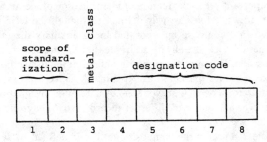

Positions 1, 2 and 3 to be alpha characters

Positions 4 through 8 to be numeric characters except for one alpha whose position and significance are to be determined by the specific Technical Committee.

FIG. 1—*Schema for ISO's proposed International Numbering System for Metals (INSM).*

TABLE 6—*Typical examples (fictitious) of the application of the INSM schema.*

INSM	Scope	Organization	Material	Organizational Code
XAA1234M	international	ISO	wrought Al alloy	ISO AlCu2Mg
XGJ99N99	regional	Pan American	cast iron	COPANT 828FG350
DESU6666	national	German	steel	DIN 1715

Current Developments

In 1984 the National Bureau of Standards and the U.S. Department of Energy initiated a cooperative program to build a prototype network system to demonstrate the power and utility of computer access to materials data. This project, Materials Information for Science and Technology (MIST) has been described in recent reports [2,5] and by McCarthy (in this publication, pp. 135-150). With the establishment in 1985 of the industry-sponsored consortium, the National Materials Properties Data Network (NMPDN), the data structures and support files developed in the R&D program of MIST were essentially adopted for broad U.S. use by NMPDN. At the present stage of development the NMPDN consists of a "network" of three independent data bases resident on a single machine. Plans are underway to add other data bases, some of which would be resident on other machines [6].

Recognizing the difficulties in materials designation and identification outlined above, the builders of the MIST demonstration system have taken a three-fold approach to metals identification:

- Composition—UNS Number (existing or requested) supplemented by an explicit description of composition.
- Key descriptors (form, processing, condition, etc.)—individual fields (in name = value format, descriptor i = value j); several usually required to differentiate materials.
- Auxiliary descriptors (source of manufacture, heat or lot number, date of production, analyzed composition, specification to which it conforms, and so forth)—nondefining, but useful for differentiating data sets, and so forth.

Composition is an important yet tricky feature of identification of metals and alloys in that it must be neither too broadly nor too narrowly defined. Although the UNS Number implies a compositional definition that could be incorporated in an auxiliary file in the computerized system, callable on demand, it will usually be desirable that the material record itself include a specific field for composition. Beyond the compositional definition itself, this field should characterize the compositional values presented as "nominal," "specification," or "analysis." If this be done, then a search for a particular type steel by a nominal designation, say "18Cr, 9Ni, $C \leq 0.08\%$" could recover a number of different records even though no exact matching of compositions is possible.

As an example, Table 7 presents the compositional limits defined by several different national standards for 304 stainless steel equivalents, a hypothesized heat analysis, and a nominal compositional description. All records containing the compositions shown in Table 7 would be retrieved by the search selection shown, although none is an exact match for the specified composition of UNS S30400. Table 7 shows that the individual national standards differ as to which elements are specified, whether a minimum exists as well as a maximum, and in the particular numeric values called for. Note that this approach permits retrieval of records for closely related materials but leaves it for the user to judge their equivalency. Experience may have shown that apparently slight variations in composition affect weldability for instance or corrosion behavior in some particular environment, hence, the value of an explicit presentation of compositional limits or analyses as an aid in materials selection. An alternate strategy is for the data base builder to select a number of aliases for a given reference composition (UNS Number), for example, the various national specification codes shown in Table 7. While this has the advantage of a more compact identification record, it substitutes the judgment of the data base builder for that of the (possibly more experienced and knowledgeable) user.

The key descriptors category also requires more explication. The more important of these descriptors include the following:

- *Form*—the geometric shape in which the material is produced.
- *Processing*—a particular process applied to the material:

    (1) synthesis processes, for example, melting, casting, powder metallurgy, and so forth.
    (2) post-synthesis processes, for example, working, joining, heat-treatment, and so forth.
    (3) surface treatment processes—Alclad, centerless ground, plated, anodized, carburized, and so forth.

- *Condition*—the metallurgical state imparted by a specific process.
- *Size*—the dimension or dimensions that critically affect properties.
- *Application Requirement*—material whose purity, property level or tightness of specification meet some defined application, for example, "aircraft quality," "bolting," "pressure vessel," and so forth.
- *Product*—the end product for which the material is manufactured, for example, for aluminum sheet: can stock, venetian blind stock, roofing, and so forth.
- *Microstructure*—feature or features of the microstructure that affect properties, for example, grain size, primary constituents, inclusion count, and so forth.
- *Property*—a property used as a controlling or defining parameter that effectively sets many other properties.

The exhibits of Figs. 2 through 7 and Tables 8 and 9, summarized in Table 10, illustrate the fact that adoption of different values of these key descriptors, singly or in combination, just as effectively define new materials as does choice of a different composition. It will further be noted that some properties will be very sensitive to the value of a certain descriptor, while others are unaffected. For the data collector and computer systems designer this poses special problems, for the result is that the (material [compositionally defined]—property—descriptor) matrix is usually very sparsely filled with values. Consideration must be given in the case of a missing descriptor as to whether it is indeterminate but inconsequential, or whether an appropriate default value must be assigned.

Each descriptor term and any other material identification code must also be provided with a list of accepted aliases if the computer system is to accommodate data sets of different origin or serve users with different backgrounds and mind-sets. Similarly the computer system must be able to cope with the collectives used in many of these categories, for example, "all wrought forms," "12 Cr steels," "flat products," "vacuum melted," and so forth. To do this, definitions must be carefully constructed, and overlaps and categorized hierarchies noted.

Again difficulties arise when an attempt is made to include data for widely different materials, from diverse sources, and to serve various applications within a single system. For example, terms, such as "product," have different definitions in different metal industries. Some would follow a definition similar to that given above; others mean a combination of application and form, for example, "hydraulic tubing," or of process and form, for example, "extruded bar." These semantic complexities must be meticulously sorted out and provided for if the system is to perform the retrieval function accurately and exhaustively.

Form	Size Section in.	Temper	Tensile Strength ksi	Yield Strength (.5% Ext. under Load) ksi	Yield Strength (.2% Offset) ksi	Elongation in 2 in. %	Rockwell Hardness F	Rockwell Hardness B	Rockwell Hardness 30T	Shear Strength ksi	Fatigue Strength ksi	Million Cycles
FLAT PRODUCTS	.040 in.	Light Anneal	62.0	30.0	....	40	–	60	57	41.0	....	....
		Quarter Hard	70.0	58.0	....	17	–	75	68	43.0	....	....
	.250 in.	Soft Anneal	58.0	25.0	....	49	–	56	55	40.0	....	....
		Light Anneal	60.0	28.0	....	45	–	58	56	41.0	....	....
	1.0 in.	As Hot Rolled	55.0	25.0	....	50	–	55	55	40.0	....	....
ROD	.250 in.	Soft Anneal	58.0	27.0	....	45	–	56	–	40.0	....	....
		Light Anneal	63.0	30.0	....	40	–	60	–	42.0	....	....
		Quarter Hard (10%)	70.0	48.0	....	25	–	80	–	43.0	....	....
		Half Hard (20%)	80.0	57.0	....	20	–	85	–	45.0	....	....
	1.0 in.	Soft Anneal	57.0	25.0	....	47	–	55	–	40.0	....	....
		Light Anneal	63.0	30.0	....	40	–	60	–	42.0	....	....
		Quarter Hard (8%)	69.0	46.0	....	27	–	78	–	43.0	....	....
		Half Hard (20%)	75.0	53.0	....	20	–	82	–	44.0	....	....
	2.0 in.	Soft Anneal	56.0	25.0	....	47	–	55	–	40.0	....	....
		Light Anneal	62.0	28.0	....	43	–	60	–	42.0	....	....
		Quarter Hard (8%)	67.0	40.0	....	35	–	75	–	43.0	....	....
TUBE	.375 in. OD X .097 in.	Hard Drawn (35%)	88.0	66.0	....	18	–	95	–	....	....	....

FIG. 2—*Effects of form, size, and temper on mechanical properties of wrought naval brass, CDA Alloys Nos. 464, 465, 466, and 467.*

Material			Impact energy	
			Mean value $\bar{x}$ in Joule	Dispersion $3\sigma_{rel} = 3\dfrac{s}{\sqrt{5'}\cdot\bar{x}}$ in %
Available RM, ISO-V-notch samples		Results after Literature	94	4.0
		Tests in Institute A	95.1	6.0
		Tests in Institute B	85.1	6.2
Normal air melted steel (N)		longitudinal specimen	118	9.1
		transverse specimen	21.8	9.0
Vacuum arc remelted steel (VAR)		longitudinal specimen	105	9.6
		transverse specimen	65.4	7.9
Electroslag remelted steel (ESR)	Steelworks A	longitudinal specimen	110	4.7
		transverse specimen	80.1	2.4
	Steelworks B	longitudinal specimen	110	6.2

FIG. 3—*Influence of steel melting practice on the impact properties of 4340 steel intended as a standard reference material (RM) as reported by the J.R.C.-Ispra.*

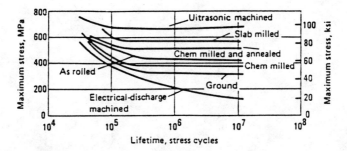

FIG. 4—*Effect of surface treatment processes on fatigue behavior of annealed titanium alloy Ti-5Al-2.5Sn (UNS R54520).*

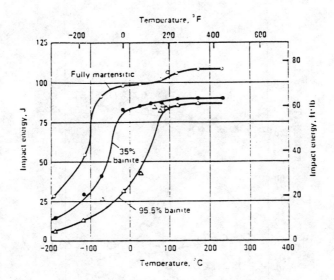

FIG. 5—*Effect of microstructure on notch toughness of 4340 steel (fully martensitic and partially bainitic microstructures were tempered to the same strength level, 150 ksi).*

Conclusion

Existing designation schemes for metals and alloys are inaccurate, ambiguous, and confusing, especially when combined. Modern computer-based materials information system demand a unique identification code coupled with a multi-faceted characterization to fully differentiate materials.

As the problems outlined in this paper have become more widely recognized, cooperative activities have been initiated, both nationally and internationally. The efforts of ASTM Committee E-49 on Computerization of Material Property Data, the European Demonstrator Program, VAMAS, and CODATA have been described in this symposium. The International Standards Organization, materials and standards bodies from the Far East and the various trade associations in the metals field need also to be involved in those standardization, definition, and

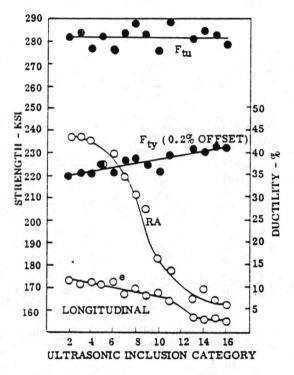

FIG. 6—*Effect of material cleanliness, as measured by inclusion incidence, on the tensile properties of 4340 steel produced from open hearth, vacuum degassed, electric furnace vacuum degassed, and vacuum arc remelt metal tempered to Ftu = 280 ksi.*

Alloy (For specification see Table 2.3.1.0(a_1) and (a_2))]	Hy-Tuf	4330 Si 4330 V	D6AC 4335 V	AISI 4340[a] D6AC	AISI 4340	98BV40[a]	300M
Form		All wrought forms			Bar, forging, tubing		
Condition		Quenched and tempered[b]			Quenched and tempered[b]		
Thickness or dia., in.		See Table 2.3.0.1(a)			See Table 2.3.0.1(a)		
Basis	S	S	S	S	S	S	S
Mechanical properties:							
F_{tu}, ksi	220	220	220	260	260[e]	280	280[c]
F_{ty}, ksi	180	186	190	215	215[e]	230	230[e]
F_{cy}, ksi	188	194	198	240	240	...	247
F_{su}, ksi	132	132	132	156	156	168	168
F_{bru}, ksi:							
(e/D = 1.5)	297	297	297	347	347	...	430[f]
(e/D = 2.0)	385	385	385	440	440	...	525[f]
F_{bry}, ksi:							
(e/D = 1.5)	260	269	274	309	309	...	360[f]
(e/D = 2.0)	286	296	302	343	343	...	396[f]
e, percent:							
L	...	...	...	10	...	...	...
LT	5[d]	5[d]	5[d]	3	5[e]	5	5[e]
E, 10^3 ksi				29.0			
E_c, 10^3 ksi				29.0			
G, 10^3 ksi				11.0			
μ				0.32			
Physical properties:							
ω, lb/in.^{3}				0.283			
C, K, and α				See Figure 2.3.1.0			

[a] Air melted steel only.
[b] Values in these columns are applicable only to steels for which the indicated F_{tu} has been substantiated through adequate quality-control inspection testing.
[c] The use of heat treatments of F_{tu} = 220 ksi or higher is subject to the specific approval of the procuring or certificating agency.
[d] Sheet 0.040 to 0.059 in., incl., 4; 0.020 to 0.039 in., incl., 2. See Table 2.3.1.0(f) for elongation applicable to consumable-electrode vacuum-melted D6AC (billets) and 4330V (all products forms).
[e] Values are for transverse direction.
[f] Bearing values are "dry pin" values per Section 1.4.7.1.

FIG. 7—*Design mechanical properties of low-alloy steels when tempered to stated levels of Ftu.*

TABLE 7—Compositional records relating to 304 stainless steels.

Record Type	Designation		Composition, % Maximum or Range						
	Country	Code	C	Mn	P	S	Si	Cr	Ni
S	USA	UNS S30400 or AISI 304	0.08	2.00	0.045	0.030	1.00	18.00-20.00	8.00-10.50
P E	France	AFNOR Z6CN18-09	0.07	2.00	0.045	0.030	1.00	17.00-19.00	8.00-11.00
C I	Italy	UNI X5CrNi18-10	0.06	2.00	0.045	0.030	1.00	17.00-19.00	8.00-11.00
F I	Japan	JIS G4303-27 -SUS27B	0.08	2.00	...	...	1.00	18.00-20.00	8.00-11.00
C A	Sweden	SIS 14-2332	0.07	2.00	0.045	0.030	1.00	17.00-19.00	8.00-11.00
T I	UK	BS 970-304S15	0.06	0.50-2.00	0.045	0.030	0.20-1.00	17.50-19.00	8.00-11.00
O N	USSR	GOST O8X18H10	0.08	2.00	0.035	0.020	0.80	17.00-19.00	9.00-11.00[a]
	West Germany	Stoff Nr. 1.4301	0.07	2.00	0.045	0.030	1.00	17.00-20.00	8.50-10.00
Analysis Nominal	heat no. 1234 search selection		0.05[b] 0.08[c]	1.50[b]	0.032[b]	0.025[b]	0.6[b]	18.2[b] 18[d]	8.7[b] 9[d]

[a] Also Ti ≤ 0.50, Cu ≤ 0.30.
[b] Actual values.
[c] Maximum values prescribed for search.
[d] Nominal values prescribed for search.

TABLE 8—*Example of same alloy and form (3003 Al tube) made by different processes for different applications.*[a]

Forming Process	Application	Available Tempers	ASTM Specifications	UNS Number
Extruded	unspecified	O, H112	B241	A 93003
Extruded, coiled	"	H112	B491	"
Drawn	"	O, H12, H14, H16 H18, H25, H113	B210, B483	"
Drawn	heat exchanger or condenser	H14, H25	B234	"
Welded	unspecified	O, H12, H14, H16 H18, H112	B313, B547	"
Unspecified	rigid electrical conduit	H12	...	"

[a]The different products and forming processes carry different dimensions, tolerances, property limits, and surface requirements as shown in both ASD [9] and the cited specifications.

TABLE 9—*Typical features of nominally 0.30C, hot-rolled steel bar and plate produced to various quality grades or application requirements.*

Quality Grade	TS, ksi	YS, ksi	El, % in 2 in.	Remarks	Typical ASTM Specifications
Merchant quality	50 to 60	25	30	looser chemicals; bendability requirement	A663
Structural quality	58 to 80	32 to 36	23	tighter chemical specifications;	A36
Reinforcing bar quality	80	50	5 to 7	inspection; internal soundness	A616
Pressure vessel quality	55 to 75	30	27	inclusion rating and other requirements	A285C
"	75 to 95	40	19	...	A299
"	70 to 90	35	15	...	A455I

TABLE 10—*Illustrative examples of key descriptors.*

Key Descriptor	Illustrative Exhibits	Reference Source
Form	Fig. 2	*CDA Standards Handbook* [7]
Synthesis process	Fig. 3	Helms and Ledworuski [8]
Post-synthesis process	Table 8	*Aluminum Standards and Data* [9]
Surface treatment	Fig. 4	*ASM Metals Handbook*, Vol. 8 [10]
Condition	Fig. 2	*CDA Standards Handbook* [7]
Size	Fig. 2	*CDA Standards Handbook* [7]
Application Requirement	Table 9	ASTM specifications as noted
Product	Table 8	*Aluminum Standards and Data* [9]
Microstructure	Fig. 5	*ASM Metals Handbook*, Vol. 1 [10]
	Fig. 6	*Structural Alloys Handbook* [11]
Property	Fig. 7	*Mil Handbook-5D* [12]

NOTE: The computer system must, in general, contain a set of validation files for the values acceptable for each of the descriptors. These values may be all-alpha strings, alpha-numeric codes or extended alpha-numeric short-hand descriptions. Illustrations of each of these from the heat treatment field are as follows:

- all alpha—annealed,
- alpha/numeric code—H38 (temper), and
- alpha/numeric shorthand description—aust 1525°F (829°C) 1/2 h, OQ + temper 2 h at 600°F (316°C).

cross-referencing activities, which can aid in metals identification for computerized systems. The activities of the MIST group and of the European Demonstrator program, as their work goes forward encompassing an ever broader scope of materials and data sources, are also valuable in identifying new problem areas of ambiguity and conflict in metals designation and identification.

APPENDIX

Listed below are some of the major reference works to metallic materials designation and identification. Not listed, but also useful, are the numerous lists of standards and specifications provided by the various trade associations and national, regional and international standards organizations. Reference *13* gives names and addresses of these bodies.

Brady, G. B. and Clauser, H. R., *1985 Materials Handbook: An Encyclopedia for Purchasing Managers, Engineers, Executives, and Foremen*, 12th ed., McGraw-Hill, New York.

Woldman's Engineering Alloys, R. C. Gibbons, Ed. American Society for Metals, Metals Park, OH, 1979, 1815 pp., 6th ed. (ISBN 0871-70086-7).

Ross, R. B., *Metallic Materials Specification Handbook*, London: E&FN Spon Ltd., 1980, 793 pp.

Simmons, W. F. and Gunia, R. B., *Compilation and Index of Trade Names, Specifications, and Producers of Stainless Alloys and Superalloys*, ASTM Publication DS45A, American Society for Testing and Materials, Philadelphia, 1972, 57 pp.

Kehler, W., *Handbook of International Alloy Compositions and Designations*, Vol. 3, ISBN 0-9911-1-6-6, Aluminum Verlag, West Germany, Heyden, 1981, 859 p.

Wahl, M. J., et al., *Handbook of Superalloys*, International Alloy, Compositions and Designations Series, ISBN 0-387-90657-6, Springer-Verlag, 1982, 258 pp.

Simmons, W. F. and Metzger, M. C., *Compilation of Chemical Compositions and Rupture Strengths of Super Strength Alloys*, ASTM Data Series DS 9e, American Society for Testing and Materials, Philadelphia, 1970.

Hucek, H. and Wahl, M., *Handbook of International Alloy Compositions and Designations*, Vol. 1, Titanium, 1976. MCIC-HB-09 (Vol. 1) Metals and Ceramics Information Center, Battelle Memorial Institute, Columbus, OH.

Unterweiser, P. M., Ed., *Worldwide Guide to Equivalent Irons and Steels*, 2nd ed., 1987, 575 pp.; and *Worldwide Guide to Equivalent Nonferrous Metals and Alloys*, 2nd ed., 1987, American Society for Metals, Metals Park, OH, 626 pp.

Hufnagel, W., Ed., *Aluminum Manual—Aluminum Materials, Standards, Compositions and Trade Marks*, ISBN 3-87017-170-7, Aluminum Vgl. Dusseldorf, F.R.G., 1983, 142 pp.

Metals and Alloys in the Unified Numbering System, 4th ed., American Society for Testing and Materials, Philadelphia, 1986, 371 pp.

World Steel Standard Specifications, English Translation from the Chinese, Foreign Technology Division, U.S. Air Force FTD-NC-23-856-74, 1974; originally published as *World Steel Designations Handbook*, Machine Industries Press, PRC, 1970.

British and Foreign Specifications for Steel Castings, Steel Castings Research and Trade Association, Sheffield, United Kingdom, 1980, 238 pp.

Index of Aerospace Materials Specifications, SAE, Warrendale, PA, 1986, 224 pp.

Wegst, C. W., *Stahlschlüssel (Key to Steel)*, Verlag Stahlschlüssel, Marbach/Neckar, FRG, 1986, 570 pp.

Arcuri, J. V. and Potts, D. L., Eds., *International Metallic Materials Cross-Reference*, 2nd ed., General Electric Co., Schenectady, NY, 1983.

British Steel Corporation, *European Committee for Iron and Steel Specifications (ECISS)*, Index of Standards PSP2, British Steel Corporation, Croydon, United Kingdom, June 1986.

Sheet Metal Industries Yearbook, "Identification of Stainless Steels," Sheet Metal Industries, Redhill, United Kingdom, 1984, pp. 71–83.

British Steel Corporation, *Iron and Steel Specifications*, 6th ed., British Steel Corporation, Croydon, United Kingdom, 1986, 335 pp.

World Standards Speedy Finder, Vol. 4: Materials, The International Technical Information Institute of Japan, Tokyo.

Standards Cross-Reference List, 2nd ed., MTS Systems Corp., Minneapolis, MN, 1977.

Department of Defense Index of Specifications and Standards, Part 1, Alphabetical Listing, Part 2, Numerical Listing, available from the Superintendent of Documents, U.S. Government Printing Office, Washington, DC 24202.

Simons, E. N., *A Dictionary of Alloys*, Hart, New York, 1970.

World Metal Index, Sheffield City Libraries, Central Library, Surrey Street, Sheffield, Yorkshire S1 1XZ United Kingdom, Telephone 0742 734714, Telex 54243, a service based on an index of over 70 000 grades of ferrous and nonferrous materials, a large collection of metallurgical and engineering journals and literature, all British Standard (BS) and U.S. standards and most European and National Standards.

"How Do I Find Out About. . . ?" No. 1 Identification of Steels; No. 3 Identification of Nonferrous Metals," Welding Institute Research Bulletin 25, 1984.

References

[1] "Metals and Alloys in the Unified Numbering System," SAE and ASTM, 4th ed., April 1986, 371 pp.

[2] Grattidge, W., Westbrook, J. H., McCarthy, J., Northrup, C. J. M., and Rumble, J. R., *Materials Information for Science and Technology (MIST): Project Overview*, NBS Special Publication 726, Department of Commerce, 1986.

[3] Promisel, N. E. et al., *Materials and Process Specifications and Standards*, NMAB Report-33, Washington, DC, 1977.

[4] "International Numbering System for Metals (INSM) International Standards Organization Report, ISO-DTR 7003, May 1986, 14 pp.

[5] Northrup, C. J. M., Jr., McCarthy, J. L., Westbrook, J. H., and Grattidge, W., "Materials Information for Science and Technology (MIST): A Prototype Materials Properties Data Network System," from *Materials Properties Data-Applications and Access*, ASME, PV&P and Computer Engineering Division, 1986, p. 167.

[6] Kaufman, J. G., "The National Materials Property Data Network, Inc., The Technical Challenges

and the Plan," from *Materials Properties Data—Applications and Access*, ASME, PV&P, and Computer Engineering Division, 1986, p. 159.

[7] *Standards Handbook, Wrought Copper and Copper Alloy Mill Products, Part 2—Alloy Data*, Copper Development Association, 1983.

[8] Helms, R. and Ledworuski, S., "Reference Material for Proof Tests of Methods and Apparatus for Notched-Bar Impact Tests," *Production and Use of Reference Materials*, B. F. Schmitt, Ed., BAM, Berlin, 1980, p. 247.

[9] *Aluminum Standards and Data*, 8th ed., The Aluminum Association, 1984.

[10] *ASM Handbook*, 9th ed., Vols. 1 and 8, American Society for Metals, Metals Park, OH, 1985.

[11] *Structural Alloys Handbook*, MCIC, Battelle-Columbus Laboratories, 505 King Avenue, Columbus, OH 43201-2693, 1986; three volumes updated periodically.

[12] *Metallic Materials and Elements for Aerospace Vehicle Structures*, MIL-HDBK-5D, 1986, updated periodically, v.p. 2 vol., available from the Superintendent of Documents, U.S. Government Printing Office, Washington, DC 10401.

[13] Westbrook, J. H., "Materials Standards and Specifications," *Encyclopedia of Chemical Technology*, 3rd ed., Kirk-Othmer, Wiley, New York, 1981, 32 pp.

Keith W. Reynard[1]

VAMAS Activities on Materials Data Banks

REFERENCE: Reynard, K. W., "**VAMAS Activities on Materials Data Banks,**" *Computerization and Networking of Materials Data Bases, ASTM STP 1017*, J. S. Glazman and J. R. Rumble, Jr., Eds., American Society for Testing and Materials, Philadelphia, 1989, pp. 43-52.

ABSTRACT: The Versailles Project for Advanced Materials and Standards (VAMAS) Task Group on Materials Data Banks has surveyed the VAMAS countries to identify areas where standards are needed to build and distribute factual materials data banks. It presented its report to the Steering Committee in May 1987.

This paper reviews the findings and recommendations of the Task Group. These have a particular relevance to the work of ASTM Committee E-49, as the input to the report related all the recommendations to the likely end result of standards for materials data and other information being stored and accessed by computer.

Many of the topics have been debated in the Workshops of Fairfield Glade, Petten, and Schluchsee in recent years. The recommendations take account of the work that has been done in several countries following these discussions. It is therefore no surprise that topics such as systems for the designation of materials, terminology, multilingual dictionaries and definitions, computerized directories for testing standards and the equivalence of materials, and methods of linking data bases and their access are given a high priority. What is different is the direct relationship between the recommendations and those producing standards, codes, and guidelines.

KEY WORDS: CEC, CODATA, computerization, data banks, designations, European Community, materials data, standards, terminology, Versailles Project for Advanced Materials and Standards (VAMAS)

What is VAMAS?

International Groups and Associations

There are many international associations of countries put together for a variety of reasons and working with various degrees of formality and levels of funding. Some associations exist to advance trade, pure science or defense, to protect the environment, or to provide aid in times of famine or catastrophe.

There are three notable groups that have expanded the range of their activities into the field of materials data. CODATA (the Commission on Data for Science and Technology of ICSU, the International Council of Scientific Unions) has some claim as the first, VAMAS (the Versailles Project on Advanced Materials and Standards) the second, and the CEC (Commission of the European Communities) the third, though by virtue of the links between the people concerned, there has been almost concurrent development.

---

[1] Principal, Wilkinson Consultancy Services, Stable Cottage, Broad Lane, Newgate, Surrey RH5 5AT, United Kingdom.

Origins of VAMAS

For many years, the heads of state and government of the seven summit countries (Canada, France, Germany, Italy, Japan, the United Kingdom, and the United States) and the CEC have met to discuss and agree on policies and programs to further international cooperative projects developing the economic, industrial, and trading health of the group.

In June 1982 at Versailles, a Working Group on Technology, Growth, and Employment was set up to prepare proposals to exploit the immense opportunities presented by the new technologies and "to remove barriers to, and promote, the development of and trade in new technologies." The recommendations for collaboration on science and technology included proposals regarding "food technology," "high speed trains," and advanced materials and standards (with Dr. E. D. Hondros, then superintendent of the Division of Materials Applications at the National Physical Laboratory or NPL and Dr. R. B. Nicholson, then chief scientist to Her Majesty's Government as the proponents). About 18 possible topics were suggested.

The significance of materials to the economies of the member countries was seen also to have the recommendation that much of the work would be of a precompetitive nature, stimulating trade through the generation of work that would provide an agreed technical basis for the generation of standards by others. Thus was born VAMAS.

There are now thirteen technical working areas in a wide range of subjects at various stages of development, some very new, others having completed their initial remits and published results.

VAMAS Technical Working Areas

The thirteen areas now active are as follows:

(1) wear test methods,
(2) surface chemical analysis,
(3) ceramics,
(4) polymer blends,
(5) polymer composites,
(6) superconducting and cryogenic structural materials,
(7) bioengineering materials,
(8) hot salt corrosion resistance,
(9) weld characteristics,
(10) materials data banks,
(11) creep crack growth,
(12) efficient test procedures for polymer properties, and
(13) low-cycle fatigue.

Outline reports on the progress of these areas are published in January and July of each year in the VAMAS Bulletin [2]. The Bulletin also lists the 19 national members of the Steering Committee, whose present chairman is Dr. Lyle Schwartz of NBS, who succeeded Dr. Ernest Hondros in May 1986 when Dr. Bruce Steiner of NBS also succeeded Dr. Tom Barry of NPL as secretary.

This paper will deal only with the working area materials data banks, listing it here with the others to show the whole range of VAMAS topics. The main activities of the other areas are developing test methods, preparing specimens, standard materials, and procedures, often by round-robin testing.

People

Each country (Table 1) is represented on the Steering Committee by two or three members. If a country wishes to participate in a particular working area, it then appoints its own national coordinator responsible for the input from and the involvement of others in that country. This means that many people and organizations are now participating in some aspect of the total project.

Funding

VAMAS has no central funds. All the work is done either as a voluntary activity within or derived from existing programs in the many organizations, or with the support of national funding. Limited government funding has been made available in several countries, but the amounts made available vary considerably and are often provided only in response to specific proposals in a particular working area or as an enabling fund, rather than covering the total cost of the work.

How Does VAMAS Fit into the International Scene?

The Pieces of the Jigsaw

Anyone outside the many materials activities might think that there is great danger of duplication of effort, of gaps not being attended to, or of a lack of coordination between the international activities. This could hamper technical progress and the fulfillment of industrial needs.

Fortunately, if this is so, it is not by reason of lack of communication between the several organizations. If there are limitations on the rate of progress in the area of materials data banks, they arise from a lack of funds or of experts with time. This is particularly so when participation in the work committed is taken on in addition to normal activities. In some instances, there is also a lack of awareness or response and participation from industry.

Other problems have not arisen because of the many informal connections between the participants. These are rarely visible to the outsider, who may be forgiven for never having heard of the work of VAMAS, CODATA, CEC, and even ASTM in the field of materials data banks, particularly if he or she works in a country not active in one or more of the groups.

There is a quite remarkable lack of information in the professional press concerning the activities of these organizations. Their own official journals provide most of the information and these do not have a very broad circulation in industry.

It can be seen from Table 1 that only the United Kingdom, France, and West Germany are at present members of all three groups. That VAMAS and the EC are closed groups does not present barriers to communication with others on general matters. There may be restrictions in the participation in their activities and in the sharing of results (for example, those from round-robin testing). However, the report on materials data banks will be an open publication.

EC—CEC

It is important to note the difference between the EC (European Community), comprising twelve countries with nine languages, and the CEC (Commission of the European Communities). The European Economic Community, the European Community for Coal and Steel, and EURATOM are based on contracts between member countries. The CEC is the joint and permanent "civil service" and administration for all three. Representatives from all twelve coun-

TABLE 1—*Group member countries.*

VAMAS	EC/CEC	CODATA
7 countries + CEC	12 countries	18 countries
Closed group	closed group	Open group
Limited national funding	contract funding	Limited funding for travel
		Australia
	Belgium	Brazil
Canada		Canada
CEC		China
	Denmark	
	Eire	
France	France	France
Federal Republic of Germany	Federal Republic of Germany	Federal Republic of Germany
		German Democratic Republic
	Greece	Hungary
		India
		Israel
Italy	Italy	Italy
Japan		Japan
	Luxembourg	
	Netherlands	Poland
	Portugal	
	Spain	South Africa
		Sweden
United Kingdom	United Kingdom	United Kingdom
United States		United States
		U.S.S.R.

tries are sent to the CEC in a fixed ratio and important documents are published in all nine official languages.

The CEC is divided into several Directorate Generals of which DG XIII, Telecommunications, Information Industries, and Innovation, is responsible for developing and administering the five-year action plan that includes materials data banks. This action plan commenced soon after the CEC Workshop at Petten 14–16 November 1984 [3]. It will be described in detail in the paper by Krockel and Steven in this volume. The CEC program differs from that of VAMAS and CODATA in having central funds allocated to accomplish the five-year plan.

The growing use of EC, the European Community, rather than EEC, recognizes the broadening purpose and scope of activities of the community as a whole.

CODATA

CODATA is concerned for the most part with data activities in pure and applied science covering a very wide range of subjects. Understandably, from its aims and origins, only a relatively small part of its work can be directly applied by engineers.

CODATA's interests in materials data stemmed from the activities of its Commission on Industrial Data, whose chairman is Dr. J. H. Westbrook. The growing interest in and general awareness of present needs in this area were fed by the workshops at Fairfield Glade 7–11 November 1982 [4], in which CODATA was one supporting organization, and the workshop CODATA held at Schluchsee 22–27 September 1985 [5]. The paper by Barrett in this volume will provide an up-to-date description of the CODATA activities related to materials data banks.

VAMAS Technical Working Area 10: Materials Data Banks

Purpose

The objective of the task group for this technical working area was to survey all standardization aspects related to materials data systems. The task group's mission was to prepare a consensus report presenting as comprehensively as possible the need for standardization related to the computerization of material property data. This included the standards for data-related information (materials, designations, properties, testing methods, reporting), data bases, networks, computer communications, access, applications, and any others having an impact on this field. The report was to be completed within 12 months, and was approved for publication by the VAMAS Steering Committee at its meeting in Pisa on 13-15 May 1987.

Method

The cochairmen of the Task Group, H. Kröckel of the CEC, JRC (Joint Research Centre) Petten and J. R. Rumble Jr. of the NBS, with K. W. Reynard, divided the collection of information and opinions and the writing of the first draft among them. Other members of the task group provided input. Each drew upon their own experience and that of those within their own countries and the international organizations with whom they had contacts.

The complete first draft was circulated to all members of the task group and to many others for comment and criticism before being rewritten for presentation to the steering committee for their consideration. Over 40 people made direct contributions or provided comments, and the knowledge and ideas of many others contributed to the final report.

Specific proposals for further work were not submitted at that time, though some possible topics arising from the recommendations were discussed and commented on at the steering committee meeting. In retrospect, this seems to have been a mistake, as there will be almost a twelve-month gap in the work in this area. However, a period for reflection and development should produce better defined and more acceptable proposals.

The Report and Recommendations

The section that follows is based on the executive summary of the report [6] as presented to the steering committee. Passages in quotes are taken from the final draft as presented to the steering committee and should differ only slightly from the published report.

To those active within VAMAS, ASTM, CODATA, and the CEC, some of the recommendations may seem rather obvious and so basic as to be almost axioms. The report will, it is hoped, pass into the hands of a much wider audience, many of whom are not familiar with what has gone before. So rightly, it seeks to present the fullest picture to those newcomers.

"The primary purpose of the VAMAS Task Group on Materials Data Banks was to identify the areas where standards development would significantly and positively impact the building and dissemination of materials data banks. Materials data banks are becoming substantial elements of the computerized flow of information on materials properties from the generation of this information to its use. Their features and operational conditions will therefore be influenced by all standards related to any phase of this information flow."

"Accordingly, the review includes the aspects of data generation, data analysis, data presentation, access to data, and the use of data; and the report presents sets of recommendations under four headings:

- basic considerations for handling data,
- material data generation and reporting,

- materials data banks, and
- access to materials data.

"They are summarized here in a condensed form and include the indication of the types of organizations to which they are addressed:

A—national standards bodies,
B—international standards bodies,
C—engineering standards bodies,
D—prestandardization activity bodies,
E—(prototype) materials data systems,
F—Versailles Project on Advanced Materials and Standards—VAMAS, and
G—Committee on Data of the International Council of Scientific Unions—CODATA."

Basic Considerations for Handling Materials Data

"Basic data handling standards impact the whole of the materials data flow, as they are related to general issues in the areas of materials designation and nomenclature systems, the conjunctive categories of terms, concepts, definitions, symbols, and abbreviations, multilingual terminology, and the system of units."

Recommendations

"Existing systems for designating engineering materials should be fully cataloged. Systems for a given material type should be brought into equivalency or developed into a more comprehensive form for selection as an international standard (A and B)."

"For material types where no materials designation system has been developed, one should be. The first step should be information gathering sessions on proprietary, informal, manufacturer, or other systems that may exist (A)."

"Each of the separate national standards activities needs to have its definitions related to materials property data listed and scrutinized to produce an agreed multilingual list of equivalent definitions (A, F, and G)."

"VAMAS should encourage the cooperation of individuals, organizations, and governments in Europe with the CEC effort to build MAT-TERM 1 (a project to produce a multilingual dictionary of terms from the Demonstrator Data Banks) and provide the links with ASTM E49.03 and the appropriate Japanese project (F)."

"Standards for creating indexes, thesauri, and related metadata guides are needed to characterize individual materials so that standard data dictionaries can be constructed (A and B)."

"The use of ISO 31 [7] is encouraged as the source of a standard set of symbols, and a special set of symbols should be developed for terminal screens and computer printers noting ISO 2955.1983 [8] (A, B, and C)."

"The use of SI units is strongly recommended in all phases of the work on materials data banks (A, C, and E)."

Much effort is being expended on terminology in information technology. To the users it sometimes seems to be a tedious and nit-picking exercise. To those who suffer financially as a result of misunderstanding, all the effort is worthwhile. The persistence of differing designation systems for specific materials is parochial and also understandable. In the long term, this situation is unjustified; in the short term, it is seemingly impossible to make reasonable progress.

Standards for Materials Data Generation and Reporting

"Recognizing that the definitions of engineering materials properties are largely conditioned by the procedures and practices that have been codified as standard tests, the standards for test methods must be considered. Computerization introduces an extension of the required standards covering conversion of test results to properties, comprehensive reporting of results (taking into account machine-readability), and the full and adequate characterization of the materials."

Recommendations

"A computerized directory of national and international test method standards should be established to identify the equivalences and differences between them for each material group and property (F)."

"A full listing of necessary metadata should be part of every materials testing standard. The lists should be standardized internationally (A and B)."

"Standards for derived data obtained by data analysis should be included in each test method standard (A)."

"To facilitate the computerization of data from publications, standards for the presentation of factual, numerical, and tabular data in the literature should be developed in parallel with data reporting standards (A and B)."

With the exception of the first recommendation in this group, there seems to be more likelihood of an international consensus, if only because virtually nothing exists or has to be replaced.

Standards for Materials Data Banks

"Standards for this subject do not cover data banks themselves, as these are rapidly evolving and should not be restricted prematurely. Instead, the standards concern the environment of interaction and user interfaces. Many standards are already available in the general area of data banks and only the requirements that are specific for materials data systems are taken up here."

Recommendations

"Metadata profiles and standard data items should be defined for specific applications for which data banks are being designed. For example, data banks for the evaluation of test results require data items describing test method, material, specimen, test condition, actual data, and data source (D)."

"When a data bank derives its metadata from an established design code (for example, American Society of Mechanical Engineers (ASME) Boiler and Pressure Vessel Code), the metadata scope and output screens should be defined with the respective application, and in important cases, adopted as a standard (C)."

"Metadata profiles, file formats, and input and output formats should allow conversion to and from an international "Materials Data Interchange Format" (A, B, and C)."

"Existing standard principles and standard elements for the format and syntax of query language should be used wherever possible (E)."

"A review should be made of the immediate need for more specific computer software standards related to building materials data banks. In particular, data-bank portability and modularity are the areas of highest concern (A and B)."

"The functional specification of a general material data-bank log-on module including disconnection recovery, security provisions, user guidance, and systems messages should be standardized (A and B)."

"Guidelines or standards for the access administration module should be developed so that user information on accounting units, time, cost, and so forth can be harmonized (A and B)."

"Guidelines for a user-friendly, harmonized design of data collection forms should be established (A and D)."

Some of these aspects are already being addressed by activities in the United States and the CEC. In particular, the first Code of Practice has now been published by the CEC as a result of the first phase of the Material Data-Banks Demonstrator Programme [9]. This Code also deals with some aspects of the next section.

Standards for Data Access

"Access to materials data is considered under the specific condition of access to computerized data systems, which has three standardization aspects:

1. standards related to data,
2. standards related to computer-human communication, and
3. standards related to computer-computer communication.

"The central issues for the material field are a suitable materials data interchange format and the features of distributed data banks (networks). The highly important issue of intelligent and user-friendly user interfaces is deliberately only briefly included in this standards review."

Recommendations

"Standards and codes of practice related to distributed data-bank systems must be in place before such systems can be built for materials data. Groups developing materials data standards should maintain close contact with computer standards development groups such as ISO TC 97 (information processing systems) and the CEC Materials Data-Banks Demonstrator Programme (A and B)."

"Groups working on standards for the computerization of materials data must maintain close working relationships with groups developing standards for graphics, computer-aided design, computer modeling, and computer-aided manufacturing (A and C)."

"Groups working on standards for the computerization of materials data must maintain close working relationships with groups working on standards for graphics, computer-aided design, computer modeling, and computer-aided manufacturing (A and C)."

"Standards must be developed for the Materials Data Interchange Format (MDIF) to allow for compatibility with other networking standards such as those being developed under the Open Systems Interconnection (OSI) model [10] (A and B)."

"Standards should be developed for well-established, verified, and widely accepted techniques for modeling materials behavior or for statistical methods applicable to materials test data to support computer-assisted engineering (A, C, and G)."

In the above, the word "standards" refers to the whole range of practical tools that can be enlisted in addition to formal or official standards. Codes of practice, guidelines, and statements of today's best practice all can assist in the achievement of better compatibility in this dynamic field. Each has its place depending upon the state of development of the particular part of the subject and the rate of change that is taking place. There is only one thing worse than the lack of standards and that is the existence of standards that inhibit progress and the exchange of information and create artificial and unnecessary barriers. Protectionism and the promotion of minority views should have no place in this work.

What Next?

The Task Group now has three tasks. First, it must ensure that the recommendations are transmitted to the appropriate organizations. Second, it must monitor progress in their implementation. Third, it has to formulate practical and acceptable proposals based on the recommendations of the report and present them in time for consideration at the next steering committee meeting in Brussels at the end of February, 1988.

There is little doubt that in addition to anything done under the VAMAS banner, other projects are likely to be started as a result of the first report of the group.

What is Acceptable Prestandardization?

There are some difficulties in this area in determining what are acceptable VAMAS proposals. These are due to rather different national attitudes to what is meant by "prestandardization." In the United Kingdom, it would not be thought to be an infringement on the work of the British Standards Institution (BSI) if some person or organization were to present something as advanced as a draft of a standard for their consideration. Such a document might in due course be taken up by the appropriate BSI Committee, who would take it through the whole rigorous process to the production and publication of a standard. Indeed, in several instances, consultants have been employed by BSI to produce a first draft. However, they do not assume that the committee will leave that draft unchanged either in technical content or in its form of presentation for the potential users of the standard.

This view of what is acceptable prestandardization activity is not held equally in other countries, nor do all other standards' bodies adopt the same method of producing standards. In the United Kingdom, BSI (that is, the organization with all its 3000 committees) is the sole maker of standards and so may feel able to operate in this way when other less unique organizations may not.

For this reason, the steering committee felt that a proposal related to the coordination of terminology was probably not something that VAMAS could do effectively and was beyond the scope of "prestandards" activity VAMAS had set for itself. The working group may still address this topic and produce an acceptable proposal, though it may be more appropriate for the CEC or the individual standards bodies to initiate the work.

Conclusion

The needs of the user in industry are paramount. They must come above those of the research worker, the computer expert, or the data-base builder or manager if the true value and contribution that these three groups make to the materials information scene is to be recognized and supported. Nevertheless, in several aspects of materials data work, scientific research and the pursuit of knowledge is important in itself. Sadly, reaction against this second component of overall development has gone too far in some places and the balance has been lost. In the development of materials data banks, there is always an end user to be considered. Therefore, it should be a lesser problem in this instance than in some others to involve the engineer, the user of the materials, in all stages of development.

References

[1] "Proposal for Multilateral Collaborative Research. Advanced Materials and Standards," Summit Working Group on Technology, Growth and Employment, National Physical Laboratory, Teddington, United Kingdom, 1983, p. 17.
[2] *VAMAS Bulletin,* Numbers 1-7, National Physical Laboratory, Teddington, United Kingdom, Jan. 1985, Jan., July 1986-1988.

[3] Kröckel, H., Reynard, K. W., and Steven, G., Eds., "Factual Material Databanks," Proceedings of a CEC workshop held at JRC Petten 14–16 Nov. 1984, published by CEC, Luxembourg, ISBN 92-825-5322-1, p. 178.

[4] Westbrook, J. H. and Rumble, J. R., Eds., "Computerised Materials Data Systems," Proceedings of a CODATA workshop held at Fairfield Glade, TN, 7–11 Nov. 1982, published by NBS, Gaithersburg, MD, p. 133.

[5] Westbrook, J. H., Behrens, H., Dathe, G. and Iwata, S., Eds., "Materials Data Systems for Engineering," Proceedings of a CODATA workshop held at Schluchsee, Federal Republic of Germany, 22–27 Sept. 1985, ISBN 3-88127-100-7, published by FIZ, Karlsruhe, FRG, p. 189.

[6] Kröckel H., Reynard, K. W., and Rumble, J. R., Eds., "Factual Materials Data Banks, the Need for Standards," the Report of a VAMAS Task Group, National Bureau of Standards, Gaithersburg, MD, July 1987.

[7] "ISO 31/0 to 31/13," TC 12 Quantities, units, symbols, conversion factors, and conversion tables, International Standards Organization, Geneva, Switzerland, 2nd Ed. (some parts), 1978–1981.

[8] "ISO 2955 Information Processing—Representation of SI and Other Units in Systems with Limited Character Sets," TC 97 information processing systems, International Standards Organisation, Geneva, Switzerland, 2nd Edition, 1983, p. 5.

[9] "Code of Practice for Use in the Material Data-Banks Demonstrator Programme," prepared by Matsel Systems Limited for the Commission of the European Communities. Document No. XIII/MDP(MAT-02)-OS-03, CEC, Luxembourg, Nov. 1986, p. 12.

[10] "ISO 7498 Information Processing Systems—Open Systems Interconnection-Basic Reference Model," TC 97 Information processing systems, International Standards Organization, Geneva, Switzerland, 1984, p. 40.

National and International Materials Data Base Activities

J. Gilbert Kaufman[1]

The National Materials Property Data Network, Inc.—A Cooperative National Approach to Reliable Performance Data

REFERENCE: Kaufman, J. G., "**The National Materials Property Data Network, Inc.—A Cooperative Approach to Reliable Performance Data,**" *Computerization and Networking of Materials Data Bases, ASTM STP 1017*, J. S. Glazman and J. R. Rumble, Jr., Eds., American Society for Testing and Materials, Philadelphia, 1989, pp. 55-62.

ABSTRACT: In response to the critical need identified by a number of expert studies, the National Materials Property Data Network, Inc. was established to provide engineers and scientists with easy on-line access to worldwide sources of reliable, well-documented, numeric/factual material property data. A pilot MPD Network from Stanford University, is now on-line. It is based upon the MIST technology developed jointly by the National Bureau of Standards (NBS) and the Department of Energy (DOE). The status of that pilot system and future plans for the MPD Network are described herein, along with a number of other data base activities that are complementary to or being developed cooperatively with the Network.

KEY WORDS: materials property data (MPD) network, data base, network, materials, properties

A Cohesive National Approach

The motivation for a cohesive national policy optimizing our ability to share data is particularly compelling at the present time [1-4]. Many of our traditional sources of data on engineering materials, such as producers and government laboratories, are either cutting back their activities or eliminating them completely. It is more important than ever that we make every effort to share data and, therefore, to store and handle it in a compatible fashion and to make as much information readily accessible on-line as possible.

Additionally, the development of sources for retaining numeric data in well-documented and machine-readable form greatly increases the value of our research dollars, as the individual data may be thus kept "alive" and accessible for ready comparison with newly developing materials research. The mechanism will also exist to resurrect "lost" data from earlier programs, the results of many of which are now resident in untraceable company or laboratory files or in the drawers of the original researchers, never to be seen again.

A concerted effort to incorporate the individual results of the more valuable research programs in searchable electronic sources, can substantially reduce the amount of new testing required to establish performance comparisons for competitive materials being considered for new applications. We can also accelerate the commercialization of new, advanced materials by assuring full assessment and analysis of all available data in minimum time.

This is not to say that there are no longer concerns about proprietary data and a lack of

---

[1] President, National Materials Property Data Network, Inc., 2540 Olentangy River Rd., P.O. Box 02224, Columbus, OH 43202.

willingness to share certain sensitive types of data; it would be unrealistic to suggest that is the case. At least two types of such concerns remain: one of private industry, unwilling to share proprietary data with its competitors when it judges those data to be part of their competitive edge; the other of government agencies concerned that the key data will fall into the hands of other governments that will use the information in unfriendly actions or make it available to others who may do so. These are real industrial and geopolitical/strategic issues and they will not change.

Nevertheless, there are tremendous volumes of numeric/factual materials data outside those sensitive areas. There is also increasing recognition of the great economic justification for being able to store and access these data in the most technically advanced and compatible fashions. The National Materials Property Data Network, Inc., known more widely as the MPD Network, is currently the major effort to build a cohesive national system.

The National Materials Property Data Network, Inc. (MPD Network)

The MPD Network is a not-for-profit corporation with the mission of providing engineers and scientists with easy access to high quality, well-documented numeric or factual materials performance data [5,6]. It was established to fill the critical need for materials selection and design decisions based upon more accurate and reliable performance information [2,3]. This need was identified in several expert studies, including one by the U.S. National Materials Advisory Board [4,7-9].

The basic model selected for the MPD Network is illustrated in Fig. 1. Users are linked with multiple sources of data via a gateway which includes: (1) a user interface to facilitate searching; (2) a metadata system to deal with the varying nomenclature, terminology, abbreviations, and other data about the data; (3) a neutral data base interface that deals with the different languages, software, and hardware of the individual, potentially geographically distributed sources; and (4) response options providing the user with some choice in the form in which the data are down-loaded. The individual sources are well-focused data bases built and managed by experts in their respective fields.

The unique features of the MPD Network include the following:

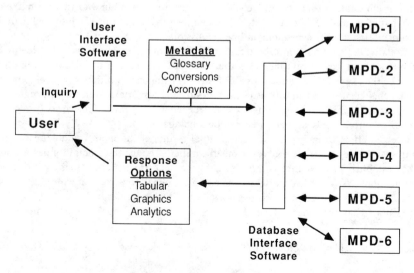

FIG. 1—*Schematic of model selected for MPD network.*

- online access to multiple sources of numeric/factual data,
- distributed sources,
- evaluated data wherever practical, and
- a metadata system.

The last feature, the "metadata" system, is a key to the general utility of the Network [10-12] in dealing with compatibility issues in numeric material property data. These include the different nomenclatures used to describe the materials, such as the diversity of terms considered acceptable for many material properties and the variations in units and abbreviations. The metadata system is a file that incorporates all of the technically acceptable material names and property terms. It is accessed upon user inquiry so that all data responsive to the inquiry are provided regardless of how they are stored in the system. Similarly, upon request, the "standard" names or terms and all acceptable aliases for any specific material or property are provided.

The assumption is that users will have their own internal data bases, just as they have their own hard-copy resources today. But they will periodically want to go on-line to update their internal source to cover new materials or new performance conditions. Therefore, it is vital that the gateway be extremely easy for occasional users to access and use in searching for information. Considerable attention is being given to that aspect of the prototype system; in fact, a new user should be able to negotiate the pilot system without reference to any written materials.

The Cooperative Aspects of the Approach

It is the intent of the MPD Network to work in a cooperative and compatible manner with industry, universities, technical societies, and government agencies and laboratories to achieve a cohesive system of accessible and interchangeable sources of performance data. When other organizations are building or maintaining data sources or both, the role of the MPD Network is to provide another avenue of access to those sources without restricting their distribution by other means in any way. The MPD Network will not build new data bases in competition with those built by other organizations, but rather, fill gaps and provide data base building competence where voids are identified. In doing so, the network will regularly contract that function to other well-qualified contributors, and does not maintain a staff for data base building functions.

The MPD Network also publicizes the data sources and data base building capabilities of other organizations, both in its MPD Newsletter and in the directory of information on other sources being made an integral part of the Network resources. Users receive guidance on sources of machine-readable data other than those that appear on the Network. They learn the content and scope of data included and how to access those other sources.

The MPD Network is developing a number of international links. It intends to continue overseas data base and networking activities wherever practical. In particular, contact is maintained with the CEC European Network Demonstrator Project [13,14]. It expects that it will eventually be possible to establish a link to that system and provide U.S. access to additional overseas sources of data. Similarly, the MPD Network is an active participant in CODATA, the international committee on technical data [15,16], and is cooperating in the development of guidelines for the quality and reliability of data and data base management systems and operations as well.

Pilot MPD Network

Two specific recent aspects of MPD Network development are worth noting. The first is the pilot system is now on-line to financial sponsors of the MPD Network. It is undergoing extensive and upgrades evaluation of its data base management system, user interface, and search strategies. The pilot network is the result of a cooperative venture with the National Bureau of Standards' Standard Reference Data Program (NBS/SRD) and the Department of Energy (DOE). The development of the pilot network is known as the MIST program (for Materials Information for Science and Technology) [17]. The role of MPD is to commercialize MIST technology while DOE uses it in compatible form within their extensive laboratory system.

The MPD/MIST system is based upon SPIRES, a general-purpose data base management system developed and supported by Stanford University. It runs in more than 30 installations in addition to Stanford's. Major contributions to system design were made by Lawrence Berkeley Laboratories [18,19] and Sci-Tech Knowledge Systems, Inc. [12,20].

SPIRES was chosen for this project because of its unique combination of capabilities for numeric as well as textual data, its proven ability to handle large numbers of simultaneous users, its flexibility and extensibility, and its history of support for innovative applications. SPIRES is considered a mature, fourth-generation data base management system (DBMS) with full data management capabilities. These include schema definition and control; report generation; full-screen form design; on-line documentation; interactive executive language; comprehensive security mechanisms for value level, data integrity, and recovery; and an extensive library of standard processing procedures for input validation, encoding, automatic indexing, query translation, and output formatting.

The pilot MPD Network runs on an IBM 3084 under STS at Stanford's Information Technology Services (ITS). Users access the system via Telenet or commercial telephone lines. Most connections are at 1200 baud, although 2400 baud is also supported, and as 9600 baud will be in the future.

In relation to the model in Fig. 1, the pilot MPD Network may be depicted as in Fig. 2. The solid-lined portions of the system are relatively well-developed, while those components shown in dashed lines have not yet received much attention (reporting options) or are still in development (the three new data sources).

There are at least four significant hurdles to be dealt with in the model MPD Network discussed above:

(1) multiple data bases of differing types of data,
(2) different material/property terminology,

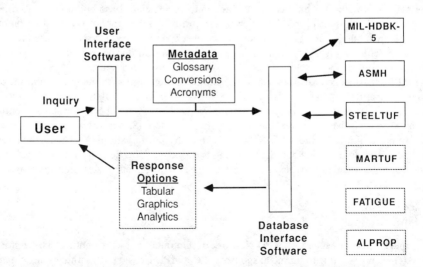

FIG. 2—*Schematic of pilot MPD network operating from Stanford University and based upon SPIRES.*

(3) different geographic locations, and
(4) different data base management systems (DBMS).

It is important to note that the existing experimental pilot network deals only with the first two of these, as a common host and common DBMS (SPIRES) was used. Differing geographic locations will not be too severe a hurdle if common DBMS are used. However, linking different DBMS is a major challenge. It remains to be seen how the economics of this problem compare to those of transferring all sources to a single DBMS.

At the present time, the pilot MPD Network provides access to three sources: (a) sections of MIL-HDBK-5, the design manual of the U.S. Aerospace industry [21], (b) a few sections of ASMH, the Aerospace Structural Metals Handbook [22], and (c) STEELTUF, a data base of more than 20,000 individual toughness tests of more than 50 steels. Three more data bases are being developed for possible addition to the MPD/MIST pilot systems, depending upon its performance in the current evaluation phase.

The user is offered four search strategies in the pilot MPD Network, including the option to search by data base, by material, or by property, and an "expert" mode wherein the experienced user may enter all of the desired search query delimiters beforehand and be led directly to the responsive data. The first three paths are heavily menu guided, while the latter assumes full knowledge of system breadth by the user.

Also included are directories to other sources of on-line data not yet on the MPD Network, and other sources of machine-readable data not yet on-line. In both cases guidance on how to access them is provided. Finally, a mechanism for direct feedback by the user to MPD Network management is provided in every access.

Linkage with STN International

The second significant development for the National Materials Property Data Network, Inc. is the recent agreement with the Chemical Abstracts Service (CAS) of the American Chemical Society (ACS) to move toward distribution of the commercial version of the Network on STN International [13]. STN International is the on-line scientific and technical information network operated cooperatively by ACS, Fachinformationszentrum Energie, Physik, Matematik Gmbh (FIZ, Karlsruhe) and the Japan Information Center of Science and Technology (JISCT), which already provides links among U.S., European, and Japanese sources. Through this affiliation, the MPD Network will move closer to its mission of providing access to worldwide sources of numeric material property data.

Also a not-for-profit service, STN International exists to promote the advancement of the scientific community worldwide by improving the ease and quality of the exchange of technical data. It was founded in 1983, and has concentrated primarily upon chemical and chemical engineering data. The U.S. technical leadership is provided by CAS, one of the world's principal centers for chemical and material science information since 1907.

ASTM Committee E-49 on Computerization of Materials Property Data

The MPD Network participates actively in the deliberations of ASTM Committee E-49 [23,24] in the development of standard classifications, guides, practices, and terminologies for building and accessing materials property data bases. Based upon the work of E 49, the MPD Network is able to use and promote guidelines to aid those individuals and organizations building data bases or intelligent knowledge systems. The network ensures that high levels of quality and reliability are maintained and that compatibility with other sources is assured. The latter is essential if we are able to readily and cost-effectively share and compare data across organization lines.

It is appropriate to note the many important standardization activities by other groups such as American National Standards Institute (ANSI), Institute of Electrical and Electronics Engineers (IEEE), International Standards Organization (ISO), and others; some are summarized in Ref 25. However, ASTM Committee E-49 is the first to focus upon the relatively complex world of numeric performance data for engineering materials.

Related/Complementary Data Base Activities

The following areas of activity by U.S. and overseas technical societies also make numeric materials property data more accessible:

The National Association of Corrosion Engineers (NACE)

NACE is building a Corrosion Data Center with NBS. Some personal computers (PC) data disks are now available on materials compatibility, and support is being sought for a much broader program with data available on disk and eventually on-line via the MPD Network.

The American Society of Metals (ASM) and the Institute of Metals (IOM)

These groups build and operate a metals data file on-line; data screened from the literature are cited. Also, ASM produces PC data disks for a variety of materials. These are available for use with ASM's MetSe12 (a derivative of MetalSelector) and EnPlot software. In addition, ASM is building a machine-readable source of alloy phase diagrams at the NBS.

The American Society of Mechanical Engineers (ASME)

With a number of other organizations, ASME is building a Computerized Tribology Information System (ACTIS) at the NBS; data will be available on disks and, eventually, on the MPD Network.

The American Welding Society (AWS) and American Welding Institute (AWI)

These organizations are building a group of expert systems and data sources covering various aspects of weld procedures, weld qualification, and the performance properties of welds. Some disks are now available. They expect to make a finished product available on the MPD Network.

The American Ceramic Society (ACerS)

The ACerS publishes Ceramic Source, a summary of numeric data for a wide range of ceramics. It plans to build a machine-readable version to be made accessible on-line.

There are also a number of private industry activities on numeric data bases. These include those by PLASPEC, CENCAD, ESDU (London), and Engineering Information (MATUS, London). We apologize for any that may have been overlooked. In addition, groups are developing a number of machine-readable numeric data bases as part of proprietary industrial or government programs or both intending limited access. These too have been omitted from discussion.

Summary

The explosion of interest in computerized materials property data bases has resulted in two significant developments in the United States—the incorporation of the National Materials

Property Data Network, Inc. (MPD Network), and the formation of ASTM Committee E 49 on the Computerization of Materials Property Data. The MPD Network provides engineers and scientists with easy on-line access to worldwide sources of reliable, well-documented, materials property data. A pilot MPD Network has been established at Stanford University based upon the MIST technology developed by NBS and DOE. It is now available on-line to financial supporters of the network. Agreement has been reached with ACS's Chemical Abstracts Service to host the MPD Network on STN International in the future.

References

[1] Ambler, E., "Engineering Property Data—A National Priority" *Standardization News,* ASTM, Philadelphia, Aug. 1985, pp. 46-50.
[2] U.S. Department of Commerce, "The Economic Effects of Fracture in the United States," National Bureau of Standards Special Publication 647-1 and 2, Washington, DC, 1983.
[3] U.S. Department of Commerce, "The Economic Effects of Metallic Corrosion in the United States," National Bureau of Standards Special Publications 511-1 and -2, Washington, DC, 1978.
[4] "Materials Data Management—Approaches to a Critical National Need," National Materials Advisory Board (NMAB) Report 405, National Research Council, National Academy Press, Washington, DC, Sept. 1983.
[5] Kaufman, J. G., "The National Materials Property Data Network: Response to a Critical Need," *Chemical Engineering Data Sources,* D. A. Jankowski and T. B. Selover, Eds., AIChE Symposium Series No. 247, Vol. 82, 1986.
[6] Kaufman, J. G., "The National Materials Property Data Network, Inc.—the Technical Challenges and the Plan," *Materials Property Data—Applications and Access,* MPD-Vol. 1, PVP-Vol. 111, American Society of Mechanical Engineers, New York, 1986, pp. 159-163.
[7] "Computerized Materials Data Systems," *Proceedings of the Workshop at Fairfield Glade,* 7-11 November 1982, J. H. Westbrook and J. Rumble, Jr., Eds., National Bureau of Standards, Washington, DC, 1983.
[8] "Material Property Data for Metals and Alloys: Status of Data Reporting, Collecting, Appraising, and Disseminating," Numeric Data Advisory Board, National Academy Press, Washington, DC, 1980.
[9] Westbrook, J. H., and McCreight, L. R., Eds., "Computerized Aerospace Materials Data," *Proceedings of a Workshop on Computerized Materials Property and Design Data for the Aerospace Industry,* June 1986, AIAA, New York, Jan. 1987.
[10] Westbrook, J. H., *Standards and Metadata Requirements for Computerization of Selected Mechanical Properties of Metallic Materials,"* NBS Special Publication 702, U.S. Department of Commerce, Washington, DC, Aug. 1985.
[11] Westbrook, J. H., "Some Considerations in the Design of Properties Files for a Computerized Materials Information System," *Proceedings of the Ninth International CODATA Conference,* North-Holland, Amsterdam, 1985.
[12] Westbrook, J. H., this publication, pp. 23-42.
[13] Krockel, H., and Steven, G., "EC Action Towards a European Pilot Network for Materials Data Information," *Materials Property Data—Applications and Access,* MPD-Vol. 1, PVP-Vol. 111, ASME, New York, 1986, pp. 175-182.
[14] Krockel, H. and Steven, G., this publication, pp. 63-74.
[15] Westbrook, J. H., Behrens, H., Dathe, G., and Iwato, S., Eds., "Materials Data Systems for Engineering," *Proceedings of a CODATA Workshop, Schluchsee, FRG,* Sept. 1985, Fachinformationszentrum, Karlsruhe, 1986.
[16] Glaeser, P., Ed., *The Role of Data in Scientific Progress,* Proceedings of the Ninth International CODATA Conference, North-Holland, Amsterdam, 1985.
[17] Grattidge, W., Westbrook, J., Northrup, C., Jr., McCarthy, J., and Rumble, J., Jr., "Materials Information for Science and Technology (MIST): Project Overview," NBS Special Publication 726, U.S. Department of Commerce, Washington, DC, Nov. 1986.
[18] McCarthy, J., this publication, pp. 135-150.
[19] Northrup, C. J. M., Jr., McCarthy, J. L., Westbrook, J. H., and Grattidge, W., "Materials Information for Science and Technology (MIST): a Prototype Materials Properties Data Network System," *Materials Property Data—Applications and Access,* MPD-Vol. 1, PVP-Vol. 111, ASME, New York, 1986, pp. 167-175.
[20] Grattidge, W., this publication, pp. 151-174.
[21] Military Standardization Handbook, Metallic Materials and Elements for Aerospace Vehicle Structures, MIL-HDBK-5E, Department of Defense, Washington, DC (updated approximately annually).

[22] Aerospace Structural Materials Handbook (ASMH), Department of Defense, Battelle-Columbus, OH (updated approximately annually).
[23] Kaufman, J. G., "Standardization for Materials Property Data Bases and Networking," *Standardization News*, Feb. 1986, pp. 28–33.
[24] Kaufman, J. G., "Towards Standards for Computerized Materials Property Data and Intelligent Knowledge Systems," *Standardization News*, March 1987, pp. 38–43.
[25] Shilling, P. E., "Standards for the Materials Property Data Network," *Standards Engineering*, Vol. 38, No. 3, June, 1986, pp. 55–57, 68.

Hermann Kröckel[1] and Günter Steven[2]

European Activities Towards the Integration and Harmonization of Materials Data Systems

REFERENCE: Kröckel, H. and Steven, G., **"European Activities Towards the Integration and Harmonization of Materials Data Systems,"** *Computerization and Networking of Materials Data Bases, ASTM STP 1017,* J. S. Glazman and J. R. Rumble, Jr., Eds., American Society for Testing and Materials, Philadelphia, 1989, pp. 63–74.

ABSTRACT: In a follow-up to three previous actions plans, the Commission of the European Communities (CEC) at present operates a five-year "Program for the Development of the Specialized Information Market in Europe" (1984–88). This program aims in particular at improving the market conditions for information services to ensure their commercial viability and at developing specialized information products and services that are innovative and unique.

For the practical execution of the program, the CEC has defined a number of priority areas, one of which is factual materials data banks. The central activity of this area, the Materials Data Banks Demonstrator Program, aims at the joint demonstration of a pilot network accessible in all member countries of the European Community (EC) linking a number of European on-line data banks. A central user guidance system, which is under development, facilitates data-base location, access, and searching, and provides translation support.

The participating data banks are, by their origin, heterogeneous in structure, operational features, and language. The demonstrator program therefore is accompanied by harmonization efforts that address operational, linguistic, and terminology issues of the system. The progress of these supporting projects is of particular relevance to the present international activities on standardization for materials data systems.

A code of practice harmonizing the operational features of data bases joining the demonstrator program has been developed and come into force. A multilingual terminology project collects terms and definitions used by the cooperating data bases to establish a multilingual dictionary covering an estimated 2000 concepts. These activities refer to existing international standards wherever possible and foster exchange and harmonization with similar activities organized in other parts of the world.

KEY WORDS: materials data, data banks, data bases, information services, data base networks, data base directory, demonstrator program, information market, standardization, harmonization, codes of practice, multilingual terminology, EURONET-DIANE, European Community

During the past few years an unprecedented growth of activities on data banks, information systems, and networks for materials and the related standardization occurred worldwide. The principal reason is the present move towards the computerization of data on materials characteristics and properties, and of many of the processes in which such data are involved.

Some drivers and trends of this move have been reviewed in Ref *1*. They can be related to two main areas of technological change:

---

[1] Head of engineering Data Bank Unit, Commission of the European Communities, Joint Research Center, Petten, The Netherlands.
[2] Researcher, Commission of European Communities, Directorate General XIII, Luxembourg.

(1) the removal of technical barriers by progress in information technology, telematics, software, low-cost equipment, and standards and

(2) the growing demand for computerized materials data caused by the computerization of engineering methods for design, analysis, and manufacturing.

These trends operate equally on the national/international and the institutional level, that is, they influence both public as well as company internal data-bank development. Most of the industrialized countries have, during the last few years, started to take account of their early materials data-base activities, which have mostly grown spontaneously and without coordination, and to build programs for better controlled and coordinated development by using and supplementing the existing basis. In the European Community (EC), two lines of action have determined the present situation:

(1) national policies and institutional activities in the EC member countries and
(2) actions taken by the Commission of the European Community (CEC).

These lines are by their nature interactive, and for some years they have seen an increasing degree of coordination. Though taking into account this interaction, the subsequent review will mainly concentrate upon the initiatives taken by the CEC.

Development of the Infrastructure for Information Systems

In the past (1960-1975), activities in the field of materials information have been only parts of more general programs for the promotion and development of scientific and technical information systems. The early phases of CEC activities have concentrated on the creation of the technical and organizational infrastructure and the realization of international coordination and collaboration. A major objective of the previous three EC action plans (1975-83) in the area of scientific and technical information has been the development and implementation of EURONET and DIANE.

EURONET is the predecessor of the present public, packet-switched telecommunications network (PSN) implemented by a Consortium of the European Post, Telephone, and Telegraph (PTT) administrations on the initiative of the CEC. DIANE is an information service network that presently provides on-line access to some 800 data bases operated on 50 host computers in all EC member countries.

DIANE is supporting the development and implementation of integrated, cooperative, user-oriented information systems and services. It has created conditions for the introduction of the common command language (CCL), for a project aiming at the interconnection of European hosts, and for the development of an intelligent interface facility.

These developments are of general utility not only for the first generation, that is, bibliographic information services for materials that have operated on EURONET-DIANE since 1972, but also for the implementation of the present, second generation, the factual data systems.

Development of Materials Data Banks for Engineering

Among the early second-generation materials data bases, that is, factual data banks presenting data on-line in public environments in the EC, were the Iron and Steel Data bank, now called Steelfacts, of Betriebsforschungsinstitut (BFI) in Germany, and the European collaboration on thermodynamic data and phase diagrams, Thermodata, which was supported by the CEC with systems developed in the United Kingdom, France, Germany, and Sweden.

A considerable increase in these activities took place between 1978 and 1980. Among them, the CEC Joint Research Center at Petten and Ispra began their work on the High Temperature Materials Databank [2]. Similar projects have been brought to life in several countries of the EC on national and private bases. Rapid development is going on and undoubtedly, more European systems will become known in the near future. About 50 projects are currently running in the member countries of the EC. In addition, many company-owned data bases not destined for public use are computerized in structures similar to the advanced, public ones. A good list of machine-readable materials data bases is shown in Appendix C of Ref 3. Most of the better known European systems are included here, and reference to more extensive and detailed listings is made.

A closer look at the ongoing activities shows that the majority of all materials data systems are in their infancy and many projects—even those that have reached the operation stage—are immature. This observation is in agreement with the general situation reported for the United States [4] as well as Japan [5].

Factual data bases differ in character from bibliographic data bases in a way that affects both the technology used to store and access the information and their marketability. Materials data have various peculiarities [1,6], for example, a single property of an engineering alloy determined by a typical material test can be shown to be influenced by several hundred parameters that have their origin in the characteristics of the material and its production processes, the test method and conditions, the testing controls and environmental parameters, the specimen characteristics, and various technical, commercial, and standard conditions.

The ways in which these parameters influence each other and the property under consideration are complex and sometimes even poorly known, in addition to the problem posed by their quantity. Unlike fundamental physical data, engineering materials data constitute a pragmatically optimized, multiparameter system. They are not only subject to physically accessible factors, but also to those relating to production technology, trade, legislation, safety, and standardization. The resulting complexity has usually made the computer implementation of engineering material data systems very demanding and expensive.

An important cost factor is the data input preceded by data capture in the format of the data bank, either paper-based or using electronic devices, disks, tapes, or on-line formats. This process needs expert knowledge and is one of the key areas in which rationalization is desirable, but difficult to achieve. A group of data-bank operators in the EC has started to discuss the harmonization of the input procedures, but obviously, there are various obstacles to be overcome.

There are also many unsolved problems at the data output end of the system where communication with materials specialists and engineering users takes place. The development of intelligent access procedures and query languages, which relieves untrained users of the need to acquire detailed system knowledge, is far from satisfactory. The resulting low usage, in combination with the high cost of the data input process, causes prohibitively poor cost-effectiveness for many systems.

Moreover, unlike bibliographic files factual data files need the care of experts for their development as well as for their operation and maintenance. Most of them are therefore rather specific and restricted in scope, and this is one of the most problematic conflicts with user needs. Users want access to data for many materials, various properties, and various types of information. This disparity will not be solved in any part of the world by a single, super data base, but rather, by setting up networks integrating distributed data bases and offering network features that satisfy the users' need for comprehensiveness, comfort, speed of access, availability, and quality of data.

The development of materials data banks in the EC as well as elsewhere therefore has to address various problem areas that need a great deal of individual effort combined with national and international cooperative programs. This is also the basis for the initiatives taken by the CEC in this field; they support and supplement the institutional and national activities in the EC member countries and develop special European actions for the integration and harmonization of a pilot network and the opening of the market for materials data via a demonstration program.

CEC Initiatives for the Integration of Materials Data Banks

The EC Information Market Development Program

A framework for initiatives on materials data services and their cooperation became available when the CEC implemented a five-year program (1984-1988) for the development of the specialized information market in Europe. This program aims at

1. improving the information environment and market conditions for the use of information products and services to ensure, as far as possible, their commercial viability and
2. creating or developing specialized information products and services that are innovative and unique, and that offer added value with a view to improving their responsiveness to the needs of a wide range of users.

Within the program, easier access to the existing data bases for the professional end-user of information is a major concern of all CEC initiatives. Appropriate projects are partly undertaken within the existing DIANE concept, for example:

(1) the development of a European Host Network (EHN) that aims at interconnecting a number of European hosts by means of a virtual interchange protocol, which will enable users to access a large number of data bases operated on different hosts by accessing one host only;

(2) the development of an "Intelligent Interface," which directs the user automatically to the appropriate data base and assists him in getting the right information without knowing the complex procedures for accessing different data bases on different hosts.

To execute the program, the CEC has defined a number of priority areas, one of which is factual materials data banks. It places particular emphasis upon materials properties data banks for the engineering community, the improvement of the operation, validation, and utilization of these data banks, their market penetration, and cooperation towards a European materials data information system.

The Petten CEC Workshop on Factual Materials Data Banks

The technical background for the materials data banks (MDB) priority area of the program is provided by the recommendations of the "CEC Workshop on Factual Materials Data Banks," which was held in 1984 at JRC Petten to structure actions in this area based on materials data and engineering expertise. By definition, the workshop addressed "machine-readable data on technically relevant properties of engineering materials." Its main objective was to review the status of materials data bases in the EC to define the goals for a cooperative approach to integration, so that recommendations and priorities could be evaluated as to how the CEC structures, funds, and means of action should be optimally used in this area.

The specific recommendations issued by the workshop are detailed in Ref 7. Their emphasis is on awareness, standardization, terminology, functional features in relation to data sources and data applications, the organization and coordination of European cooperation, and the need to demonstrate an integrated information system. To coordinate CEC actions with related projects carried out at the national level, the participants recommended the creation of a "European Task Force on Materials Data Banks." The Petten recommendations have been quite closely followed in the subsequently defined European work program.

Survey of the European Work Program

In consultation with the Task Force, which started its work in February 1985, the CEC has established a medium-term work program with the following action plan for progress on materials data systems (Fig. 1):

1. central actions,
    (a) directory of data information sources (inventory/directory),
    (b) demonstrator program, and
    (c) defining a European system,

2. enabling actions,
    (a) multilingual dictionary of technical terms and definitions,
    (b) guidelines for the functional specifications of software and standards for uniformity and user friendliness in access, and
    (c) standardization of test methods and test result presentation,

3. supporting actions,
    (a) analysis of economic and cost-benefit aspects,
    (b) strategy for coping with a variety of user needs,
    (c) educational aspects, and
    (d) confidentiality and restrictions on data access.

The central actions (A) of the program are those actions that constitute the progress route towards the projected European system of materials data information services.

The enabling actions (B) of the program are harmonizing-standardizing in nature and aim at the generation of references, guidelines, and standards, which are indispensable prerequisites for the development of a homogeneous European system.

The functions of the supporting actions (C) are important constituents of the MDB Program

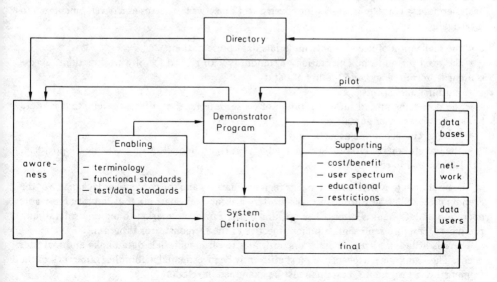

FIG. 1—*Action plan of the European Community for a pilot network of materials data banks.*

Area, but they can only be achieved by stimulating, supporting, and evaluating work in these fields and applying the resulting conclusions and concepts to the guidance of the program.

Interaction with International Activities

Many options for international cooperation are contained in action levels B and C. Coincident with CODATA projects implemented by the Schluchsee Workshop "Materials Data Systems for Engineering" (September 1985) [3], the CEC work program addresses terminology, data source directory formation, economics, educational aspects, data reporting, and standards.

Another activity of considerable interest is the data-bank standards activity carried out under the auspices of the "Versailles Project for Advanced Materials and Standards" (VAMAS) to which the CEC has strongly contributed [8]. Many of its recommendations address subjects of vital importance to the progress of the EC work program.

The cooperation with CODATA and VAMAS has also produced close contacts between the EC Work Program and corresponding activities in the United States and Japan, which go back to the Fairfield Glade Workshop on Computerized Materials Data Systems in 1982 [9] and continue in contacts with the U.S. National Materials Properties Data Network [10] and ASTM Committee E49 on the Computerization of Materials Property Data [11].

The Demonstrator Program

Objectives

The Demonstrator Program is one of the essential recommendations of the CEC Workshop. It is the action of the European Work Program to which the main effort and the majority of the planned resources are being addressed at present. Its function and objectives are therefore reviewed in some detail.

Its goal is to establish a multipurpose test phase and, simultaneously, a first step and pilot concept for the development of a European system of materials data information services. It has, therefore, a correspondingly wide spectrum of objectives. It serves as a development tool to investigate

- the realization of user-friendly materials data presentation;
- the need for and implementation of standards with regard to software functions, access standards, terminology, and quality of data;
- multilingual aspects;
- the development and implementation of a user interface at different levels, that is, data banks, networks, user guidance;
- commercial-economic aspects; and
- the integration of different service media, that is, on-line, electronic off-line (floppy disk, CD-ROM), and hard-copy (handbooks).

The limitation to restricted national or regional markets and the consequent lack of use of the economies of scale the European market offers are the main barriers that hamper the viable operation of data-bank services. The opening of the European market to materials data-bank producers therefore represents a major objective of the Demonstrator Program.

There is little knowledge about users' attitudes towards materials data banks and real user needs. These unknown aspects can most efficiently be investigated within the framework of the Demonstrator Program. Questions must be examined in relation to

- the quality and type of data needed,

- the applications for which the data are needed,
- the user friendliness of data banks, and
- the commercial aspects of materials data banks.

Demonstrating the advantages and potential of materials data banks based on a coordinated and joint approach to be adopted by a group of data base producers is the most important task. This should be supported by awareness actions such as:

- the organization of user seminars,
- public relations and advertising, and
- the publication of newsletters, pamphlets, and special articles in magazines.

Apart from the positive effect of enlarging the market for the participating data banks, the program should also avoid unnecessary and wasteful duplication in researching materials data and creating materials data banks.

General Concept

To achieve these objectives by demonstrating a pilot network, it is necessary to make it capable of running a realistic, operational, on-line exercise over a longer period in genuine dialogue with a "typical" population of engineering users. Indeed, a prototype of a full-size materials data information network is needed to demonstrate those functions that allow users to test and evaluate the system and its use. These functions are

- the features of the individual data base,
- the features of the network,
- the access and guidance of the user,
- the qualitative and quantitative sufficiency of data,
- the administrative and invoicing system, and
- the international and multilingual environment.

A review of the availability and maturity of European data banks gave rise to an estimate of the potential for participation of approximately ten systems. This was rather well confirmed in the first approach. However, during the preparation phase in 1986 and 1987, some fluctuations and changes took place.

The networking concept must take into account the heterogeneity of the data banks offered, which are mainly from the four major EC countries France, Germany, Italy, and United Kingdom and represent diverse data-base management systems (DBMSs), host computers, access procedures, user interfaces and, of course, data scopes and structures. While the data bases are in four languages, the user environment represent a maximum of nine languages in twelve countries.

Since these countries have access to the linked packet switched network, which has satisfactory technical standards, the conceptual effort is not so much concentrated upon technical but on logical and organizational networking aspects.

Participation

Although the initial demonstrator program plan suggested running it in three parts for on-line, offline, and "novel" features, a number of practical considerations regarding the concentration of efforts have led to the adoption of on-line operation only as a first approach. Data banks capable for on-line operation have been selected based on a public call for a declaration of interest. Others will be integrated at a later stage of the program.

Not considering unpredictable changes, the overall data scope will cover the following items:

1. materials

    a. iron, steels, ferrous alloys,
    b. nonferrous alloys,
    c. plastics,
    d. ceramics, glass, carbon, graphite,
    e. composites, and
    f. powders,

2. properties

    a. mechanical,
    b. physical, chemical,
    c. thermodynamic, phase diagrammes,
    d. technical, machining, and
    e. corrosion, hydrogen influence,

3. character

    a. unevaluated and evaluated test data,
    b. data extracted from literature,
    c. handbook type data,
    d. supplier data, and
    e. standard data.

Some of the data banks are associated with existing bibliographic files and some are developed as "knowledge-base systems" offering extensive modeling algorithms for practical engineering topics or materials selection procedures.

The eleven data banks that are at present part of the demonstrator program are shown in Table 1. More details about the features of most of these systems can be found in Appendix D of Ref 7.

Common Data Bank Features: The EC Code of Practice

The implementation of a number of common features by all data banks of the Demonstrator Program is a prerequisite for the realization of a pilot network that integrates a heterogeneous spectrum of individual data banks. Extensive analysis and evaluation of the candidate data banks, which was jointly achieved in a number of meetings in 1986, has led to the definition of a consensus Code of Practice, [12] which sets rules for common features at the following four levels:

(1) the access level,
(2) the host operations level,
(3) the system operations level, and
(4) the content level.

On these four levels, the following main features are controlled by the code of practice:

1. access level

    a. telecommunication connections,
    b. user manual,
    c. user contract,
    d. terminal type,
    e. user languages, and
    f. availability,

TABLE 1—*EC demonstrator program, data banks.*

Data Bank	Full Name	Institution	Subject
HDATA	Hydrogène Data	Ecole Nationale Supérieure de Chimie de Paris (France)	hydrogen-materials interactions
THERMODATA	THERMODATA	THERMODATA, Association, Grenoble (France)	thermodynamic properties, complex equilibrium calculations in inorganic materials
CETIM	Banque de données matériaux au CETIM	CETIM, Senlis (France)	common engineering material properties of metals, plastics, composites, adhesives for design offices in material selection
MDF/I	Metals Data File	The Institute of Metals, London (UK)	mechanical/physical properties of ferrous/nonferrous alloys
PERITUS	PERITUS	Matsel Systems Ltd., Liverpool (UK)	selection of metal and plastics materials and processes for design
MATUS	Materials User Service	Engineering Information Company Ltd., London (UK)	mechanical, thermal, and electrical properties and processibility of plastics, metals, ceramics, glasses and composites
COMETA	Sistema Informativo Componenti e Materiali	ENEA, Roma (Italy)	engineering properties of metallic materials and components for design
INFOS	Informationssystem für Schnittwerte	EXAPT-Verein, Aachen (FRG)	cutting value properties of metallic materials for turning, drilling and milling
POLYMAT	DKI Plastics Data Bank	Deutsches Kunststoff-Institut Darmstadt (FRG)	characteristics, properties, processing, and application aspects of plastic materials
ISTMA/ SOLLMA	Database of Metallic Materials for Pressure Vessels Construction	Rheinisch-Westfälischer Technischer Überwachungs-Verein e.V., Essen (FRG)	specified engineering properties and test values of ferrous and nonferrous alloys for pressure vessels
HTM-DB	High-Temperature Materials Data Bank	CEC, DG XII— JRC Petten (Netherlands)	mechanical and corrosion properties of high-temperature materials

2. host operations level

   a. prompts and error messages,
   b. host welcome,
   c. log-on procedures, and
   d. host messages,

3. system operations level

   a. charging information,
   b. system welcome,
   c. help functions,
   d. help desk,
   e. error messages,

f. session recording,
   g. local storage, and
   h. common command language (CCL), and

4. content level

   a. terminology,
   b. scope,
   c. completeness, and
   d. units, abbreviations.

The Code of Practice identifies two categories of features: mandatory and optional (recommended). The implementation of the mandatory features is a minimum requirement for the participation of data banks in the demonstrator program.

User Guidance System

Two other aspects are important in operating an information service network that consists of individual data banks:

1. the availability of a common identity for service providers, and
2. the existence of a unique and multilingual user reference to the total scope of network services.

For this purpose, a User Guidance System has been developed that is suited to fulfill a number of functions:

1. information on data banks

   a. scope, structure, size,
   b. services, products,
   c. access procedures,
   d. usage cost,
   e. language(s),
   f. retrieval methods, and
   g. addresses (host, owner) and

2. selection aids

   a. standard menu,
   b. reference dictionary,
   c. common command language (CCL),
   d. cross-referencing,
   e. descriptor index,
   f. problem-oriented retrieval, and
   g. multilingual features.

The system is designed to be initially operated as a centralized, toll-free referral service on the European Community Host (ECHO) and may, at a later phase, be integrated into the Intelligent Interface Facility, which will provide for:

(1) automatic log-on procedures,
(2) user support functions (help desk), and
(3) user administration (contracts, charging).

The pilot system in the English language has been recently implemented on ECHO and is being demonstrated.

Common Reference Terminology

There is an interest that the vocabulary of data banks be based on international standards that make the used terminology system-independent and allow universally understandable communication. The need for standardized terminology increases when different factual data banks are linked in networks. Moreover, the overall scope of the required vocabulary will then be wider.

If the network of factual data banks is international, it raises additional requirements in relation to data-bank intercommunication and databank-to-user communication, which are both becoming multilingual. The need here is for multilingual equivalence in two linguistic scopes:

(1) the languages of the data banks in the network and
(2) the languages of the users.

In the demonstrator program, the present data banks are in four languages: English, French, German, and Italian. The users represent a community of nine languages.

The requirements for the short-term development of an agreed upon common reference terminology for materials information are based on the scope of the demonstrator program. This will lead to a complex set of terms, since several layers of industrial practice are involved as well as several languages, each with established national habits in the use of terms. Two aids to communication and standardization are therefore required:

(1) a thesaurus, to provide synonyms of terms and scope notes and
(2) a dictionary, to provide definitions and multilingual equivalents.

These short-term aids need to be established in all nine EC languages to provide links to the four languages used in the data banks. They are to be developed on the basis of about 200 to 300 of the most-used technical terms of each of the data banks. They will result in a common, multilingual working dictionary of about 2000 terms for the use of the demonstrator program, which will be implemented as a subscope of the CEC terminology data bank EURODICAUTOM, accessible on the ECHO host. In establishing this dictionary, it is desirable that

(1) it be supported by existing multilingual thesauri, which should function as reference thesauri and
(2) it should endeavor to use existing international standards, or to take into account international standardization efforts through ISO, CODATA, VAMAS, and ASTM.

Further Activities

While testing how well the system meets its purpose can commence before the end of 1987, the main part of the Demonstrator Program is expected to begin in 1988. It will concentrate upon an awareness program in which the system will be presented and demonstrated to the research and industrial-user community in all member countries of the European Community. Feedback from users will be evaluated to assess the need for improvement and further development.

Providing information to and dialogue with the user community are the principal features of the program. Among the activities accompanying and supporting the demonstrator program is the development of a directory of materials data information sources that will become available as printed and on-line versions. The printed version will be combined with directories from other countries into a world directory by CODATA. The EC on-line directory will be accessible on the ECHO host before the end of 1987. The on-line user can then access a system of referral to materials information sources supplementary to the User Guidance System leading him to the data bases of the demonstrator program.

References

[1] Kröckel, H. and Westbrook, J. H., "Computerized Materials-Information Systems," *Philosophical Transactions of the Royal Society London* A323, 1987, pp. 373-391.

[2] Commission of the European Communities, Joint Research Center, High Temperature Materials Program, Data Bank Project: First Demonstration Report on the High Temperature Materials Data Bank of JRC, EUR 8817, Sept. 1983.

[3] Westbrook, J. H., Behrens, H., Dathe, G., and Iwata, S., Eds., "Materials Data Systems for Engineering," Proceedings of a CODATA Workshop held at Schluchsee, Fachinformationszentrum Energie, Physik, Mathematik, GmbH, Karlsruhe, West Germany, Sept. 1985.

[4] Westbrook, J. H., "American Developments in Computer Access to Materials Data," in *Factual Material Data Banks*, CEC Workshop-JRC Petten EUR 9768 EN, 1985, pp. 70-81.

[5] Iwata, S., "Materials Data Activities in Japan," in *Factual Material Data Banks*, CEC Workshop-JRC Petten EUR 9768 EN, 1985, pp. 54-69.

[6] Dathe, G., "Peculiarities and Problems of Materials Engineering Data," in *The Role of Data in Scientific Progress, Proceedings of the 9th International CODATA Conference*, Jerusalem, 1985, pp. 233-241.

[7] Kröckel, H., Reynard, K. W., Steven, G., Eds., *Factual Materials Data Banks*, Proceedings of the CEC Workshop on Materials Data Information Systems in Europe held at JRC Petten, 14-16 Nov. 1984.

[8] Kröckel, H., Reynard, K. W., Rumble, J. R., Eds., *Factual Materials Databanks—The Need for Standards*, The Report of a VAMAS Task Group, Versailles Project on Advanced Materials and Standards (VAMAS), in press, 1987.

[9] Westbrook, J. H. and Rumble, J. R., Eds, *Computerized Materials Data Systems*, Proceedings of a Workshop held at Fairfield Glade, TN, 7-11 Nov. 1982.

[10] Kaufman, J. G., "The National Materials Property Data Network, Inc., The Technical Challenges and the Plan," in *Materials Property Data: Applications and Access*, MPD-Vol. 1/PVP-Vol. 111, ASME, 1986, pp. 159-166.

[11] Kaufman, J. G., "Towards Standards for Computerized Materials Property Data and Intelligent Knowledge Systems," in *Standardization News*, Vol. 15, No. 3, 1987, pp. 38-43.

[12] *The Operation of Materials Property Data Systems in the EC, A Code of Practice for Use in the Materials Data Base Demonstrator Program*, CEC, 1987. Available from G. Steven, CEC, DG XIII-B, Building Jean Monnet, Luxembourg.

Yunwen Lu[1] and Shousan Fan[1]

Materials Data Activities in China

REFERENCE: Lu, Y. and Fan, S., **"Materials Data Activities in China,"** *Computerization and Networking of Materials Data Bases, ASTM STP 1017,* J. S. Glazman and J. R. Rumble, Jr., Eds., American Society for Testing and Materials, Philadelphia, 1989, pp. 75–79.

ABSTRACT: Current materials data activities in China are described. Some features are summarized and some applications are listed. A variety of materials data bases have been introduced. Some of them are described in detail. In addition, the organization of materials data bases in China is reviewed and its future predicted.

KEY WORDS: data bases, materials data, alloys

Since the end of 1970s, several data base actions have taken place in China. In 1978, the First National Conference on Data Bases was held, the second in 1982, and the third in 1984. These conferences have promoted the building of data bases in China [1]. Until now, based on incomplete figures, 154 science and technology numerical data bases have been built, including 32 materials data bases; most of these are still under construction. Owing to the expanding need for data, more and more industries, research institutes, and universities are planning to establish new materials data bases.

The applications of these materials data bases cover the following areas:

- selection and substitution of materials,
- failure analysis of materials,
- optimization of processing parameters, and
- materials design.

The current status of materials data activities in China has some key features:

- Data activities in China are still in the formative stage. Most data bases were founded in recent years; only a few of them are in active use. Both experimental data and handbook data are included in the data bases. Generally, they are small.
- A number of data bases are based on the personal computer (IBM PC/XT or IBM PC/AT) and use dbase II or dbase III as the data base management system (DBMS). Some of them will be installed on a minicomputer.
- As of now, many materials data bases are distributed and have not yet been united to a network. A plan of organization is in progress.

Existing Data Bases Relating to Materials

Most of the Chinese materials data bases are listed in Tables 1 to 7. They can be divided into seven areas:

---

[1] Associate professor and lecturer, respectively, Research Institute of Materials Science, Tsinghua University, Beijing, China.

TABLE 1—*Fundamental data (nuclear, atomic, molecular, and crystal structure).*

Data Base	Developer
Chinese Evaluated Nuclear Data Library	Research Institute of Atomic Energy
Atomic and Radiation Data	Research Institute of the Ministry of Nuclear Industry
Nuclear Data	Shanghai Institute of Atomic Nuclei
Atomic and Molecular Data	Beijing Normal University
Crystallography	Beijing Institute of Chemical Metallurgy
X-ray Diffraction	Beijing Institute of Physics

TABLE 2—*Data bases concerning chemical and thermal properties.*

Data Base	Developer
Thermochemistry	Beijing Institute of Chemical Metallurgy
	Zhengzhou Institute of Technology
	Sichuan University
	Beijing Institute of Chemistry
Thermophysical Properties	Chengdu Institute of Measurement and Testing Technology
Physical Properties for chemical engineering	Ministry of Chemical Industry
	Huadong Institute of Chemical Engineering
	Dalian Institute of Technology
	Chinese General Chemicals, Inc.
	Beijing Institute of Chemical Engineering
	Nanjing Institute of Chemical Engineering
Phase Diagrams of Alloys	Beijing University of Iron and Steel
Metallurgical thermodynamics	Beijing University of Iron and Steel
Phase equilibria	Zhejiang University

TABLE 3—*Spectra.*

Data Base	Developer
Mass Spectra	Beijing Institute of Chemistry
	Dalian Institute of Chemical Physics
Infrared Spectra	Shanghai Institute of Organic Chemistry
	Jinxi Institute of Chemical Engineering
NMR	Changchun Institute of Applied Chemistry
	Dalian Institute of Chemical Physics
	Dalian Chromatography Center
Chromatography	Zhenzhou Institute of Technology
	Dalian Chromatography Center
Ultraviolet Ray Spectra	Jinxi Institute of Chemical Engineering

- fundamental nuclear, atomic, molecular, and crystallographic data,
- data bases concerning chemical and thermal properties,
- spectral data,
- corrosion and tribology,
- metallic materials,
- rare earth materials, and
- nonmetallic materials.

TABLE 4—*Corrosion and tribology.*

Data Base	Developer
Atmospheric corrosion of materials	Beijing University of Iron and Steel Wuhan Institute for the Protection of Materials Sichuan Institute No. 54 Guangzhou Institute of Electrical Equipment Guangzhou Institute of Degradation
Soil corrosion data base	Shenyang Institute of Corrosion and Protection of Metals
Marine corrosion at Chinese beaches	Dalian Institute of Technology
Tribology	Wuhan Institute for the Protection of Materials

TABLE 5—*Metallic materials.*

Data Base	Developer
Alloy Steels	Beijing Institute of Iron and Steel
Nonferrous Alloys	Beijing Institute of Nonferrous Metals
Aeromaterials	Ministry of the Aircraft Industry
Mechanical engineering materials	Shanghai Institute of Materials
Cutting tools	Chengdu Institute of Cutting Tools
Strength of structures	Zhengzhou Institute of Mechanical Engineering
Failure analysis	Beijing Institute of Aeronautics

TABLE 6—*Rare earths materials.*

Data Base	Developer
Materials from Rare Earths	Changchun Institute of Applied Chemistry
Physical and Chemical Properties of Rare Earths	Changchun Institute of Applied Chemistry
Extraction of Rare Earths	Changchun Institute of Applied Chemistry

TABLE 7—*Nonmetallic materials.*

Data Base	Developer
Fine ceramics	Tsinghua University
Nonmetallic materials	Wuhan University of Technology
Amorphous materials	Shanghai Institute of Optics and Fine Mechanics
Silicon epitaxy parameters	Hebei Institute of Technology

Among these, the following materials data bases are detailed:

The Alloy Steel Data Base

This data base was developed by the Beijing Institute of Iron and Steel. The data for 110 grades of stainless steels have been stored. The data for nearly 1500 grades of alloy steels will be collected and stored in the near future. The materials properties include steel standards, chemical composition, physical properties, mechanical properties, relaxation data, corrosion resistance data, and so forth. The data base is mainly applied in the selection of materials. An expert system for materials selection is being developed.

Nonferrous Alloys

This data base was developed by the Beijing Institute of Nonferrous Metals. It contains the data for 360 grades of aluminium alloys, including 160 grades from the United States, 140 grades from the Soviet Union, and 80 Chinese grades. Data for copper alloys and for some rare metals are being collected and stored. The data base system also includes several computation programs for the optimization of the technology for deformation forming, such as drawing, extrusion, and forging. Therefore, the mechanical property data of alloys can be directly applied in alloy processing.

Aeromaterial Data Base System

This data base system was established in 1984 by the Ministry of the Aircraft Industry. The system includes 11 data bases, which contain two million data points for 600 grades of metals and 400 grades of nonmetallic materials. The data on ten properties for every type of material, such as chemical composition, physical and chemical properties, mechanical properties, and so forth, are collected and stored. This data base system is applied in material selection for aircraft design. It is still a personal computer based system with graphic capability.

Ceramics Data Base

This data base was developed by the Research Institute of Materials Science of Tsinghua University. The materials include oxides, nitrides, and carbides, and so forth. The data for 50 properties and 16 processing parameters for each material are collected and stored. This data base is applied in developing high-temperature ceramics for engines and new functional ceramics for electronic devices.

Corrosion Data Base System

An atmospheric corrosion data base system for materials has been developed by the Beijing University of Iron and Steel and the Wahan Institute for the Protection of Materials. The data base system collects corrosion data, such as the corrosion rate, average corrosion thickness, change in mechanical properties, and so forth from nine atmospheric corrosion experimental stations. These stations are located in Beijing, Qindao, Wuhan, Sichuan, Guangzhou, and Hainan Island. Therefore the data reflect the corrosion properties of materials in different geographic environments. Nearly 100 000 samples for 320 grades of steel, nonferrous alloys, coatings, and plastics have been tested at the stations.

In addition to the atmospheric corrosion data base, the Institute of Corrosion and Protection of Metals has developed a soil corrosion data base for metals. It is based on data collected from 4284 metal samples buried in 203 sample pits in 29 locations monitored by 19 soil corrosion stations. From these data, a national soil-corrosion diagram will be prepared. Besides these, a marine corrosion data base has been developed by the Dalian Institute of Technology. It can be applied in the shipbuilding industry, in marine petroleum prospecting, and in oceaneering.

Rare Earth Data Base

China ranks first among all countries in rare earth resources. Therefore, research on the rare earths is a very active area in China. Based on this research, Changchun Institute of Applied

Chemistry has developed a rare earths data base, which collects and distributes the data concerning materials based on rare earths, the physical and chemical properties of rare earths, and the extraction of rare earths [2].

Organization and Future Plan

Under the leadership of the Chinese National Committee for CODATA, a national workshop on materials property data bases was held in Beijing on 23-25 October 1986. Forty participants from research institutes, universities, and industries were present. The workshop emphasized the critical importance of both developing more materials property data bases in China and organizing most of these materials data bases into an effective interacting group. As a result, a subcommittee on material data bases that belongs to the Chinese National Committee for CODATA was established. It includes four working groups, as shown below:

- engineering structural materials,
- corrosion of materials,
- functional materials, and
- rare earth materials.

A national conference on material data bases will be held every 2 years and a publication, "Materials Data Base Letter," will be published [3].

Furthermore, a proposal for a National Data Network for Advanced Materials was submitted to the State Science and Technology Commission (SSTC), which is the highest science and technology policy-making body in China. The establishment of this network has been designated as one of the highest priority projects. It covers the following areas of advanced materials:

- Advanced metals
- Polymers
- Composites
- Fine Ceramics
- Electronic materials

The network will be a distributed system, including several workstations based on personal computers. Its purpose is to develop new materials in China. At present, the critical tasks for the project are:

- standardization, uniformity, and international compatibility in data presentation and the structure of the data bases,
- developing a distributed data system and networking, and
- international cooperation.

We hope that this symposium will help us to solve such problems.

References

[1] Division of Data Base Creation, "The current status of scientific databases in China," unpublished.
[2] Kwahk, M. and Xu, Zhihong, "Overview of China's Scientific Data Bases," *Proceedings of 10th CODATA Conference,* in press.
[3] Lu, Y., "The Development of Materials Data Bases" *Electrotechnical Journal of China,* Vol. 5, 1987, pp. 16-18.1.

Satoshi Nishijima,[1] *Yoshio Monma,*[1] *and Masao Kanao*[2]

Japanese Progress in Materials Data Bases

REFERENCE: Nishijima, S., Monma, Y., and Kanao, M., **"Japanese Progress in Materials Data Bases,"** *Computerization and Networking of Materials Data Bases, ASTM STP 1017,* J. S. Glazman and J. R. Rumble, Jr., Eds., American Society for Testing and Materials, Philadelphia, 1989, pp. 80–91.

ABSTRACT: Recent trends in the development of materials data systems in Japan are reviewed. As the building of materials data bases is a major concern of industry, academic societies, and government laboratories and the sharing of materials information is a common need among interested parties, the government is supporting the building of materials data bases and networking with the Open System Interconnection (OSI) model. Data base activities under such programs are briefly discussed. However, a more detailed discussion is given to a comprehensive materials data base on the materials strength of engineering steels and alloys that has been developed by the National Research Institute for Metals (NRIM) and Japan Information Center of Science and Technology (JICST). The importance of incorporating the evaluation procedure is emphasized. In addition, some specifics on Japanese language in the system environment are discussed.

KEY WORDS: alloy, creep, data base, fatigue, materials, network, steel, strength

Nomenclature

CSJ	The Ceramics Society of Japan
GRIK	Government Research Institute of Kyushu, AIST
GRIN	Government Research Institute of Nagoya, AIST
IPRI	Industrial Products Research Institute, AIST
JAICI	Japan Association of International Chemical Information
JFCA	Japan Fine Ceramics Association
JFCC	Japan Fine Ceramics Center
JKTC	Japan Key Technology Center
MEL	Mechanical Engineering Laboratory, AIST
MSSJ	Mass Spectroscopy Society of Japan
NCLI	National Chemical Laboratory for Industry, AIST
NIRIM	National Institute for Research in Inorganic Materials, STA
P-S-N	probability-stress-number of cycles to failure
RIPD	Research Institute for Production Development
RIPT	Research Institute for Polymers and Textiles, AIST
SCEJ	Society of Chemical Engineers, Japan
SSRT	Slow Strain Rate Test

---

[1] Director, failure physics division, and head, first subgroup, 5th research group, respectively, National Research Institute for Metals, 2-3-12 Nakameguro, Meguro-ku, Tokyo 153, Japan.
[2] Formerly deputy director-general, National Research Institute for Metals.

Introduction

The development of the computerization of materials data in Japan was started in the late 1970s. The activities supporting materials data systems in Japan until 1983 were reviewed by Iwata [1]. Since then, a growing number of materials data systems have been developed. Today, data generation activities that lead to the computerization of materials data are very common. Typically they are linked to a computer-aided experimental system using microcomputers. Although there are some sophisticated data systems on specific classes of materials, they have been mostly developed in industry and oriented for in-house, private use.

The collection and compilation of materials data have been traditionally conducted by academic and engineering societies. Many data books based on such activities are available. In view of the importance of information processing and management in science and technology, the Japanese Government has given materials data systems, R&D major emphasis. However, there are still a very limited number of factual data systems available to the general public in Japan [2].

The present paper deals with the status quo of materials data systems being developed mainly under the support of the Japanese Government. We begin with a discussion of three specific materials data base efforts: Those of the Society of Materials Science, Japan (SMSJ), the Science and Technology Agency, and the National Research Institute of Metals.

Fatigue Data File of the Society of Materials Science, Japan

In the past, a great deal of time and effort have been spent by SMSJ to obtain the fatigue properties of structural materials [3–5]. The effect of a corrosive environment on fatigue properties has especially attracted much attention. Since the fracture of materials caused by fatigue is intrinsically probabilistic, it is essential to have as much data as possible for design analysis and materials development. The output of nationwide cooperative projects by the Society of Materials Science, Japan has been published in two compilations on fatigue data and one on the corrosion-related properties of steels and alloys (supported by a grant-in-aid from the Ministry of Education). Nine volumes of data in three series (about 6000 print-out pages in total) are available. They are also offered as machine-readable magnetic tape. Table 1 summarizes the voluminous data.

TABLE 1—*Summary of SMSJ fatigue data files.*

Material Type	Series A [3]		Series B [4]		Series C [5]
	Number of Alloys	Number of S-N	Number of Alloys	Number of da/dn-ΔK	Data Size[a]
Structural steels	67	272	34	272	450
Low alloy steels	50	1230	28	237	1350
Stainless steels	30	302	11	101	2750
Cast and forged steels	31	227	14	25	...
Others	26	126	2	22	...
Aluminum alloys	45	187	6	72	600
Copper alloys	7	29	4	8	200
Titanium alloys	2	7	2	8	150
Cast and others	15	44	1	2	700
Total	273	2424	102	747	6200

[a] Sum of data sets for S-N, S-t, da/dn-ΔK, da/dt-t, and SSRT under a corrosive environment.

Integrated Materials Information System of the Science and Technology Agency

The Science and Technology Agency (STA) of the Japanese Government has been supporting the concept of an Integrated Chemical Database System [6], (Fig. 1). This system combines ten chemical data bases with networking to share information on chemical substances developed by government laboratories [7,8]. The most important product of this project is a system called STARS (Stereochemically Accurate Registry of Substances), similar to but more comprehensive than the Chemical Abstracts Service (CAS) Registry, with 3-D structural information and Japanese language capability. The Japanese Information Center for Science and Technology (JICST) plays a central role in the project in cooperation with various professional institutes and data centers. They already have put a mass spectroscopy system (MASS) [9,10] into their on-line service, JOIS (JICST On-line Information System), and soon will add more numeric data bases. Table 2 summarizes the components of the integrated chemical and materials data system (see Nomenclature). The total concept in Fig. 1 has not yet been completely implemented because of problems of incompatibility among data base designs and in networking, but steady efforts towards improvement are being made.

NRIM-JICST Data Base on Engineering Steels and Alloys

Since 1966 the National Research Institute for Metals (NRIM) has conducted extensive materials testing programs on the creep and fatigue properties of steels and alloys. The Data Sheet Programs are being undertaken to obtain the original data on multiheat samples of engineering steels and alloys from commercial stock so that the data can be used in a realistic evaluation in design analysis, life prediction, materials development, and so forth. Well-balanced data with least scatter should serve as the standard reference for many situations in the assessment of materials strength. So far, more than 8000 stress-rupture specimens with accumulated rupture times of 1.1×10^8 h (12,780 years) and 4×10^{11} cycles of fatigue tests have been obtained.

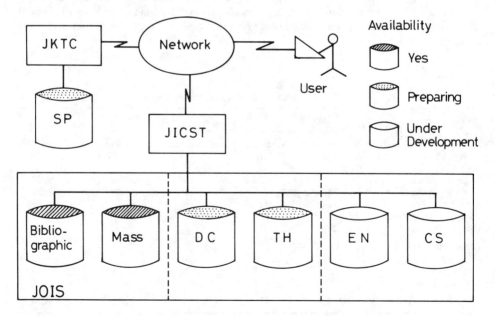

FIG. 1—*Concept of computer networking for materials information.*

TABLE 2—*Components of integrated materials information systems in Japan (Fig. 1).*

Data Base	Abbreviation	Organization	Typical Data Items	Availability
Chemical substances	DC	JICST JAICI	DD of substances, data base locators	Jan. 1988
Spectrum	SP	NCLI JKTC	IR and NMR spectral data of standard substances	Jan. 1988
Mass spectrum	Mass	JICST MSSJ	mass spectral peaks for elements and molecules	Since Mar. 1987
Thermophysical, thermochemical properties	TH	JICST RIPD SCEJ	basic constants, PVT, phase equilibrium, thermodynamics	Jan. 1988
Engineering steels and alloys	EN	NRIM JICST	mechanical properties, creep properties, fatigue properties	Target April 1989
Crystal structures	CS	JICST NIRIM	crystal types, lattice parameters	Target April 1989

In 1985, NRIM and JICST signed a contract to work jointly to build a comprehensive materials data base on engineering steels and alloys using the data sheets as the core. The primary emphasis of the data base is on the material information for structural applications. The public on-line service is scheduled to be available early in 1989 (Phase I). Ultimately, this data base will be put into the worldwide service network. In Phase II, NRIM will expand coverage to include nonferrous materials and other properties.

In August 1986, NRIM and JICST conducted a survey of some 600 potential users of the data base in Japan. According to the results of the survey, 50 percent need such data more than once a month. The materials strength data of engineering steels and alloys are used for materials selection (60%), remaining life prediction (50%), and structural analysis in design (44%). The common problems that concern users are the whereabouts of the data (37%), the time and cost that must be expended to get the data (35%), and the reliability of the data (23%). Although 63% of them are the producers of materials strength data, about two-thirds of the data produced are unpublished.

There are three categories of materials properties data included in the data base. The primary one is strength-related properties and the secondary one is the physical property data. Data covering areas other than materials properties are also provided to meet users' needs. The major properties given for each heat/cast of a steel/alloy are:

- hardness, impact values
- tensile properties (at room and high temperatures)
- creep properties (stress-rupture, creep strain-time)
- fatigue properties (room and high-temperature rotating bending, torsion, axial loading, welded joints, high-temperature low-cycle, creep-fatigue interaction)

The physical properties given for each material type as the representative values are:

- density
- melting point (range), heat of melting
- specific heat
- thermal expansion coefficient
- thermal conductivity
- elastic moduli (Young's modulus, rigidity, Poisson's ratio)

Supplemental items include:

- materials specification (standard, chemical composition, heat treatment)
- processing history (product form, dimension)
- testing/measurement environment (specimen, environment)
- nature of the data (raw or evaluated data)
- data source (reference, test lab)

The list of the metadata of the data base consists of about 250 record items. Figure 2 shows the structure of files and records of the data base. The major properties and materials covered are shown in Tables 3 and 4.

Importance of Data Evaluation

One of the most important features of a factual data base is the possibility of data evaluation of the retrieved data set. In building the NRIM-JICST system, some representative methods of evaluation will be implemented. The evaluation of data to produce standard reference data is very important when the data are added to the data base from outside. This is because the

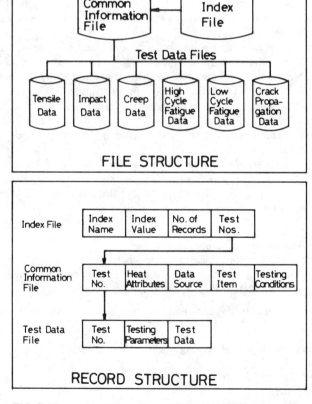

FIG. 2—*Structure of files and records in NRIM-JICST data base.*

TABLE 3—*Summary of NRIM creep data sheets.*

Material Type	Material	Form	Number of Heats
Carbon steels	0.2C	tube	9
	0.3C	plate	8
	1.25Mn	tube	2
Low-alloy steels	UTS 600 MPa	plate	21
	0.5Mo	tube	12
	1.3Mn-0.5Ni-0.5Mo	plate	5
	0.5Cr-0.5Mo	tube	9
	1Cr-0.5Mo	tube	11
	1Cr-0.5Mo, NT	plate	8
	1Cr-1Mo-0.25V	forging	11
	1Cr-1Mo-0.25V	cast	9
	1.25Cr-0.5Mo-Si	tube	10
	1.25Cr-0.5Mo-Si, NT	plate	13
	2.25Cr-1Mo	tube	13
	2.25Cr-1Mo, NT	plate	6
	2.25Cr-1Mo, QT	plate	7
High-alloy steels	5Cr-0.5Mo	tube	9
	9Cr-1Mo	tube	11
	12Cr	bar	9
	12Cr-1Mo-1W-0.3V	bar	9
	304H	tube	10
	304	weld	39
	316H	tube	9
	316	bar	6
	316	plate	2
	321H	tube	9
	347H	tube	9
	800H	tube	6
	800H	plate	6
	0.4C-25Cr-12Ni	cast	5
	0.4C-25Cr-20Ni	cast	14
	0.4C-25Cr-35Ni	cast	7
Heat-resisting alloys	660	forging	3
	S590	bar	3
	N155	cast	9
	600H	tube	5
	750	bar	4
	713C	cast	8
	U500	cast	11
	700	bar	2
	X45	cast	4
Total	41		363

validity of materials' properties modeling can be only checked with the data with the least scatter. NRIM data is especially helpful in this regard because it provides well-balanced and comprehensive data sets from multiple lots.

Typical examples of data evaluation using NRIM data are shown. Figure 3 compares three distinct behaviors of stress-rupture in 304H steel. When the aluminum content was more than 0.03%, a detrimental effect on long-time strength has been reported [11]. Figure 4 shows an example of the curve fitting of creep strain-time data for 304 steel. The validity of various

TABLE 4—*Summary of NRIM fatigue data sheets.*

Fatigue Properties	Material Type	Material	Form
Room temperature, high and low cycle properties	carbon steels	0.25C	bar
		0.35C	bar
		0.45C	bar
		0.55C	bar
	low-alloy steels	0.38C-1.5Mn	bar
		0.43C-1.5Mn	bar
		0.20C-1Cr	bar
		0.40C-1Cr	bar
		0.20C-1Cr-0.2Mo	bar
		0.35C-1Cr-0.2Mo	bar
		0.40C-1Cr-0.2Mo	bar
		0.31C-2.7Ni-0.8Cr	bar
		0.20C-0.5Ni-0.5Cr-0.2Mo	bar
		0.20C-1.8Ni-0.5Cr-0.2Mo	bar
		0.40C-1.8Ni-0.8Cr-0.2Mo	bar
		0.47C-1.8Ni-0.8Cr-0.2Mo	bar
Elevated temperature, high and low cycle, creep/fatigue properties	high-alloy steels	304	bar
		430	bar
	carbon steels	0.30C	plate
		0.45C	bar
	low-alloy steels	0.35C-1Cr-0.2Mo	bar
		1.25C-0.5Mo-Si, NT	plate
		2.25Cr-1Mo, NT	plate
		1Cr-1Mo-0.25V	forging
	high-alloy steels	403	bar
		304	plate
		316	plate
		616	bar
		800H	bar
High and low cycle, crack growth properties	0.2C	plate, base metal, welded joint	
	UTS 500 MPa	plate, base metal, welded joint	
	YS 500 MPa	plate, base metal, welded joint	
	UTS 600 MPa	plate, base metal, welded joint	
	UTS 800 MPa	plate, base metal, welded joint	
	304	plate, base metal, welded joint	

models of the creep constitutive equation must be established in the design analysis for a fast-breeder reactor [12]. Heat-to-heat variations in fatigue strength are illustrated in Fig. 5, in which the scatter of a JIS (Japanese Industrial Standards) grade steel (14 heats) with standard heat treatment reflects the variation in chemical composition [13]. Figure 6 shows an example of the analysis of the fatigue crack propagation rate (1259 data points) for manual, gas-metal arc, and submerged arc welded joints with various heat input conditions. Since the effects of welding methods and heat of input are rather small, with a statistical treatment, the evaluated curves could be used as the basis for in-service inspection intervals and life prediction [14].

Data Bases on Advanced Materials

The STA is also promoting the development of factual data bases on thin films and very fine particles. This is part of the effort to create new materials based on hybridization in microstructural design [15].

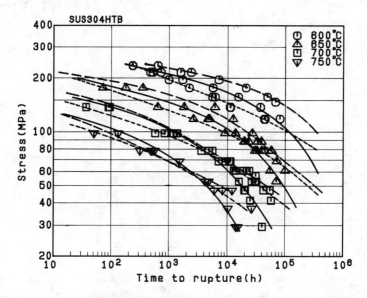

FIG. 3—*Example of heat-to-heat variation in stress-rupture properties of 304H steel.*

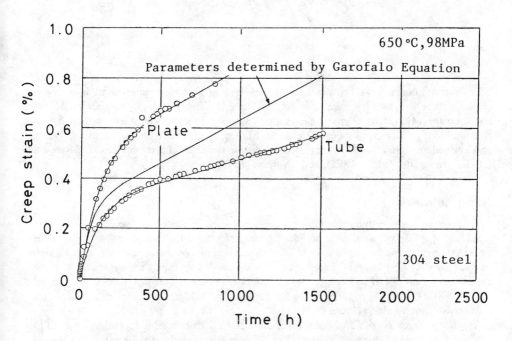

FIG. 4—*Example of curve fitting to creep strain-time data for 304 steel.*

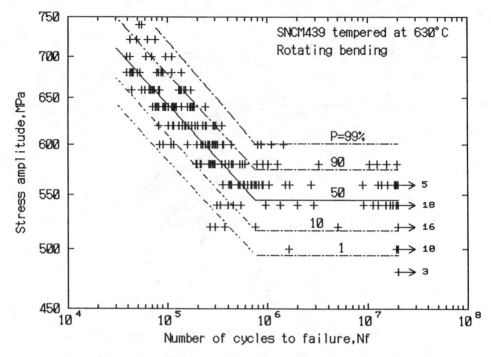

FIG. 5—*Example of P-S-N analysis for fatigue strength of a low-alloy steel.*

The recent interest in and demand for factual data on advanced materials such as fine ceramics, composites, and polymers are being addressed by the Ministry of International Trade and Industry (MITI) [16]. Figure 7 shows a schematic of integrated data base project for new ceramics. Many government laboratories under the auspices of the Agency of Industrial Science and Technology (AIST) of MITI are jointly working in the fields of ceramics (JFCC, GRIN, and GRIK), composites (MEL, IPRI), and polymers (RIPT).

System Environment

Until recently, using Japanese on computers has been burdensome, because of the extra graphics capability required. The standard character set for Japanese is composed of about 6400 characters in two levels of two-byte *kanji* (Japanese graphic character) code. At least 16 by 16 pixels on CRT and 24 by 24 pixels on printers are needed to display a *kanji* character. However, the progress of LSI technology allows the storage of the character images on a read-only memory (ROM) chip. Also, today, good Japanese lexical analyzers are available. They can at least transform syllable-to-clause efficiently.

Another problem is the *kanji* code: there are a few different codes, depending on both the machine type and the operating system. The differences among them are greater than that between MS-DOS and UNIX. This means that the software for the data file transfer needs to support the differences in the different *kanji* codes. Direct conversion between a data file containing *kanji* code and that containing one-byte ASCII code is extremely difficult. Thus, many

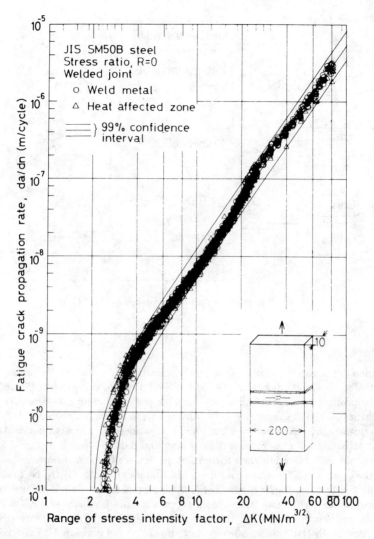

FIG. 6—*Example of analysis of fatigue crack growth for various welding conditions.*

systems, including NRIM-JICST's, employ the bilingual (Japanese and English) mode directly in the user interface.

The data communication network available in Japan is the public telephone and DDX-P. The standardization of networking along the OSI model is under way in cooperation with the interested parties. Two major activities are underway to produce the OSI networking model in Japan. One is a seven-year project sponsored since 1985, by MITI. This "Interoperability Database System" is being carried out jointly by the Electrotechnical Laboratory (ETL) of AIST and the Interoperability Technology Association for Information Processing [17]. Distributed database systems technology and networking based on OSI are among the targets being pursued. The other project is industrial and is directed by the Promoting Conference for Open Systems Interconnection, which consists of seven major computer firms.

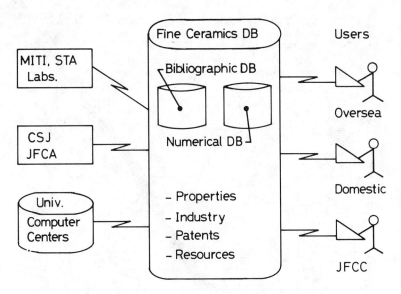

FIG. 7—*Fine ceramics data base proposed by MITI* [16].

Conclusion

In addition to the above activities of various organizations, those of the Versailles Project on Advanced Materials and Standards' (VAMAS) Technical Working Area "Materials Data Banks" are having an impact on many national laboratories building data bases. The international activities of ASTM Committee E-49 and the CODATA Task Group on Materials Data Base Management are attracting the attention of many concerned people in Japan. An effort to coordinate activities in the computerization of materials data in Japan was started recently by the Subcommittee on Materials Data Systems of the Science Council of Japan.

Under the support of STA, three government laboratories (NRIM, GRIN, NCLI) and JICST are working jointly to investigate common ground for mutual access in the distributed (network) environment of materials' data bases. Metals, ceramics and problems related to the system environment are being investigated. The central subject in this study is the Materials Data Interchange Format (MDIF), which allows data to be exchanged between data systems [18].

Acknowledgments

The authors are grateful to Messrs. N. Onodera and K. Shimura of JICST, who kindly provided information on their activities. They also would like to thank the Japanese members of the VAMAS Working Group on Materials Data Banks for their discussion of the subject. The computer graphics work done by A. Miyazaki and A. Ishii of NRIM is also appreciated.

References

[1] Iwata, S., "Materials Data Activities in Japan," in *Factual Materials Data Banks*, H. Kröckel, K. Reynard, and G. Steven, Eds., EUR9768EN, CEC-JRC Petten, 1985, pp. 54–69.
[2] Onodera, N. and Atago, T., *Tetsu-to-Hagane (Journal of the Iron and Steel Institute of Japan)*, Vol. 71, No. 15, 1985, pp. 1726–1733.
[3] Tanaka, T., Editor-in-Chief, "Data Book on Fatigue Strength of Metallic Materials," 3 Vols., Society of Materials Science, Japan, 1982.

[4] Nakazawa, H., Editor-in-Chief, "Data Book on Fatigue Crack Growth Rates of Metallic Materials," 2 Vols., Society of Materials Science, Japan, 1983.
[5] Komai, K., Editor-in-Chief, "Data Book on Stress Corrosion and Corrosion Fatigue Properties of Metallic Materials," 4 Vols., Society of Materials Science, Japan, 1987.
[6] Okimura, N., *Joho Kanri (The Journal of Information Processing and Management)*, Vol. 25, 1982, pp. 71-79.
[7] Fujiwara, S., Uchida, H., Fujiwara, Y., Tikizane, S., Kudo, Y., Araki, K., Komuro, S., and Tominaga, I., *Joho Kanri (The Journal of Information Processing and Management)*, Vol. 25, 1983, pp. 840-859.
[8] Tajima, M., Nakadate, M., Mizoguchi, T., Nagayama, D., Oshima, T., Hiraishi, J., and Atago, T., *Joho Kanri (The Journal of Information Processing and Management)*, Vol. 25, 1983, pp. 929-948.
[9] Osugi, U., *Proceedings of the 9th CODATA Conference*, 1984, pp. 165-168.
[10] Onodera, N., Atago, T., and Uchida, H., "The JICST Thermophysical and Thermochemical Property Data Base System," JICST Technical Note No. 6, JICST, Jan. 1985.
[11] Shinya, N., Kyono, J., Tanaka, H., Murata, M., and Yokoi, S., *Tetsu-to-Hagane (Journal of the Iron and Steel Institute of Japan)*, Vol. 69, No. 14, 1983, pp. 1668-1675.
[12] Monma, Y., Yokoi, S., and Yamazaki, M., *Proceedings of the Fifth International Conference on Pressure Vessel Technology*, Vol. 2, American Society of Mechanical Engineers, 1984, pp. 1366-1392.
[13] Nishijima, S. and Ishii, A., *Transactions of NRIM*, Vol. 24, No. 4, Dec. 1986, pp. 277-286.
[14] Ohta, A., Soya, I., Nishijima, S., and Kosuge, M., *Engineering Fracture Mechanics*, Vol. 24, No. 6, 1986, pp. 789-802.
[15] Yamada, A., *Kogyozairyo*, Vol. 31, No. 11, 1983, pp. 18-24.
[16] Tomita, I., *Ceramics*, Vol. 22, No. 8, 1987, pp. 626-642.
[17] Tabata, K., OSI, Japan Standards Association, 1987.
[18] Kröckel, H., Reynard, K., and Rumble, J. Eds., "Factual Materials Databanks—The Need for Standards," The Report of a VAMAS Task Group, July 1987.

Claude Bathias[1] *and Benard Marx*[2]

Use of Materials Data Bases in France

REFERENCE: Bathias, C. and Marx, B., "**Use of Materials Data Bases in France,**" *Computerization and Networking of Materials Data Bases, ASTM STP 1017,* J. S. Glazman and J. R. Rumble, Jr., Eds., American Society for Testing and Materials, Philadelphia, 1989, pp. 92-98.

ABSTRACT: In order to know the interest and the use for materials data bases in France, a questionnaire was distributed, 5000 copies by the end of 1986. The answers obtained are summed up in this paper. They show what types of organizations are interested in data bases, what their goals are, and what the market is for data bases. In conclusion, we point out the important need of an international cooperation to develop the materials data bases.

KEY WORDS: data bases, materials data bases, metals, polymers, composites

The main findings from a survey organized by CODATA France and DBMIST/Ministère de la Recherche et de l'Enseignement Superieur [1] are presented in this paper. The survey was intended to provide answers to the following questions:

(1) Is the French scientific and technical community interested in materials data bases?
(2) How are existing materials data bases used?

About 5000 questionnaires were distributed and 363 complete answers were received. The results follow.

Type of Organization (Table 1)

Thirty-three percent (33%) of the answers came from industrial companies. A similar number of answers came from scientific researchers (in higher education, engineering schools and CNRS laboratories) and from engineers (in other public research centers, industrial companies and nonprofit making organizations). The percentage of answers from documentation centers is higher in industry than in universities. In industry, the documentation center is a more frequent data base user than it is in universities and CNRS laboratories.

Regions (Table 2)

The results are very imbalanced. The total number of answers from two regions (Ile de France and Rhône Alps) is about half of the general total. The various regional administrations should introduce programs to promote data base use by scientific researchers and engineers and encourage them to participate in data base production.

---

[1]Professor, University of Technology of Compiègne, BP 649, Compiègne, Cedex 60206, France. President of the CODATA France Materials Working Group.
[2]DBMIST, French Ministry for Education, 3 Boulvevard Pasteur, Paris 75015, France.

TABLE 1—*Types of organizations.*

Status	Number of Responses	%	From Documentation Centers	%
Higher education				
universities	84	23	...	...
engineering schools	67	19	...	...
others	2	0	...	...
Total	153	42	7	3
CNRS laboratories	22	6	1	5
Other public organizations	55	15	1	2
Nonprofit organizations	20	6	3	15
Industrial societies	113	31	20	18
Total	363	100	32	43

TABLE 2—*Regional distribution of replies.*

Region	Number of Responses	%
Alsace	6	2
Aquitaine	14	4
Auvergne	3	1
Bourgogne	4	1
Bretagne	11	3
Champagne-Ardennes	2	1
Centre	11	3
Corse	1	...
Franche-Comté	3	1
Ile-de-France	108	30
Limousin	4	1
Lorraine	20	5
Languedoc-Roussillon	11	3
Midi-Pyrénées	14	4
Nord-Pas-de-Calais	16	4
Basse-Normandie	1	...
Haute-Normandie	13	4
Picardie	6	2
Poitou-Charentes	4	1
Pays de la Loire	12	3
Provence-Alpes-Côte-d'Azur	24	7
Rhône-Alpes	66	18
Départements d'Outre-Mer	4	1
Foreign or nonspecific	5	1
Total	363	100

Activity Sector (Table 3)

Chemistry is the most strongly represented sector. This is a classical situation in France, given the importance of chemistry in metallurgy and in polymers. It is a national problem where materials are concerned. The community of chemical scientists is more frequently addressed than that of mechanical engineers. This is unfortunate since there is as great a need for mechanics and physics in materials science and engineering as in chemistry.

TABLE 3—*Distribution of replies by sector of activity.*

Sectors	Number of Responses	%
Chemistry	82	23
Mechanics	44	12
Physics	42	12
Electrical, electronic	34	9
Information	34	9
Construction, civil	29	8
Energy	23	6
Earth sciences	17	5
Transportation	15	12
Other	43	12
Total	363	100

Materials Type (Table 4)

The results show that to meet the major user requirements, data bases must be produced in the following sectors: metals, polymers, and composites. The biggest market is for these types of materials.

Data Base Use Objective (Table 5)

There has been a considerable increase in the trend towards the use of data bases for education (training, research, and development). Even in industry, data bases are most often used as educational tools. This trend is perhaps too strong considering that only 15% of respondents use data bases to create new materials and the obvious advantages of data bases for research departments. However, these results also show that only 50% of the engineers interested in data bases for conceptualization actually use them whereas this percentage drops to 30% in other data base applications.

Seventy-three percent (73%) of the engineers are interested in data bases for production (Table 6). This application is less common in higher education, which is to be expected.

Only 20% of the respondents in technical research centers use data bases in production. This is a national weakness, especially in iron and steel metallurgy. Fifty-three percent (53%) of the answers from engineers state that data bases are useful for training, research, and development.

TABLE 4—*Types of materials.*

Material	Number of Responses	%
Biotechnology, materials	39	11
Metals	174	48
Polymers	150	41
Composites	134	37
Ceramics	111	30
Construction	81	22
Electronics	101	28
Glasses	58	16
Wood and natural materials	39	11
Total	887	111

TABLE 5—*The use made of data bases.*

Action	Number of Responses	, %
Research, development, training	285	79
Conceptualization	54	15
Processing, production, manufacturing	124	34
Application	93	26
Totals	556	154

TABLE 6—*Use made of data bases by the different types of organizations.*

Action	Higher Education	Laboratory	Other Public Organizations	Nonprofit Organizations	Technical Societies	Average
Research, development, training	94	95	80	80	53	79
Conceptualization	9	9	20	10	15	15
Processing, production, manufacturing	14	18	22	20	73	34
Application	6	5	37	40	38	26

This is a lower percentage than in the replies from universities and the CNRS, but it is nevertheless important. It may be thought that these replies relate to the types of materials, but this is not the case (Table 7). Production represents between 20 and 30% of data utilization for all types of materials just as research and development represent between 75 and 88%. The type of material covered is not a determining factor in data base utilization; the application has a greater influence on the decision to use a data base.

Data Base Use

There is a marked difference between bibliographic data base use (81% of the answers) and factual data base use (36%) (Table 8). External data bases are more frequently used than internal data bases (either bibliographic or factual).

TABLE 7—*Use of data bases in relation to the type of material.*

Materials	Research, Development, Training	Conceptualization	Processing, Production, Manufacturing	Application
Metals	75	10	34	30
Polymers	74	19	32	28
Composites	81	16	32	31
Ceramics	83	16	25	28
Electronic	76	18	29	29
Construction	88	15	25	26
Glass	84	14	19	22
Wood, natural materials	79	16	26	31
Biotechnology	87	23	13	28

TABLE 8—*Use of data bases.*

Application	Factual Data Bases	, %	Bibliographic Data Bases	, %
External	98	27	265	78
Internal	55	15	85	23
Total	132	36	293	81

When comparing data base use with the type of organization to which the user belongs (Table 9), it becomes apparent that higher education does not frequently use data bases. However, industries use data bases more frequently (48%). The percentages for bibliographic data bases and online suppliers (Table 10) show that the PASCAL, CAS, and METADEX data bases are cited in more than half the answers. QUESTEL, ESA-IRS, DIALOG, and CEDOCAR represent 80% of the hosts cited.

The percentages for bibliographic data bases and online suppliers (Table 10) show that the PASCAL, CAS, and METADEX data bases are cited in more than half the answers. QUESTEL, ESA-IRS, DIALOG, and CEDOCAR represent 80% of the hosts cited.

The importance of THERMODATA should be noted in the results concerning external factual data bases use (Table 11).

Budgets for Data Base Use (Table 12)

Fifty-nine percent (59%) of the answers indicate an annual expenditure of less than 10 000 francs ($1500) and only 9% indicate an expenditure of over 100 000 francs ($15 000). Average annual expenditure is about 20 000 francs ($3000). The market for materials data bases is not very large considering the work involved in producing and maintaining a data base. It is not possible for companies to make profits with these data bases. Although it is possible that the situation will change and in the next few years there will be a rapid and important increase in materials data base use, at present it seems difficult for a private company to make money with data bases alone.

TABLE 9—*Data base use of different types of organization.*

	Application of Factual Data Bases					
	External		Internal		Total	
Status	Number	%	Number	%	Number	%
Higher education	27	18	11	7	33	22
CNRS laboratories	9	41	1	5	9	41
Other public organizations	18	33	16	20	28	51
Nonprofit organizations	7	35	2	10	8	40
Industrial societies	37	33	25	22	54	48
Total	98	27[a]	55	15[a]	132	36[a]

[a]Average.

TABLE 10—*Use of external bibliographic data bases.*

Responses, %	BD biblio.	Host	Responses, %
25	PASCAL	QUESTEL	35
16	CHEM. ABS	ESA-IRS	24
11	METADEX		
6	INSPEC	DIALOG	11
4	INPI	CEDOCAR	8
4	COMPENDEX		
4	NTIS	INFOLINE	5
3	CETIM	SDC	5
3	WPI		
3	NASA	STN	4
2	RAPRA	DATA-STAR	2
2	EDF-DOC		
2	NORIANE	GCAM	1
1	FIESTA		
1	ELODIS	THERMODATA	1
1	LABINFO		
1	THERMDOC	ECHO	1
1	CETIF		
10	Other data bases	Other hosts	3
100			100

We must request the French government to encourage different types of organizations to produce data bases. Producers must consider the potential market and appropriate size of the data base.

From the results of the survey on expenditure for data base use, it seems difficult to set up data bases for broad industrial sectors. The solution is to produce a number of small highly specialized data bases, with the backing of the public authorities. A network must be organized to link these data bases. Regional and national governments should cooperate in the field of data bases (with the European Community). Given the size of the task with which we are faced, collaboration should exist on all levels, be they national, European or international.

TABLE 11—*Results concerning external factual data bases use.*

Name	Producer	Domain	Number of Responses Naming Data Bases
ADEP	Université de Limoges	ceramics	1
ARTEMISE	CRRC, Nancy	geochemistry	2
BSS	BRGM, Orleans	geology	5
CINDAS	Purdue University	technology	1
CORROSION	Marcel Dekker	corrosion	1
EMIS	IEE, G.B.	electronics	3
FIMAC	CETIM, Nantes	composites	1
G3F	INSA, Lyon	composites	1
GAPHYOR	University of Paris Sus, Oresay	physics, chemistry	2
HEILBRON	Chapman and Hall, G.B.	chemistry	2
HYDROGENE DATA	ENSWCP, Paris	hydrogen in materials	1
JANSSEN	Janssen Chemnica, Bcigique	chemistry	1
MDF	ASM	metals	4
MERCK	Merck	chemistry	1
MICROPLAST	EAPH, Strasbourg	plastics	1
NIH/EPA	NBS	mass spectrometry	1
PPDS	ICE, G.B.	physical chemistry	3
PROCESS	Simulation Science	thermodynamics	1
PROCOP	CISIGRAPH	plastics	2
THERMODATA	SGTE, Grenoble	chemical metallurgy	48
TOXDATA	NLM	chemistry	1
VULCAIN BDM	CT-DEC, Cluses	mechanics	3
Nonprofit			28
Total			114

TABLE 12—*Annual budget for data base use.*

Annual Budget	Number of Responses	Percent
>100 000 F ($17 500)[a]	31	9
10 000 F to 100 000 F ($1750 to $17 500)	63	17
1000 F to 10 000 F ($175 to $1750)	131	36
<1000 F ($175)	83	15
Nonprofit	55	15
Total	303	100

[a]May 1988.

Reference

[1] Utilisation des banques de données sur les matériaux, resultats d'enquête, Oct. 1987, CODATA FRANCE et DBMIST/MRES, 33 pp.

Anthony J. Barrett[1]

CODATA Activities on Materials Data

REFERENCE: Barrett, A. J., **"CODATA Activities on Materials Data,"** *Computerization and Networking of Materials Data Bases, ASTM STP 1017,* J. S. Glazman and J. R. Rumble, Jr., Eds., American Society for Testing and Materials, Philadelphia, 1989, pp. 99-106.

ABSTRACT: The International Council of Scientific Unions Committee on Data for Science and Technology, (CODATA), is broadly representative of the industrialized nations. Because of this wide international constitution, CODATA has served as an effective catalyst in bringing together those who are involved in producing, managing, and using materials data systems and in arranging collaborative efforts between them. This role is currently being fulfilled by the identification of national and multinational materials data projects and of the experts who are involved in them; the promotion of awareness and communication; influencing the preparation of standards, guidelines, and consistent terminology; and the dissemination of knowledge and practices in the technology of building and using materials data bases. In addition, CODATA aims to assist the materials data base manager by drawing attention to and providing information on such issues as liability, cost-benefits, and user-friendliness.

KEY WORDS: CODATA, cost-benefits, international cooperation, liability, materials, materials data bases, user-friendliness

CODATA was founded as an interdisciplinary scientific committee of the International Council of Scientific Unions (ICSU) in 1966 and seeks to improve the quality, reliability, management, and accessibility of data in all fields of science and technology. It aims to facilitate international cooperation among those collecting, organizing, and using data. These objectives are addressed through the preparation of key data sets, for which consistent international use is desirable, the coordination of multinational projects, and the establishment of formal standards to promote the compatibility of data bases. CODATA's objectives are also met by providing guidelines for the presentation of data, supplying information on the sources of reliable data, offering education and training, and organizing conferences and workshops.

CODATA has a long history of involvement with data related to, for example, thermodynamic properties for use in chemical engineering. Within the scope of the present symposium, that is, data relating essentially to the mechanical properties of engineering materials, the involvement is of more recent origin. A number of papers have been presented at the biennial International Conferences over the last 20 years. At the most recent such conference, one in 1986 on the Computer Handling and Dissemination of Data [1], a substantial proportion of the contributions were related to materials data. Attention has been directed to the specific issues of the computerization and networking of engineering materials property data bases in specialized workshops.

---

[1]Consultant, ESDU International Ltd., 251-259 Regent Street, London, W1R 8ES, United Kingdom, and chairman, CODATA Task Group on Materials Data Base Management.

International Workshops on Computerized Data Systems

At the initiative of its Commission on Industrial Data (founded originally as an ad hoc Working Group in 1974), CODATA has jointly sponsored two major international workshops. The first of these has had a substantial influence on the initiation of national and multinational enterprises that are building comprehensive materials data systems. Some of these systems have now reached the demonstrator stage and are the subject of papers at the present symposium. As an introduction to more recent CODATA activities, it is worthwhile recalling briefly some of the achievements of these carefully focused workshops.

Fairfield Glade Workshop

A workshop devoted to the discussion of problems confronting the development of computerized materials data systems was organized at Fairfield Glade, TN, in Nov., 1982. Financial support and planning was provided by CODATA in association with several other official and commercial bodies in the United States and from the Federal Republic of Germany.

Participants from some seven countries took part. Among the significant results of this workshop [2] were the conclusions that there were no significant technical barriers to the development of a computerized system for materials data and that, without prompt cooperative action, either a large unsatisfactory system or uncoordinated, independent data bases would evolve. The major problems foreseen included those of raising the necessary financial resources, data validation and establishment of data quality, user-friendliness, the lack of studies of the economic value of such systems, and standardizing and cross referencing materials properties and test methods. Subsequently, several of those problems have been addressed by national and international groups and others are being given further attention by CODATA, as we shall see later.

The system concept recommended by the Fairfield Glade workshop was a coordinated, distributed system of independent data bases, each of limited scope, connected by a common gateway computer in such a manner that the user gains access to all data bases with a single phone call. The gateway was also foreseen as providing support, on-line directories, and other services. This concept, in essence, is that which has been implemented by the Material Properties Data Network, (MPD), in the United States. The problems relating to standards and the cross referencing of materials that were identified at Fairfield Glade have subsequently been addressed by ASTM Committee E-49 on the Computerization of Materials Property Data, on which J. G. Kaufman is reporting at this symposium.

In the wider international context, the Fairfield Glade workshop acted as a stimulus and pattern for other national or multinational studies. Notable among these was the workshop on factual materials data banks organized by the Commission of the European Communities (CEC) at Petten, the Netherlands, in Nov. 1984 [3]. The activities following this initiative are being reported at this Symposium by H. Kröckel and G. Steven. There are several important differences between the systems concept of the demonstrator being activated as part of the CEC program and that of the MPD. Notably, the CEC demonstrator directs the user to individual data bases rather than giving access to all data bases via a single gateway connection as conceived at the Fairfield Glade workshop. A range of different approaches is represented among the national and multinational systems reported at the 10th International CODATA Conference [1].

Schluchsee Workshop

In September, 1985, CODATA originated a workshop at Schluchsee in the Federal Republic of Germany on materials data systems for engineering. It was jointly sponsored and funded by CODATA with a number of German information centers. The Fairfield Glade workshop had identified international participation in the development of computerized systems as a missing factor in the provision of access to broad files of materials properties. So among the goals set by CODATA for this new workshop were the identification and assessment of conditions for international cooperation in building computerized data systems for engineering materials.

The five-day workshop involved participants from 13 different countries. It also attempted to obtain a reasonable balance among those present with respect to their backgrounds—system users (design engineers), system developers (materials engineers and computer technologists), system marketers, and government managers of technical information. The final report [4] defined a number of desirable projects relating to data base building, management, and support. It recommended actions relating to the provision of widely acceptable guidelines for data base building, the establishment of liaisons between CODATA's Commission on Industrial Data and a number of national and international agencies, and liaison with national and other bodies producing directories of materials data resources. The report also proposed the establishment of an international group to provide communication between data base managers.

In effect, the Schluchsee workshop recognized that the national systems then evolving went some way towards avoiding the proliferation of uncoordinated independent data bases that the Fairfield Glade workshop had warned against. However, this evolution was still uncoordinated at the international level. Though many of the individuals working on the various national and multinational projects were frequently in personal contact and attended the same national and international meetings, the group of professionals involved in these projects as managers or as users was widening rapidly. Among other things, there was an urgent need to build bridges between users and data base managers, and for collaboration between data base managers, on an international scale.

From this point, the activities of CODATA with respect to engineering materials have become more focused on a number of specific projects we now describe.

Directories

In the past decade, the traditional attitudes towards engineering materials have been changing in the industrialized and many other nations. Gone are the days when designers could rest assured of secure sources of supply or could try to cover the shortcomings of design by the profligate use of relatively cheap materials. The expanding ranges of new materials and processes have also placed demanding commercial premiums on the selection of materials both for traditional and for new applications.

This has been reflected in the concern shown by governments and their support of legislation and studies as typified by the passing of the Materials Act of 1984 in the United States and the publication of the "Collyear Report" on new materials and processes [5] in the United Kingdom. Unfortunately, in the modern world, government concern is not always accompanied by the funding to apply to the problems at the root of that concern. Nonetheless, it is gratifying that official pronouncements such as those mentioned have highlighted the importance of information systems as an integral part of the technological resources of a nation.

Once the importance of information and data resources has been recognized and publicized by official bodies, it is not unusual to discover that many of those who should be using these resources are not doing so; some are even unaware of their existence! This has resulted in a number of projects to set up directories of materials information sources. Typical of these are national directory projects in the United Kingdom and in the Federal Republic of Germany and a multinational initiative in the European Community. At the time this article was written, production of these directories had reached or was nearing completion.

At the international level over the past 20 years, CODATA has listed data sources in its Directories of Data Sources for Science and Technology. These embrace such subjects as crystallography, chemical kinetics, and nuclear and elementary particle physics. Technical data for engineering materials have more recently been addressed and directory chapters covering mechanical properties and corrosion are under preparation. Surveys of these areas, which cover countries outside the European Community, have been made by the National Bureau of Standards (NBS) on behalf of CODATA. The information contained in the European Community directory—which now embraces that in the previously mentioned United Kingdom and Federal Republic of Germany directories—will be merged with that collected by the NBS. The end result will be comprehensive international CODATA directories of materials data resources on mechanical properties and on corrosion.

The CODATA directories include information on many sources, some of which do not offer computerized data bases. In a further initiative, the International Council for Scientific and Technical Information (ICSTI), which is a scientific associate of ICSU, has set up a Numeric Data Group with the collaboration of CODATA. The interest of this Group includes international cooperation in data compilations. One project to this end relates to the identification of numeric data bases worldwide. This project will produce a directory of machine searchable data bases. In due course the ICSTI directory will include a section on engineering materials, which will in effect be that subset of the CODATA directory that refers solely to machine readable data bases on materials.

Materials Data Base Management

Based on some of the recommendations of the Schluchsee workshop, CODATA has instituted a Task Group on Materials Data Base Management. The objectives of this group are to provide an international vehicle for communication between, and coordination of, builders, managers, and users of machine readable engineering materials data bases. Areas of action for the Task Group include

- promoting communication and awareness,
- influencing the preparation of standards, guidelines, and terminology,
- providing education in the technology of building and using materials data bases, and
- promoting knowledge and practices to improve the bridge between users and managers of materials data bases.

The constitution of this group had to embrace a wide range of national interests as well as establish a positive connection with at least the major national and multinational initiatives in materials data base systems, standardization, and other activities relating to those systems. A further boundary condition was to do this with a core group of experts small enough to ensure positive commitment and participation in determining and implementing the group's actions.

The task group comprises members and corresponding members from China (People's Republic), France, Germany (Federal Republic), Japan, Sweden, the United Kingdom, the United States, and the U.S.S.R. The same members, by reason of their personal affiliations, also provide direct links with several major national and multinational activities, which include the following:

ASTM Committee E-49,
Chinese national programs,
COMECON collaborative materials data base programs,
the European Community collaborative program,
Japanese national programs,
the National Materials Property Data Network of United States,
U.S.S.R. national programs, and
the Versailles Project on Advanced Materials and Standards.

A report on VAMAS, the last of those listed above, is being given by K. Reynard at the present symposium.

Beyond its relatively small formal membership, the task group is identifying those professionals worldwide who are involved in materials data base building and management and who can be associated with the group's projects. Thus the core group of members and corresponding members can apply international leverage on a scale that is probably unique to CODATA.

How is this task group proposing to act in response to the needs it was set up to serve? This may best be answered by reference to some of the initial projects and issues that have been illuminated in the group's discussions since it first met in June, 1987.

Register of Materials Data Base Managers

The directories previously described will give the names of contacts appropriate to each data source, but these contacts will not necessarily be data base managers. Furthermore, many private and company data bases will not appear in the directories. All data bases, no matter what their ownership, have similar problems in their professional management. Because it is the intention of the CODATA task group to serve all data base managers, it is desirable that as many of them as possible be located. This is the purpose of a new register now being compiled.

Newsletter

In addition to identifying the principal actors on the international materials data base scene, it is necessary to establish a regular channel of communication with them. A newsletter has been produced to provide a central resource of regular, authoritative information on works on standards, network developments, new data bases, government programs, and short notes on issues that need to be brought to the attention of data base managers and principal users. The newsletter is being distributed to individuals in the 17 CODATA countries through their national committees. Other associates of the group who would not normally receive information by this route will receive separate mailings. In addition, appropriate technical societies will be cooperating by reproducing this newsletter, wholly or in part, in their own journals.

Guidelines for Data Base Building

There are a number of studies on the principles of data base building and some standards and codes. It seems important to summarize the most widely accepted practices in a concise, easy-to-read document that will be of particular value to managers new to the field of materials data bases. The objective of these guidelines is to promote good practice in such matters as using standards, consistently ensuring ease of use, making provision for hot lines, establishing internal quality assurance, and controlling access to data (for example, to protect data bases against contamination through unauthorized access). A set of such guidelines is under preparation. It will be made widely available to data base managers in the form of a CODATA Guideline Bulletin.

Liability

Legislation regarding the liability of producers of products and services and consumer protection is changing in many countries. Numeric data bases, such as those on materials, contain information that will be used directly in technological applications. Moreover, the trail back from the catastrophic failure of a technological artifact to the numerical information on which its construction was based may be highly visible.

This exposure must be understood by the data base manager in relation to the legislation applicable in the community he is serving. For example, the defense traditionally thought to be provided in some countries by disclaimers may not actually exist in law and may now have to be augmented or replaced by provisions in the contract between the data base provider and the user. The task group is examining the different national and international developments in this area to see what, if any, real commonality exists. Its main, immediate action, however, is to alert materials data base managers to the need to inform themselves on the specific legislation appropriate to the communities in which their data bases will be used.

Cost-Benefits Studies

Managers of information services frequently find themselves spending a great deal of time and effort in trying to convince governments or commercial customers of the benefits those services can bestow in relation to the price paid for them. Price is related to the cost of setting up, maintaining, and operating systems. The task group agrees that these cost factors are tractable and could, if necessary, be studied on an international basis. However, there is much greater difficulty in demonstrating the benefits side of the equation in sufficiently effective and incontrovertible terms.

In the past estimates have been made of potential national losses or of company losses that might result from a failure to apply known information. Also, case histories have been collected on actual losses that have been traced back to a failure to apply known information. With the possible exception of a few instances, these efforts seem to have made little impression on the governments of nations or on the directors of companies that might be vulnerable to such losses. Possible explanations for this paradoxical situation have been offered in Ref 6.

One industrial view is that if the development of materials data bases is properly market driven, then the selling of them, as with other products, will not be a problem. The benefits will be self evident and price will be a secondary consideration. While there is general agreement that better knowledge of their markets is vital to all data base managers, it also seems to be general experience that the traditional procedures employed by professional marketing consultants are not effective in producing this knowledge. Conventional market surveys are conducted against a background of definable product cost-benefits, which, as has already been noted, are difficult to formulate for materials data bases.

The CODATA task group is not alone in trying to address this problem, of course, and a number of studies have been made in several nations and multinational groups. These are being reviewed in preparation for an attempt to specify a new approach to assist the materials data base manager in gaining a better understanding of his market.

User-Friendliness Issues

In planning his response to an industrial market as discussed above, the data base manager is largely attempting to accommodate national and commercial desiderata not of his making. By contrast, when he addresses the difficulties the customer encounters when using his system, the data base manager is attempting to deal with problems which, in most cases, are very much of his own making. Such nonprice factors are as vital to the customer's acceptance of a system as are the benefits the customer may hope to obtain from reliable data at a competitive unit cost.

Members of the CODATA task group who are from an industrial background maintained in a recent discussion that from their point of view the user-friendliness of data base systems is of *primary* importance. They are not alone in their view. A recent announcement for Interact '87, an authoritative international conference on human-computer interaction, observes:

"People-problems" are now widely recognized as the major issues which determine success or failure in the effective and profitable use of information systems. Therefore designers, manufacturers, purchasers and users are asking how systems can be made safe, more useful, less tiring, as well as easier to learn and use.

Rumble and Westbrook [7] comment that though the term "user-friendly" is overworked, in relation to the materials data base system it has a specific meaning: that the system can be used without having to turn constantly to manuals or telephone hot lines and without having to adopt new nomenclature, but be able to do simple searches easily, to make more difficult searches straightforwardly, and to manipulate data without hassle.

Rumble and Westbrook's "definition" must be coupled with the objective of retrieving numeric data with minimum probability of errors resulting from faulty communication. In the case of bibliographic data bases, the absence of user-friendliness may result in unnecessary cost, frustration and, at worst, a breakdown in communication. Even so, the consequences of error in this case are effectively insulated from the decision-making process. In the case of the numeric data base, however, that which is retrieved will be used immediately in deciding a course of action or in making a technical calculation. So while user-friendliness may do little more than offer convenience to the user of a bibliographic system, it is of crucial importance to the user of numeric data. This reflects not only on the commercial attractiveness of what the data base manager has to offer, but it could also be pivotal to his defense in a legal liability suit.

Considerations such as those outlined above have caused both the CODATA Task Group on Materials Database Management and the ICSTI Numeric Data Group to prioritize user-friendliness as a topic on their agendas. Both are aware of the considerable amount of research currently in progress in such areas as the "man-machine interface" and "human-computer interactions." While this research is applauded, there is an urgent need to put together a straightforward, practical aid package that will crystallize the principal features and practices employed in those softwares and systems that do exhibit a reasonable degree of user-friendliness. Both groups must promote the idea that systems and software designers must pay more than lip service to the need for user-friendliness.

It is my personal experience that the majority of the failings of current softwares and systems in ease and reliability of use by engineers (as distinct from computer experts) can be rectified by applying the same rules that ensure legibility and understanding during communication via long-established media. Rules that are unique to electronic data manipulation systems are probably a small proportion of those that need to be followed to achieve an attractive and safe level of user-friendliness.

Conclusion

CODATA activities on materials, within the scope represented by the present symposium, first related to bringing qualified, international opinion and expertise to bear on identifying the conditions under which materials data base systems could be made feasible and the problems that had to be overcome before those systems could become practicable. More recently, the attention of CODATA has turned to problems associated with the management of the systems that have now begun to emerge and to issues that will enable those systems to be economically viable whether within individual, national, or international economic domains. Several of those problems and issues that need to be made visible to the managers and users of these systems are already receiving attention with a view to the production of guidelines influencing the production of standards and the initiation of cooperative actions that can be supported internationally. Other issues, such as copyrights, other proprietary interest questions, and data base security, will receive attention later.

The development of materials data base systems is a complex and expensive business. CODATA, by virtue of its wide international constitution, is well placed to assist the beneficial exploitation of the global resource of materials information by all users of individual systems and of national and multinational networks. These benefits will be obtained through the sharing of experience and expertise amongst practitioners in many nations. They will also derive from the dissemination of the results of their achievements and those of other experts who are willing to associate themselves with the same objectives. All who feel they have a contribution to make to this work are invited to make their interests known to the CODATA Task Group on Materials Data Base Management and to participate in the group's communications and projects.

References

[1] Glaeser, P., Ed., "Computer Handling and Dissemination of Data," *Proceedings of the 10th International CODATA Conference,* Elsevier Science Publishers, Amsterdam, 1987.
[2] Westbrook, J. H. and Rumble, J., Eds., "Computerized Materials Data Systems," Steering Committee of the Computerized Materials Data Workshop, Fairfield Glade, TN, Nov. 1982.
[3] Kröckel, H. et al., Eds., "Factual Material Data Banks," Compilation of discussion reports compiled by the Commission of the European Communities Workshop, Petten, EUR 9768 EN, Luxembourg, 1985.
[4] Westbrook, J. H. et al., Eds., "Materials Data System for Engineering," *Proceedings of a CODATA Workshop, Schluchsee,* CODATA Karlschruhe, Germany, Sept. 1985.
[5] Collyear, J. et al., "A Program for the Wider Application of New and Improved Materials and Processes," Report of the Materials Advisory Group, Department of Trade and Industry, HMSO, London, Feb. 1985.
[6] Barrett, A. J., "The Costs of Not Having Refined Information," in *The Value of Information as an Integral Part of Aerospace and Defence R&D Programmes,* Advisory Group for Aerospace Research and Development (AGARD) Proceedings, Vol. CP 385, AGARD, Paris, 1985, pp. 5.1–5.9.
[7] Rumble, J. and Westbrook, J. H., Eds., "Computerizing Materials Data—A Workshop for the Nuclear Power Industry," National Bureau of Standards Special Publication 689, Washington, DC, Jan. 1985.

Emerging Issues

Stanley Y. W. Su[1] *and Abdullah Alashqur*[1]

Uniform Treatment of Integrated CAD/CAM Data and Metadata

REFERENCE: Su, S. Y. W. and Alashqur, A., "**Uniform Treatment of Integrated CAD/CAM Data and Metadata,**" *Computerization and Networking of Materials Data Bases, ASTM STP 1017*, J. S. Glazman and J. R. Rumble, Jr., Eds., American Society for Testing and Materials, Philadelphia, 1989, pp. 109–125.

ABSTRACT: An integrated data base is the heart of an integrated manufacturing system. It serves as the glue to bind together heterogeneous component systems, which are often developed independently for supporting product design, process planning, manufacturing, product inspection and testing, and product service and maintenance, into a fully integrated system. To establish and use such a data base, a data model that is very rich in semantics is necessary to model the complex structural properties, operations, and constraints associated with the data objects commonly found in computer aided design (CAD)/computer aided manufacturing (CAM) applications. Furthermore, it would be necessary to have a data dictionary facility to allow the users of various component systems to define, display, and modify the schema of the integrated data base. This paper presents a semantic association model called SAM*, which contains six general constructs for modeling the complex relationships (or associations) among data objects. These constructs can be used in an embedded or recursive fashion to allow very complex objects to be explicitly defined in a schema that contains the metadata of an integrated data base. In this work, the same semantic model is also used to model the contents of the schema so that the metadata, as well as the application data of an integrated data base, can be uniformly defined, stored, displayed, modified, and manipulated by the users. This paper also describes an interactive data dictionary system implemented for the definition and manipulation of metadata and application data modeled by the semantic model.

KEY WORDS: data base, schema, metadata, data dictionary, computer aided design (CAD)/computer aided manufacturing (CAM)

Factory automation has become a new technology pursued by all industrial countries. The main objectives of factory automation are to (1) increase the productivity of factories, (2) improve the quality of products, (3) relieve factory workers from working in unpleasant, hazardous environments, (4) minimize production delays, and (5) increase resource use. For a manufacturer to economically compete with other manufacturers, it is necessary to integrate the various control processes and data used in the design, manufacturing, sale, and service of products so that feedback from any of the processes can be used to effect better design, planning, control, and production. Computer Integrated Manufacturing (CIM) can achieve such an integration by the use of computers and computational techniques in design, planning, and manufacturing processes. Central to a CIM system is a data base management system (DBMS) that facilitates the sharing of data among the component systems.

Future CIM systems will most likely be networks of heterogeneous computer systems, each of which controls and supports the operations of numerical control (NC) machines, robots, transport vehicles, machine tools, and so forth. This is because (1) low level manufacturing equipment is likely to be produced by different manufacturers who use different computer hardware

[1]Professor and researcher, respectively, Database Systems Research and Development Center, University of Florida, Gainesville, FL 32611.

and software for the control and support of this equipment, and (2) the factories moving towards automation are likely to add new but different equipment and computer facilities to the existing facilities rather than entirely replacing them for economic reasons. Component systems in a factory network will have their own data stored in those structures most suitable for supporting their different processing needs and will have quite different data management facilities, which may range from full scaled DBMSs, to file management systems, and to simple application programs. The problem of integrating heterogeneous data base systems is discussed in Refs 1 and 2.

The heterogeneous nature of a CIM system introduces a number of data base requirements. First, the data used to control and support the design, manufacturing, sale, and service of products will be physically stored at and processed by the component systems. It is necessary to have a common data model, that can be used to explicitly define the structures, constraints, and operations associated with the data. These structures, constraints, and operations represent the semantic properties of the data. Second, the integration of the diverse activities in a factory requires that the data model be rich in semantics and capable of defining the diverse engineering, statistical, scientific, and business data bases necessary for the operation and management of modern factories. Third, the data model should provide a strong data typing capability to allow the definition and processing of the complex data types found in design, manufacturing, and materials property data.

The existing data models, such as the relational, network, and hierarchical models, and the commercially available DBMSs built based on these data models, are not entirely suitable for managing the data bases in the CIM environment [3,4]. This is because these data models and systems are mainly designed for managing business oriented data bases rather than scientific and technical data bases. Many data types and semantic properties useful in computer aided design (CAD)/computer aided manufacturing (CAM) applications are not captured by the constructs of the available models. The available models are quite inadequate in modeling even some of the business oriented data bases.

Motivated by the limitations of the existing data models for CAD/CAM applications, we have developed a Semantic Association Model (SAM*) [5,6] for modeling the CAD/CAM data and for storing data in a heterogeneous CIM environment. The model is rich in semantics in comparison with the other models. It can be used to model not only the data of a wide variety of applications, but also the metadata that define the application data. SAM* has been adopted by the National Bureau of Standards under a project called the Automated Manufacturing Research Facility (AMRF). Under the support of the National Bureau of Standards, we at the University of Florida have designed and implemented a prototype data dictionary system for computer integrated manufacturing data bases. The main feature of this dictionary is that the contents of the dictionary are modeled using the same model (that is, SAM*) used to model the application data. Consequently, the metadata and the data can both be defined, stored, and processed homogeneously using the same data base creation and data manipulation facilities of DBMSs.

This paper describes a refined version of SAM* based on its original version [5] and on our recent experience in modeling CAD/CAM data (see the following section). The conceptual schema for the data dictionary as modeled by SAM* is then presented, followed by a description of an implemented interactive dictionary system and some concluding remarks.

The Semantic Association Model SAM*

An Overview of SAM*

In SAM*, a data base is modeled as a network of interrelated object classes. An object class contains a set of objects that share the same structural properties, operational characteristics, and semantic constraints. (The terms "concept type" and "concept occurrence" are used in Ref 5 instead of "object class" and "object," used here. The latter are more suitable for the CAD/CAM data addressed in this paper.) Objects and object classes are defined and represented in terms of other objects and object classes. An object class to be defined is called a "defined class" and the classes used to define a defined class are called "constituent classes." A defined class can be defined by grouping either a set of similar objects (for example, a set of integers that represents an object class "Item-number") or a set of dissimilar constituent classes whose objects are used to represent the objects of the defined class (for example, each object in a defined class "Part" is defined in terms of a part number of a constituent class "Part-number" and a part name of a constituent class "Part-name"). These different types of groupings are called associations. Six association types (that is, six different ways to group objects and object classes) are provided by SAM*. They are Membership (M), Aggregation (A), Interaction (I), Generalization (G), Composition (C), and Crossproduct (X). (The seventh association type, Summarization, presented in Ref 5 has been dropped from this version of the model. The semantics captured by that association type is now incorporated in the association types "Composition" and "Crossproduct".) Each association type has a different set of structural properties, operational characteristics, and semantic constraints. Some of these properties, characteristics, and constraints may overlap with those of the other association types. An object class may have a number of associations of different types with other object classes. Its objects inherit all the properties, characteristics, and constraints of these association types. One of these association types is considered dominating. It is used to label the object class, as will be shown later.

Since an object class can be defined in terms of a number of constituent classes and itself can serve as a constituent class of many other classes, it is necessary to have a mechanism to name each interrelationship between a defined class and a constituent class. In SAM*, attributes are used to model and name the interrelationships among object classes. An object class is defined by a set of attributes. Each attribute is a mapping from the defined class to a constituent class. It draws values from the constituent class. A set of values corresponding to the attributes of a defined class forms an *instance* of that class. Thus, an instance is a representation of an object in a class. In SAM*, an object can belong to many classes and have different representations (that is, have different instances) in these classes. This is distinctly different from the concept of object and object class hierarchy in the traditional object-oriented programming languages like Small-talk [7], in which an object can belong to a single class.

Two kinds of attributes are distinguished in SAM*: descriptive attributes and structural attributes. An object class can have either descriptive attributes or structural attributes or both. The values of the descriptive attributes of a defined class describe or characterize the objects in the class and the values of structural attributes specify how the objects in the defined class are formed in terms of the constituent classes that these attributes are associated with. The association between the defined class and its constituent classes as represented by descriptive attributes is called the Aggregation association. The structural attributes, on the other hand, represent either the Generalization, Interaction, Composition, or Crossproduct association that a defined class has with its constituent classes. Before explaining the different types of SAM* associations, three representations of a SAM* data base are presented below.

1. Data base Schema: This is a textual representation of a data base using a data definition language (DDL). The schema contains a set of declarative statements posed in the DDL and defines the attributes (representing the structural properties), the procedures and functions (representing the operational characteristics), and the production rules (representing the semantic constraints) associated with the object classes in a data base.

2. Semantic Diagram (S-diagram): This is a graphic representation of object classes and their associations. In an S-diagram, nodes represent object classes and links represent attributes. A node is labeled by one of the characters M, A, I, G, C, or X to show the prominent

association type it has with other classes. The S-diagram gives a visual description of the data base contents, which provides a more convenient representation than a schema from the users' point of view. It is an abstraction of the schema in which some of the details represented in the textual representation are omitted from the S-diagram. It is possible to provide several levels of abstraction using the S-diagram depending on the amount of detail to be shown at each level. In the lowest level of abstraction, which is above the schema representation, object classes, their attributes, and some of the commonly used constraints can be shown in the S-diagram. Details about the operations, rules, and other constraints are hidden at this level. An S-diagram at a higher level of abstraction may show only the object classes that involve certain types of association. Providing such a layered graphical representations facilitates the top-down browsing of a complex data base by the data base user.

3. Generalized Relation (G-relation): This is a tabular representation of the instances of an object class similar to the notion of relation in the relational data model. A G-relation is a nonnormalized relation defined over a set of attributes, each of which is associated with either a system-defined object class or a user-defined object class. A system-defined object class can be either a simple class (for example, integer, real, character, boolean) or a complex class (for example, set, ordered-set, bag, vector, matrix, date, and so forth, which are in turn defined over other object classes). The user-defined classes are those defined by the user (or DBA) for an application. Thus, a tuple of a G-relation is a hierarchical structure of attribute values drawn from the related object classes. A G-relation is the logical structure for storing and manipulating the instances of an object class. Each tuple contains a globally unique object identifier (OID). Objects refer to other objects using their OIDs. Unlike the relational model in which tuples of one relation are related to tuples of another relation using the concept of primary key, candidate key, and foreign key, the tuples in a G-relation are related to the tuples of another G-relation using their OIDs. One of the advantages of using OIDs instead of keys is that changes to the values of the primary key or candidate keys will not affect the identity of an object and its associations with other objects, thus simplifying the enforcement of referential integrity [8].

In the remainder of this section, the different association types of SAM* are described. Examples are used to illustrate the use of these association types to model structural relationships among objects and object classes found in CAD/CAM data.

Membership Association

An object class of the membership association type is the grouping of a set of homogeneous objects to form a domain from which other object classes can draw values. The instances of a type-M object class are considered atomic, meaning that they cannot be further decomposed into more primitive components from the DBMS point of view. Figure 1 shows some examples of type-M object classes. Part# and Color are type-M object classes whose instances are integers and strings, respectively. A constraint can be specified with a type-M object class to limit the range of its legitimate instances. For example, a constraint on the object class Part# can be specified to limit the range of legitimate part#'s to 1 to 1,000. Instances of a type-M object class are legitimate values that can occur in a data base, rather than the actual values entered by the user into the data base. They do not have to be explicitly stored in the data base. A discussion of the role of constraints in data base systems is provided in Refs 9 and 10.

An object class of association type M forms a static domain whose instances are fixed at the time the data base is created and from which attributes of other object classes can draw values. Object classes of other association types form dynamic (time-variant) domains. A user can insert or delete objects from such object classes.

In addition to the simple object classes such as Integer, String, and Boolean, a type-M object

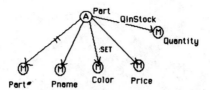

FIG. 1—*Object classes of association types M and A.*

class can be defined over a complex object class such as Set, Ordered-set, Duplicate-set, Vector, Matrix, Time, Date, Dollar, or Rule.

Aggregation Association

A set of object classes (the constituent classes) can be aggregated to define another object class (the defined class) in such a way that the objects of the defined class are characterized or described by the instances of the constituent classes. The object class "Part" in Figure 1 is an example of an object class with an aggregation association type. It is therefore labeled by an A for aggregation. An instance of the type-A object class Part is a high-level object that is described by an aggregation of instances taken from the constituent object classes Part#, Pname, Color, Price, and Quantity. The attributes that map instances of Part to those of the constituent classes are descriptive attributes. Inter- and intra-instance constraints can be specified for the objects of a type-A object class.

By default, a descriptive attribute has the name of its domain object class. An attribute can also have a different name. In this case, the link that represents the attribute is labeled with the new name (for example, the attribute QInStock that draws values from its domain object class Quantity). A descriptive attribute can be defined over a structured object class (for example, the Set, Vector, and so forth of an underlying object class that is constructed from the instances of its underlying object class). In this case, the link that represents the attribute is labeled with the structured type. As an example, the domain of the attribute "Color" of part is a set of colors. Individual elements of such an attribute whose domain is structured can be accessed by the user. That is, it is considered nonatomic or decomposable. (This is different from a type-M complex object class. In that case, it is considered atomic.)

One (or more) of the attributes of a type-A object class can be designated by the data base designer or the user as the key attribute(s). The key attribute, which is graphically represented by a link crossed by double dashes, can be used to uniquely identify an instance within an object class to input or output data associated with the object. It is different from the system-wide, unique OID of the object.

Interaction Association

An object class can be defined for representing a set of facts, each of which records an interaction or relationship between the objects of two or more constituent classes. In this case, the defined class has an interaction association with the constituent classes. For example, the object class Shipment in Figure 2 is a class with an interaction association. It models the relationship between the objects in Manufacturer and the objects in Part. Each instance of Shipment represents the fact that a manufacturer ships a part. The relationship between Shipment and the constituent classes Manufacturer and Part is defined by the structural attributes, which are distinguished from descriptive attributes in the S-diagram by using solid arrows. In addition to the structural attributes, Shipment is also defined by the descriptive attribute Quantity-shipped

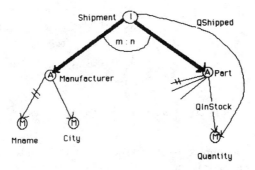

FIG. 2—*Interaction association.*

(Qshipped), which specifies an aggregation association between Shipment and Quantity. Since the interaction association is the main association that constructs the object class Shipment, Shipment is labeled by I for Interaction association even though it also has an aggregation association.

Many types of semantic constraints are inherent semantic properties of type-I object classes. A constraint rule can be specified with the schema definition to state that if a new shipment instance is to be inserted, its Qshipped value must not be more than the QInStock value of the part to be shipped. Another rule can be specified stating that if a new shipment instance is inserted, then update the QInStock value for the part to be shipped to the difference between the current value of QInStock and the value of Qshipped for the new shipment instance. The attribute Cost of Shipment can be derived or computed. A formula can be used to compute the cost of a shipment at retrieval time as the price of each part times the quantity shipped plus 10% for shipping and handling or

$$\text{Cost} = \text{Qshipped} \times \text{Price} \times 1.1$$

The mapping constraint n:m is specified between the structural attributes Manufacturer and Part. In general, any of the mapping constraints 1:1, 1:n, and n:m can be specified between any two structural attributes of a type-I object class.

Figures 3, 4, and 5 show sample G-relations for the object classes Part, Manufacturer, and Shipment, respectively. The unique object identifiers' OIDs for Part and Manufacturer are used in Shipment instead of using the key attributes. Also, instances of the Shipment object class have their own OIDs.

Part

OID	Part*	Pname	Color:Set	Price	QInStock
01	55	Engine	{Black,White}	1950.90	20
02	23	Screw	{White}	2.40	95
03	18	Switch	{yellow}	25.00	34

FIG. 3—*G-relation of the part object class.*

Manufacturer

OID	Mname	City
020	John	Gainesville
021	william	Miami
022	GM	Detroit

FIG. 4—*G-relation for the manufacturer object class.*

Shipment

OID	Manufacturer[OID]	Part[OID]	QShipped
100	020	01	10
101	020	03	20
102	022	02	35

FIG. 5—*G-relation for the shipment object class.*

Generalization Association

Objects of different object classes can be grouped into a more generic object class. This type of grouping or association is called the generalization association. The set of instances of the defined object class is a superset of the union of the objects of its constituent classes. The defined class can in turn be a subclass of another more general class, thus forming a generalization hierarchy, or be a subclass of a number of classes forming a generalization lattice. The objects of a subclass inherit all the structural properties, operational characteristics, and semantic constraints of all the superclasses in the hierarchy or lattice.

Constraints, such as Set-exclusion (SX), Set-subset (ST-SS), Set-intersection (SI), and Set-equality (SE), can be specified among the subclasses of the defined class. Figure 6 shows an example of the type-G object class. Foreign-parts and Domestic-parts are subclasses of the object class Parts. Structural attribute links are used to connect the defined class to its subclasses. The descriptive attributes Weight and Part-id are inherited by the objects in the two subclasses of Parts. This association type allows new object classes to be defined by naming their superclasses without repeating all the semantic properties that can be inherited from the superclasses.

Composition Association

The time-varying set of objects in an object class can be treated as a single object and be characterized by a number of descriptive attributes. Sets of objects of different object classes

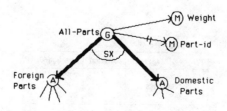

FIG. 6—*Generalization association.*

can be characterized by the same set of descriptive attributes. This semantics is modeled in SAM* by a Composition association.

An object class having a composition association with other object classes is defined by a number of structural attributes, each of which corresponds to a constituent class whose set of instances is treated as a single instance of the defined class. For example, in Fig. 7, the defined class Shop-floor-items has three instances, that is, Parts, Machines, and Robots. Each of these instances is connected to the type-C object class Shop-floor-items via a structural attribute link and is described by a value of the descriptive attribute Average-cost and another value of the descriptive attribute Heaviest. Figure 8 shows the G-relation for the object class Shop-floor-items. Descriptive data about the set of instances of an object class are generally considered as the metadata and are not part of a data model. In SAM*, metadata of this type are explicitly modeled by the composition association.

The composition association type is used in designing the conceptual schema for the data dictionary system, which will be presented later. In the schema to be shown, all object classes of the same association type are treated as instances of a type-C object class whose descriptive attributes characterize the properties of that association type.

Crossproduct Association

In CAD/CAM as well as statistical data base applications, we are often interested in recording data on a category of objects rather than the individual objects. For example, we may want to record data about the design characteristics of a category of objects called Bolt. The actual physical objects (that is, the individual bolts) are indistinguishable in a data base. They are referred to as copies in Ref *11*. Or in a statistical data base for population data, we may want to classify population by such attributes as Race, Sex, Age, State, and County and record the descriptive data of each category using descriptive attributes such as population count (Pcount) and average income (Ave-income) as shown in Figure 9. The Crossproduct association provides the mechanism for modeling the semantics explicitly. The set of instances of a type-X object class is the result of taking the Crossproduct among the sets of instances of the constituent classes connected to the defined class by structural attribute links. Figure 10 shows the corre-

FIG. 7—*Composition association.*

Shop-floor-items

OID	Object-class	Heaviest	Average-Cost
10	Parts	P5	20
11	Machines	M2	300
12	Robots	R7	9000

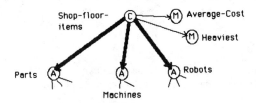

FIG. 8—*Shop-floor-items G-relation.*

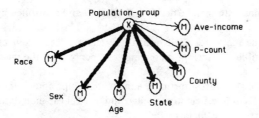

FIG. 9—*Cross product association.*

Population-group

OID	Race	Sex	Age	State	County	Pcount	Ave-income
101	Black	M	25	FL	Alachua	34000	20K
102	White	F	40	FL	Orange	24000	30K

FIG. 10—*G-relation for the population group object class.*

sponding G-relation containing two sample tuples. The descriptive data of these categories are connected to the defined class by the descriptive attribute links as shown in the figure. Alteration of object categories by increasing or decreasing the number of structural attributes will entail changes to the descriptive data of these categories. For example, categorizing population data by race and sex alone will entail the recomputation of Pcount and Ave-income values for the new categories. The so-called statistical aggregation and disaggregation operations are only meaningful to object classes of this association type. The data models used by most of the existing data base management systems do not distinguish descriptive data regarding individual objects from those regarding categories of objects.

The association types described above are six general semantic constructs used to define complex data objects and object classes. Since an object class is defined in terms of its association(s) with other object classes and the object class itself can be used to define other object classes, the complex data structures found in CAD/CAM data can be defined by nesting these association types in a hierarchical or recursive fashion or both. For example, a type-A object class can use a type-G object class as the domain of a descriptive attribute, thus allowing the attribute to draw values from a set of dissimilar constituent classes of the type-G object class. In another example, the objects of a type-I object class can be the constituent class of another high-level, type-I object class, thus allowing facts about facts to be explicitly modeled. A data base management system that uses SAM* as the underlying model recognizes the semantic distinctions of the association types and thus can automatically enforce the semantic constraints and perform the operations that are meaningful to the object classes with these association types. Therefore, the user does not have to enforce the semantics through his application programs.

The Data Dictionary Model

Approaches to Designing a Data Dictionary System

To design and manage the data of an enterprise, we first need to define the structures, formats, and relationships among these data. This information about the enterprise's data is referred to as metadata. The function of a data model is to establish well-defined and agreed-upon metadata. The data can then be defined, managed, and controlled by the DBMS via their metadata.

In a similar way, the metadata itself need to be managed and controlled. The part of the DBMS that is responsible for storing, managing, and controlling the metadata is the data dictionary system (DDS) [4,12,13].

As with the data, the structures, formats, and relationships among the metadata need to be defined to enable the DDS to efficiently manage and control the metadata. The data about the metadata can be referred to as meta-metadata.

The DDS is required to allow the metadata to be shared among different users and among the DBMS facilities. Also, the DDS should provide access mechanisms and security restrictions in addition to maintaining the consistency of the metadata. The DDS should also have the capability to accept and interpret a data definition language to be used in manipulating the metadata. An expensive approach in designing a DDS is to make it an independent part of the DBMS. In this case, the DDS should be enhanced with all the above capabilities. At the storage level, the metadata are stored in a metadata base that can be accessed only via the DDS. The data base and the metadata base are two different data bases, each having its own structures and formats for the data to be stored in it.

The second and more efficient approach in designing a DDS is to make it an integral part of the DBMS. The DDS performs functions on the metadata that are similar to the functions performed by the DBMS on the data. Allowing the DDS to use these existing functions is more efficient. For this to be successful, the application data and the metadata must be treated uniformly. That is, the metadata must be modeled by the same data model that is used to model the application data. The data model used must be rich in semantics to be able to capture the semantic properties of the metadata. This section describes the data dictionary system as modeled by SAM*. One major advantage of this approach is that the query language used for retrieving and manipulating the data can also be used for the metadata.

Dictionary Schema

Figure 11 shows the data dictionary modeled by SAM*. The System object class is a generalization of the six subdictionary object classes. Each of these subdictionary nodes models one of

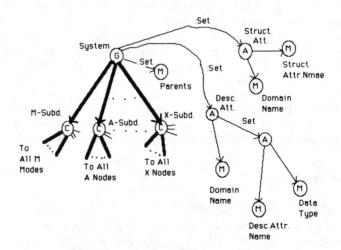

FIG. 11—*The dictionary schema.*

the six association types of SAM*. The descriptive attributes of the System node model the common properties needed to define an object class of any association type. These descriptive attributes are inherited by each of the subdictionary nodes.

Three descriptive attributes are associated with the object class System to characterize or describe a node (object class) in the application schema. These are

(1) Parents, to define the set of parent object classes of a node,

(2) Desc-attr, to specify the descriptive attributes of a node. Each node in the application schema can have a set of underlying domains from which its descriptive attributes draw their values. One or more descriptive attributes of an object class can be defined on each such domain. Each descriptive attribute can have a data type that is different from the data type of its underlying domain (for example, if the data type of the underlying domain is Integer, the data type of the descriptive attribute can be Set of Integer). This data type specification in effect converts the domain of the attribute from a simple object class to a structured object class, and

(3) a set of structural attributes, each connecting the defined class to a constituent object class.

Each of the subdictionary nodes is a type-C object class and describes the properties that are specific to one of the association types of SAM*. Figures 12 to 17 show the details of the subdic-

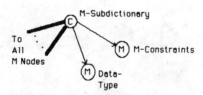

FIG. 12—*M-subdictionary schema.*

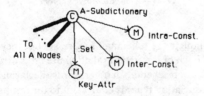

FIG. 13—*A-subdictionary schema.*

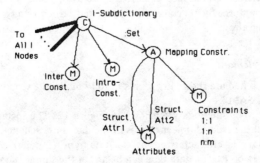

FIG. 14—*I-subdictionary schema.*

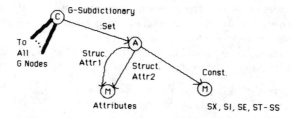

FIG. 15—*G-subdictionary schema.*

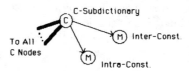

FIG. 16—*C-subdictionary schema.*

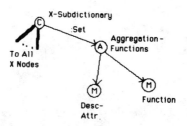

FIG. 17—*X-subdictionary schema.*

tionaries' schemata (they complete the schema of Figure 11, but they are drawn separately owing to the space limitation). Each of the structural attribute links of a subdictionary node go to one of the nodes in the application schema. In other words, declaring a node in the application schema to be of one of the association types is equivalent to (or results in) connecting that node via a structural attribute link to the subdictionary node that models that association type.

The dictionary S-diagram presented here does not show the operations and constraints associated with its object classes, which further describe the association types. For example, the operations or rules that enforce the set constraints presented earlier for a type-G object class are not shown in the S-diagram.

The Subdictionaries' schemata

The subdictionaries' schemata shown in Figures 12 to 17 model the semantic properties of the six association types of SAM*.

The schema for the M-subdictionary is shown in Figure 12. The instances of the C node M-subdictionary are all the type-M object classes in the data base schema. An instance of the C node M-subdictionary is characterized by its data type (D-type) and the constraints (M-Constraints) imposed on it. The data type can be any of the simple data types such as Integer, Real, Boolean, and Character or a complex data type such as Vector, Set, Ordered-set,

Date, and so forth. The function of the constraints that are specified for an M node is to limit its instances to those which satisfy the constraint. The M-subdictionary node inherits the descriptive attributes of the System node.

Figure 13 illustrates the subdictionary schema for the aggregation association. An instance of the C node A-subdictionary (which is a type-A object class that appears in the application schema) is characterized by a set of key attributes, intrainstance constraints (Intra-cons), and interinstance constraints (Inter-cons).

The G-subdictionary schema is shown in Figure 14. The G-constraints specify the relationship between every two subclasses of a G node. The possible relationships are set exclusive (SX), set-subset (ST-SS), set-equality (SE), and set-intersection (SI). These constraints are enforced by operations or rules that are defined in the dictionary schema (not shown in the S-diagram).

The I-subdictionary schema of Fig. 15 shows that interinstance and intrainstance constraints can be specified for an I node [5]. Also, the mapping constraints can be specified for an I node to control the relationships between the instances of every two constituent object classes. The possible mapping constraints are one-to-one (1:1), one-to-many (1:n), and many-to-many (m:n). These constraints can be enforced by operations defined with the I-subdictionary object class.

For a type-C object class (Figure 16), it is possible to specify interinstance and intrainstance constraints. Since the instances of a type-C object class are themselves object classes, the interinstance constraints specified for it are effectively interclass constraints. Similarly the intrainstance constraints are effectively intraclass constraints.

Figure 17 shows the schema for the Crossproduct association type. Statistical aggregation and disaggregation functions can be defined for a type-X object class.

Prototype Data Dictionary System

A prototype data dictionary system has been implemented on a VAX/UNIX system using the C programming language. Another version of the prototype dictionary system runs on the VAX/VMS. When a user logs into the system, s/he sees the menu of Fig. 18. This is the root of the menu tree of the system. The selection of any item from this menu invokes some other menus or programs or both.

Schema Definition Mode

The selection of item 1 from the System menu starts the schema definition mode and the user sees the menu of Fig. 19. By selecting from this menu, the user can start defining nodes of the appropriate association types. The system will prompt the user with further questions and menus that correspond to the association type s/he chooses. As an example, we show here what happens when a user decides to define an M node. The system will prompt him with the following command:

Enter node name:

After typing the name of the type-M object class he wants to define, the user sees the menu of Fig. 20, from which he selects a data type for it. After that, he gets the following prompt asking him to enter constraints on this object class:

Enter constraints:

Similarly, other menus and prompts are used for the other association types. This removes the memory burden from the user because he does not have to remember all the details.

System Menu

1. Schema definition menu
2. Schema display menu
3. Schema modification menu
4. Delete an existing schema
5. Create database
6. Data manipulation
7. Go to a UNIX sub-shell
8. Exit from system

SELECTION:

FIG. 18—*The root of the menu tree of the system.*

DB Definition Menu

1. Define M nodes
2. Define A nodes
3. Define G nodes
4. Define I nodes
5. Define C nodes
6. Define X nodes
7. End DB definition mode

SELECTION:

FIG. 19—*Schema definition menu.*

Schema Display Mode

By selecting the second item from the system menu, the user enters the schema display mode. Consequently the menu of Fig. 21 will be displayed. By selecting an item from the schema display menu, the user can get information about any node of the selected association type. This information is displayed in a tabular form that includes the node name, the children and attributes of that node, and the constraints specified for that node.

```
Select a Data Type:

 1. Integer            2. Real
 3. Double-precision   4. Character
 5. Boolean            6. Set
 7. Ordered set        8. Duplicate set
 9. Vector            10. String
11. Matrix            12. Time series
13. Time              14. Date
15. Currency          16. Compute
17. Rule

SELECTION:
```

FIG. 20—*Possible data types of an M-mode.*

```
1. Display M nodes

2. Display A nodes

3. Display G nodes

4. Display I nodes

5. Display C nodes

6. Display X nodes

7. End display mode

SELECTION:
```

FIG. 21—*Schema display menu.*

Schema Modification Mode

The selection of the third item from the System menu invokes the menu of Fig. 22. By selecting from this menu, the user can modify the schema. This can be done by adding new nodes to the schema, deleting existing nodes, modifying the constraints specified for a node, or adding or deleting attributes for an existing node.

The selection of item 4 from the System menu allows the user to delete an existing schema as a whole.

The selection of item 5 invokes a program that creates G-relations for the application data according to the contents of the dictionary, that is, according to the schema. After creating a data base, the user can start manipulating data by selecting item 6 of the System menu. Simple data manipulation operations like inserting new objects or displaying existing objects can be performed.

Item 7 of the System menu allows the user to temporarily exit from the system and execute any command at the UNIX shell level.

```
                    1. Modify M nodes

                    2. Modify A nodes

                    3. Modify G nodes

                    4. Modify I nodes

                    5. Modify C nodes

                    6. Modify X nodes

                    7. End schema modification mode
           SELECTION:
```

FIG. 22—*The schema modification menu.*

Conclusion

The need for a powerful data model to explicitly model the CAD/CAM data and metadata found in an integrated manufacturing system has been established. A semantic association model, that provides six general semantic associations for modeling complex data objects in a nested or recursive fashion or both has been described. The structural properties and semantic constraints and operations associated with the six association types can be used by a data base management system to process objects and object classes intelligently based on the association types in which these object classes participate. The operational data as well as the metadata of a data base can be uniformly modeled by the same semantic model. Thus, they can be processed and manipulated by the same data base management system without the traditional separation between a data dictionary system and data base management components. As a result, more flexibility and expressive power in modeling and more functionality in processing data and metadata can be gained.

Acknowledgment

This research is supported by the Navy Manufacturing Technology Program through National Bureau of Standards grant number 60NANB4D0017, and also by Florida High Technology and Industry Council grant number UPN 85100316.

References

[1] Gligor, V. and Luckenbaugh, G., "Interconnecting Heterogeneous Database Management Systems," IEEE Computer, pp. 33–43, Jan. 1984.
[2] Spaccapietra, S., "Heterogeneous Database Distribution," in *Distributed Data Bases*, I. Draffan and F. Poole, Eds., Cambridge University Press, New York, 1980.
[3] Maier, D. and Price, D., "Data Model Requirements for Engineering Applications," Proceedings of the First International Workshop on Expert Database Systems, Kiawah Island, South Carolina, 24–27 October 1984.
[4] Cammarata, S. and Melkanoff, M., "An Interactive Data Dictionary Facility for CAD/CAM Data Bases," Proceedings of the First International Workshop on Expert Database Systems, Kiawah Island, South Carolina, 24–27 October 1984.

[5] Su, S., *SAM*: A Semantic Association Model for Corporate and Scientific-Statistical Databases,* Information Sciences 29, Elsevier Science Publishing Co. Inc., New York, 1983.

[6] Su, S., "Modeling Integrated Manufacturing Data with SAM*," IEEE Computer, Vol. 19, No. 1, pp. 34-49.

[7] Goldberg, A. and Rotson, D., Smalltalk-80: The language and its implementation, Addison Wesley, Reading, MA, 1983.

[8] Date, C., An Introduction to Database Systems, Fourth Edition, Addison-Wesley Publishing Co., Reading, MA, 1986.

[9] Morgenstern, M., "The Role of Constraints in Databases, Expert Systems, and Knowledge Representation," Proceedings of the First International Workshop on Expert Database Systems, Kiawah Island, South Carolina, 24-27 October 1984.

[10] Shepherd, A. and Kerschberg, L., "Constraint Management in Expert Database Systems," Proceedings of the First International Workshop on Expert Database Systems, Kiawah Island, South Carolina, 24-27 October 1984.

[11] Batory, D. and Kim, W., "Modeling Concepts for VLSI CAD Objects," *ACM Transactions on Database Systems,* Vol. 10, No. 3, pp. 322-346.

[12] Allen, F., Loomis, M., and Mannino, M., "The Integrated Dictionary/Directory System," *Computing Surveys,* Vol. 14, No. 2, pp. 245-286.

[13] Sibley, E., "An Expert Database System Architecture Based on an Active and Extensible Dictionary System," Proceedings of the First International Workshop on Expert Database Systems, Kiawah Island, South Carolina, 24-27 October 1984.

Cita M. Furlani,[1] Don Libes,[1] Edward J. Barkmeyer,[1] and Mary J. Mitchell[1]

Distributed Data Bases on the Factory Floor*

REFERENCE: Furlani, C. M., Libes, D., Barkmeyer, E. J., and Mitchell, M. J., **"Distributed Data Bases on the Factory Floor,"** *Computerization and Networking of Materials Data Bases, ASTM STP 1017*, J. S. Glazman and J. R. Rumble, Jr., Eds., American Society for Testing and Materials, Philadelphia, 1989, pp. 126–134.

ABSTRACT: A major facility for manufacturing research exists at the National Bureau of Standards (NBS). The Automated Manufacturing Research Facility (AMRF) serves as a test bed and demonstration facility in support of research by workers from NBS, industry, academia, and other government agencies. The AMRF has been designed as a "data driven" control system. This permits it to handle a broad range of parts for automated manufacturing, but requires an effective interface between the data generated in a manufacturing system and the control modules that use the data. To meet this requirement, a distributed data system, the Integrated Manufacturing Data Administration System (IMDAS), has been developed. It is designed to provide the control systems of the AMRF access to the data necessary to support the design, planning, manufacturing, and inspection of parts.

KEY WORDS: distributed, distributed heterogeneous, data base management system (DBMS), data administration, automated manufacturing, computer integrated manufacturing (CIM), Integrated Manufacturing Data Administration System (IMDAS), Automated Manufacturing Research Facility (AMRF)

The National Institute of Standards and Technology (NIST) is addressing the measurement and standards needs for the automation of the small-batch, discrete-parts manufacturing industries. The Automated Manufacturing Research Facility (AMRF) is funded by NIST and the Navy Manufacturing Technology Program. It has been developed to serve as a test bed and demonstration facility in support of research by workers from NIST, industry, academia, and other government agencies.

Automated manufacturing requires sharing data among control, sensory, planning, and administrative processes. These processes are invariably distributed over many different computer systems. Unlike most existing data systems, a manufacturing enterprise requires support for diverse computer systems, data systems, and data bases in addition to real-time data access. In addition, it is necessary to provide for the integration of new systems into an operating shop without interrupting its operations. Thus, a distributed data system is necessary.

Increasing industrial automation results in factory floors containing large numbers of computer systems controlling and monitoring physical processes. The critical element in such a complex is the ability of all the associated programs and users to share information.

The AMRF is designed as a "data driven" control system. This permits it to handle a broad range of parts for automated manufacturing, but requires an effective interface between the

---

*NOTE: Contribution of the National Institute for Standards and Technology. Not subject to copyright.
[1]Leader, Integrated Systems Group, computer scientist, computer scientist, and computer systems analyst, respectively, National Institute for Standards and Technology, Gaithersburg, MD 20899.

data generated in a manufacturing system and the control modules that use the data. To meet this requirement, the Integrated Manufacturing Data Administration System (IMDAS) has been developed. It is designed to provide the control systems of the AMRF access to the data necessary to support the design, planning, manufacturing, and inspection of parts. The IMDAS handles the administrative tasks of storing new data, accepting requests for old data, locating and updating that data, and transferring the results to control systems and engineering systems. User systems in the facility need not know anything about how or where the information is stored, only how to make the request to the IMDAS.

Automated Manufacturing Research Facility

The AMRF is the NIST facility for manufacturing research. It is significantly supported by industry through donations or loans of major components and through cooperative research programs. One goal of the project is to identify and exercise potential interface standards between existing and future components of small-batch manufacturing systems. Others are to provide a laboratory for the development of factory-floor metrology in an automated environment, and to develop new ways of making precisely machined parts. Commercially available products are used in the facility wherever possible to expedite the transfer of research results to the private sector.

To provide a real test bed for interface standards, the AMRF is intentionally composed of manufacturing and computing equipment from many vendors, thereby making its construction a major integration effort [1,2]. Shop floor equipment includes computerized numerical control (CNC) machines, a coordinate measuring machine, robots, a vision system, robot carts, automated storage and retrieval systems, cleaning and deburring devices, and part fixturing and robot gripper systems. The configuration is structured around several self-contained workstations, each capable of executing a well-defined set of manufacturing functions. Each workstation is able to operate either as an independent manufacturing unit under the control of a local operator, or as an element of a multiworkstation manufacturing system under the control of a higher-level process. A typical machining workstation consists of a CNC machine tool, a robot, a materials transfer station, and local buffer areas for tools and workpieces.

The intelligence structure of each workstation includes the robot control system and machine tool control system, sophisticated sensor systems, and a workstation control system to coordinate the activities. Above the workstation level, batch manufacturing coordinators, or cells, and a floor manager, or shop controller, provide higher levels of control. The highest level of control, the facility controller, implements the "front office" functions that are typically found in small manufacturing facilities. All of these control and sensory processes are software systems that reside on interconnected computer systems, making the AMRF a distributed computing system [3]. Many different computer languages and types of computer systems make up the computing environment of the AMRF. In addition to the shop floor activities, manufacturing data preparation activities, including part design, geometry modeling, group technology classification, process planning, and off line control programming, are performed on "engineering" computer systems linked into the factory floor network. These types of data together form the global shared data base of the manufacturing facility (Fig. 1).

In the AMRF, as well as in most automated factories, data resources are physically distributed across a network of heterogeneous hardware and software systems acquired from numerous vendors. Such a distributed system requires a method of transferring information that is fast, accurate, and reliable, has consistent representation, and is independent of the actual physical location of the machines.

The AMRF uses the concept of computer "mailboxes," areas of memory in various computers to which all of the machines in a particular group have access through the network com-

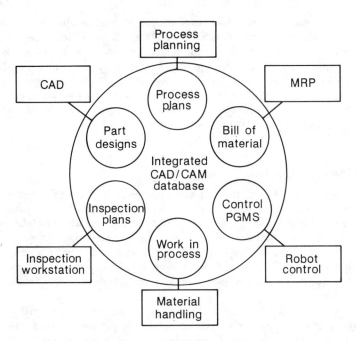

FIG. 1—*Global shared data base.*

munications system, subject to strict rules of protocol. Control processes can leave "messages" for each other and stop to read their own "mail" at opportune times without interrupting each other.

Currently, the AMRF communications network, the backbone of the distributed data system, uses an Applitek[2] broadband token bus and a combination of older computer communications protocols, including RS 232 and Ethernet systems. Work is underway to upgrade the AMRF network to one based on the principles of the Manufacturing Automation Protocol (MAP) network proposed by General Motors and others.

Computer integrated manufacturing (CIM) refers to the integration of diverse systems into an automated production complex closely coupled to the engineering and administrative systems that support and drive it. Rarely does the same kind of computer system perform engineering support, real-time control, and administrative applications. These component systems range in data management capabilities from simple shared memory managers to special-purpose software that optimize data access for a particular function to general-purpose data base management systems with a full range of supporting tools. The software system that integrates the diverse and distributed data resources of the factory environment must mask the differences inherent to these conditions. The critical element in such a complex is the ability of all the associated programs and users to share data. "Data—its generation, processing, storage and use in the implementation and control of the manufacturing enterprise—is the essence of CIM, the NBS AMRF, and the fully automated factory of the future [4]."

---

[2]Certain commercial products are identified in this paper in order to adequately describe the IMDAS. Such identification does not imply recommendation or endorsement by NIST.

Data in a Distributed Environment

A significant part of the AMRF distributed data system research has focused on the identification and modeling of factory data and relationships. A critical step in accomplishing integration in an environment of diverse applications is the definition of a common logical model of shared information and meanings. In the CIM environment, almost all data is significant to more than one application, but the way in which it is best organized for each application area may be different. For example, Production wants to keep track of component inventories by PART-TYPE and LOCATION, while Purchasing wants to organize the components by SUPPLIER, SHIPMENT, and ORDER and Accounting wants to track them by PART-TYPE, INVOICE, and PRODUCT. Naturally, having been organized for different applications, they have different schemas and often use entirely different data base management systems. Because there are interactions of the data, it becomes necessary to integrate data bases. This almost always requires an external mechanism to keep them consistent, to define interrelationships, and to describe their interaction.

Controllers retrieve and modify through views of data structured to meet the needs of the particular application. These logical views are relations whose objects and attributes may or may not coincide directly with the fields of a physical record. A distributed architecture demands that a data dictionary and directory system exist at each communications node to index and define the data sets that reside at that node. Data include the translation of logical names associated with data structures and elements and the definition of actual schemas or physical structures in terms of the local data management system. The information in the dictionary and directory system is active in that data sets are created and deleted while the test bed is operating. In addition, a common conceptual model of the entire data base complex must be maintained at some generally available location. This central model contains information on the distribution and the logical structure of the data sets, as well as the relationships between records in the data sets that span multiple nodes or are replicated at them.

The researchers within the AMRF have used existing data modeling methodologies for the discovery, concept formation, and validation of information shared by the shop floor and data preparation activities. Individual components of the AMRF are modeled and then merged to develop an integrated data model. The techniques used, Information Analysis [5] and IDEF1X [6], emphasize the discovery of the meaning and relationships between information units, in addition to recording representation forms and deriving convenient storage structures.

Commercially available software tools have been used to support these modeling activities. Analysts use these products to describe the relationships and constraints discovered in English sentences, such as: "a storage_device has one or more storage_areas associated with it," "a storage_device is uniquely identified by an equipment_id and unit_id," and "a tool is a type of reusable_resource_item." Then the control system engineers validate and refine the evolving integrated data model by deciding whether they agree or disagree with these statements.

The integration data model being used in the AMRF is SAM*, a semantic data model [7,8]. This approach was accepted over others because of the model's expressive power to represent arbitrarily complex objects and constraints. This model is the heart of the AMRF dictionary system, which makes use of the available semantics to control the distribution of the data and its complex interrelationships and constraints.

Managing Data in a Distributed Environment

As a "data driven" control system, the AMRF is capable of handling a broad range of parts for automated manufacturing. But an effective interface is required between the data generated in a manufacturing system and the control modules that use the data. Control processes and factory personnel must be able to specify requests for data services in an environmentally neutral form, a common data manipulation language (DML). The integrated data service provider must translate requests in this language into commands that can execute at the site or sites that manage the data resources. In addition, individual data units may have to be translated to resolve the representation differences across hardware boundaries. The integrated data services system may be required to assemble results from several sites and to perform user specified formatting of the resultant data. Finally, a problem characteristic of, but not unique to, the manufacturing floor is real-time access to data.

A manufacturing facility is comprised of a large number of dissimilar computer systems, data systems, and data bases. The need for sharing data among these systems results in overlapping data bases with representational inconsistencies. We have argued that we need to build a common data system to solve the integration problem. This system would make knowledge about shared data resources available for all applications. In addition, we assert that such a system must support real-time use. How does one go about constructing such a system? The simplest approach to the development of a common data system is the centralized system, a single common data base in a central computer system with communication paths to all client systems. The inherent simplicity of this architecture, particularly with regard to management and maintenance, makes it highly desirable if it can be made practical. With current computer technology, it is possible to create a central system with sufficient redundancy so that a total failure is extremely unlikely. However, the ability of such a system to handle all of the data base transactions for the whole manufacturing enterprise without bottlenecks or unacceptable delays is very questionable. In all but the smallest organizations, performance and cost considerations will dictate some distribution of function and data.

Another possibility is maintaining distributed overlapping data bases with conversion of data between them. This is viable if the number of data bases is small and the interactions between them are carefully timed and controlled. The number of conversion programs will grow proportionally to the square of the number of data bases. Any two data bases must be idle for a certain period of time for information to be transferred between them. Moreover, such data bases will normally be inconsistent and be brought into alignment only at the times of transfer between them.

The alternative, modifying programs to access data from multiple data bases, can be overwhelming. It avoids the consistency problems inherent in the data base conversion approach, but it is simply impractical in an environment of numerous systems and regular changes.

By using a common interface between programs and data bases, this complexity can be avoided. Each new program has only one interface to all data. A change to the architecture of one data base requires modification of only its interface to the common service, not to any of the applications. Most enterprises will benefit from a system of distributed data bases with common interfaces. This will enable the integration of new and existing data systems and relieve programs and programmers of network, access, and conversion problems.

The remaining choice is whether to provide the common interface from a single server with interfaces to each of the distributed data bases, or to distribute the common interface service over several systems with interfaces to each other. Again, the centrally controlled system is simpler and currently feasible, but its ability to handle all of the transactions, even when it can distribute the actual data manipulations, must be in doubt. It also acts as a single point of failure. On the other hand, the protocol problems that result from trying to do distributed data management by committee have been the subject of much academic discourse and few, if any, sound solutions.

Integrated Manufacturing Data Administration System

Our approach to providing a common interface to distributed data is the Integrated Manufacturing Data Administration System (IMDAS) [9]. IMDAS is characterized by (1) a common interface to user programs, (2) a common interface to underlying data bases, and (3) a hierarchical architecture, providing both centralized and distributed services.

Application, or control programs, communicate with IMDAS using a standard language, referencing data names from a common dictionary. The IMDAS Data Manipulation Language (DML) was closely modeled after ANSI standard SQL [10]. Extensions to SQL were made to allow the projection and selection of elements in complex objects. The DML also allows programs to specify files or memory buffers as the sources and destinations of data.

On the other side, IMDAS has a common interface to underlying data repositories, such as commercial data base systems. This minimizes the work needed to incorporate new data bases into the common data system while allowing existing systems to continue operating without change.

This set of common interfaces affords users (AMRF control processes) a generalized view of data access. They see data manipulation as operations on information units, not data bases, and are not concerned with what system or machine has the data. The result is conceptually simple. The user sees a single common data base managed by the IMDAS (Fig. 2).

The internal architecture of IMDAS is a four-level hierarchy (Fig. 3). The levels are distinguished primarily by scope of responsibility for data management. The higher the level, the more data is administered. At the bottom level of the IMDAS are the data repositories, some of which are commercial data base management systems (DBMSs). Also included are other repositories of sharable information, such as file systems, common memory, and locally developed, application-specific data managers.

BDAS—The Basic Data Administration System

The Basic Data Administration System (BDAS) resides on each computer system in the AMRF. Each BDAS integrates its local data repositories into the IMDAS, and with its associated DBMSs, executes all manipulations on that data. The BDAS must convert the IMDAS internal form of a transaction to that accepted by the DBMS that has to execute it, pass it to the DBMS, and interpret the status that comes back.

The BDAS must also convert any data involved between the DBMS dependent form and the IMDAS interchange form. Because capabilities and access techniques differ dramatically from DBMS to DBMS and can be further complicated by local operating system conventions, the interface to a particular DBMS is encapsulated in a separate process called the Command Translator/Data Translator (CT/DT) (Fig. 4). In addition, the BDAS must access local user data areas and convert between the declared user representation and the IMDAS interchange form.

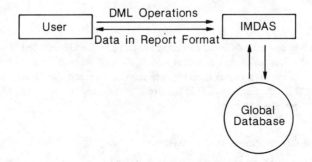

FIG. 2—*IMDAS concept.*

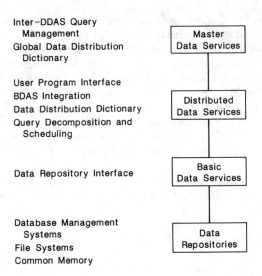

FIG. 3—*The IMDAS hierarchy.*

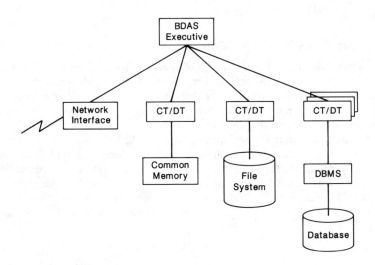

FIG. 4—*Typical BDAS.*

The BDAS receives commands from and returns status to the Distributed Data Administration System (DDAS) that supervises it, but it also deals with other BDASs as network peers to deliver data. In this way, data moves directly between the user and the data repositories rather than following the IMDAS hierarchy. This feature is mandatory for achieving the performance required in a real-time environment.

DDAS—Distributed Data Administration System

At the Distributed Data Administration System (DDAS), the collection of data repositories managed by a group of BDASs is logically integrated into a segment of the global data base, using a dictionary describing the distribution of the data. The DDAS becomes the data manager for that segment and supervises all manipulations of it. In addition, each DDAS provides the IMDAS with an interface to some set of user programs, as a DDAS is available on each computer system that is capable of supporting its activities.

User programs issue transactions to the IMDAS, represented by the DDAS, which accepts them, oversees their execution, and returns status to the user. User transactions are stated in the DML. These transactions are converted into an IMDAS standard internal form and then modified to reflect the differences between the user's external view and the IMDAS global conceptual view.

The DDAS attempts to map the transaction into a set of operations on elements of the global data base, which are managed by individual DBMSs. To do this, it consults a dictionary that describes how data is distributed over the integrated data bases. The result is a set of tasks to be executed by specific DBMSs. If any of the data is outside the segment managed by this DDAS, the whole transaction is sent to the Master Data Administration System (MDAS).

MDAS—Master Data Administration System

To integrate segments managed by separate DDASs and to execute user transactions that require this level of integration, a single system is designated the Master Data Administration System (MDAS). The MDAS is an optional component of the IMDAS, which is made necessary by having more than one DDAS. So, from our point of view, the issue of centralized versus distributed control of the distributed data bases does not have to be resolved at the outset. If a single system can manage all data activity in an enterprise, one installs the sole DDAS there and makes its controlled segment the whole global data base. But when more than one DDAS becomes necessary (as must inevitably occur), rather than trying to solve the crossover problems by committee, we appoint a Master DAS. The MDAS supervises all DDASs in a limited way. The choice of MDAS is made by the system manager when the system is started or resumed after a crash, thus avoiding the single point of failure.

IMDAS Implementation

An IMDAS prototype exists within the AMRF. The current implementation consists of one DDAS and a number of BDASs. With user programs, the IMDAS communicates over the AMRF network. The IMDAS software is written in C and Pascal and runs in both UNIX and DEC/VMS environments. The user programs have no language constraints and currently include C, Pascal, Lisp, BASIC, and Forth. User systems include Symbolics, Sun Microsystems, HP, VAX, and numerous Intel 80×86 and Motorola MC68000 controller systems.

ASN.1 and ISO 8825 [11] are used to encode all data units for interchange within the IMDAS. This machine-independent representation supports primitive types (e.g., integers, strings) as well as complex user-defined types (e.g., relations, query trees). This enhances portability with minimal expense of time and space, providing data coherency across a distributed network of heterogeneous machines.

At the BDAS level, interfaces to a variety of data base management systems exist, including RTI/Ingres and BCS/RIM. In addition to commercial DBMSs, there also exist interfaces to the AMRF Geometry Modeling System [12], to the AMRF Process Planning System [13], to common memory [14], and to file systems for UNIX and VMS.

Existing subsystems of IMDAS are undergoing continuous testing and refinement. A number of efforts to improve performance and reliability will become important as the project is transferred from the research environment to actual users. NIST researchers have been working jointly with the Database Systems Research Development Center at the University of Florida in the development of IMDAS. Their SAM* model has recently been extended to OSAM* or object-oriented SAM*. We intend to incorporate object-oriented DBMSs into future extensions of IMDAS—in particular, those developed by Servio Logic and Ontologic, provided to us through the NIST Research Agreement Program. Research Agreements with DACOM, CDC, and QINT Data Systems have provided us with the necessary tools to model the information resources in the AMRF. In addition, we work closely with Research Associates from Bendix, General Dynamics, and Computervision.

Summary

Automated manufacturing requires sharing data among control, sensory, and administrative processes. Since these processes are invariably distributed over many different computer systems, a distributed data system is necessary. Unlike most experimental distributed data systems, a manufacturing enterprise requires support for diverse computer systems, existing data systems, and data bases, as well as real-time data access. The Integrated Manufacturing Data Administration System is an experimental software system that is being used to provide update and retrieval services over preexisting, distributed, heterogeneous files and data bases to support the design, planning, manufacturing, and inspection of products in the NIST Automated Manufacturing Research Facility.

References

[1] Simpson, J. A., Hocken, R. J., and Albus, J. S., "The Automated Manufacturing Research Facility of the National Bureau of Standards," *Journal of Manufacturing Systems*, Vol. 1, No. 1, 1982.
[2] Nanzetta, P., "Update: NBS Research Facility Addresses Problems in Setups for Small Batch Manufacturing," *Industrial Engineering*, June 1984, pp. 68-73.
[3] McLean, C. R. and Brown, P. F., "The Automated Manufacturing Research Facility at NBS," *Proceedings of IFIP W.G. 5.7 Working Conference on New Technologies for Production Management Systems*, Tokyo, Japan, October 1986.
[4] Swyt, D. A., "CIM, Data, and Standardization within the NBS AMRF," *Proceedings of International Conference, Communications and Standardization in CIM*, Zurich, Switzerland, November 1986.
[5] Griethuysen, J. J. Van, "Concepts and Terminology for Conceptual Schema and the Information Base," Report ISO/TC97/SC5/WG3 N 695, International Organization for Standardization, 15 March 1982.
[6] Loomis, M., "Data Modeling—the IDEF1X Technique," Proceedings of IEEE Conference on Computers and Communications, Phoenix, AZ, March 1986, pp. 146-151.
[7] Su, S. Y. W., "Modeling Integrated Manufacturing Data Using SAM*," *IEEE Computer*, Jan. 1986, pp. 34-49.
[8] Krishamurthy, V., Su, S. Y. W., Lam, H., Mitchell, M. J., and Barkmeyer, E. J., "A Distributed Data Base Architecture for an Integrated Manufacturing Facility," *Proceedings of Conference on Data and Knowledge Systems for Engineering and Manufacturing*, Hartford, CT, Oct. 1987.
[9] Barkmeyer, E. J., Mitchell, M. J., Mikkilineni, K. P., Su, S. Y. W., and Lam, H., "An Architecture for Distributed Data Management in Computer Integrated Manufacturing," NBSIR 86-3312, Jan. 1986.
[10] "American National Standard Data Base Language SQL," ANSI X3.135-1986, American National Standards Institute, Inc., New York, Dec. 1986.
[11] International Organization for Standardization, "Information Processing—Open Systems Interconnection—Basic Encoding Rules for Abstract Syntax Notation One (ASN.1), ISO/TC 97/SC 21 N25," July 1984.
[12] Tu, J. S., Hopp, T. H., "Part Geometry Data in the AMRF," NBSIR 87-3551, April 1987.
[13] Brown, P. F., McLean, C. R., "Interactive Process Planning in the AMRF," Proceedings of Symposium on Knowledge-Based Expert Systems for Manufacturing at ASME Winter Annual Meeting, Dec. 1986.
[14] Libes, D., "User-Level Shared Variables (in a Hierarchical Control Environment)," *Proceedings of the Summer 1985 USENIX Conference*, Portland, OR, June 1985.

John L. McCarthy[1]

Information Systems Design for Material Properties Data

REFERENCE: McCarthy, J. L., "**Information Systems Design for Material Properties Data,**" *Computerization and Networking of Materials Data Bases, ASTM STP 1017,* J. S. Glazman and J. R. Rumble, Jr., Eds., American Society for Testing and Materials, Philadelphia, 1989, pp. 135–150.

ABSTRACT: Material properties data present numerous challenges that current data base management technology does not address. Major problems include data complexity, naming conventions, measurement units, summary levels of abstraction, and data sparseness. Networks and gateways to multiple material information systems compound these challenges and add others.

Experience from projects such as the Materials Information for Science and Technology (MIST) system prototype suggests several general approaches that address these problems, including modular system architecture; a three-tiered approach to naming materials, properties, and variables; an active data thesaurus; material and variable class hierarchies; structured representation of tables and graphs; and existence tables to indicate data availability by the cross product of materials, properties, and independent variables for any given data base. It also suggests that we need to go beyond the simple relational model to solve our problems and implement solutions.

KEY WORDS: information systems, information retrieval, thesauri, data description, tables (data), units of measurement, distributed data bases, electronic publishing, dictionaries, directories, handbooks, nomenclature

This paper discusses design considerations for materials property information systems, including the networking of multiple data bases. It is based largely on experience with the development of an experimental prototype system called Materials Information for Science and Technology (MIST) [1,2]. The first section of the paper outlines some major issues and problems of material properties, most of which are common to scientific data in general. The second section suggests specific design approaches that address issues discussed in the first section.

Issues and Problems of Material Properties Data

Material properties, like many other types of scientific data, present challenges for information management that most current data base management systems do not address [3]. Materials data users need many of the same features in their information systems as other scientists and engineers do, but they also have some unique requirements that have implications for systems design. Networking of heterogeneous material properties data bases compounds those challenges along with the standard (and formidable) problems of distributed data base management.

This section is further subdivided into the following topics:

---

[1] Computer scientist, Computer Science Research Department, MS: 50B/3238, Lawrence Berkeley Laboratory, University of California, Berkeley, CA 94720.

- Data Structure Complexity,
- Naming and Measurement Units,
- Summary and Abstraction Levels,
- Sparse and Diverse Data, and
- Distributed Data Base Issues.

Data Structure Complexity

Like many other types of scientific and engineering information, material properties data frequently requires more than the standard relational data model, with its simple, single-valued fields. Although some types of materials property information—notably raw, experimental data—can be represented in flat (relational) files, many others defy straightforward relational representation.

For example, Fig. 1 shows a simplified hypothetical example of material property information typically found in an engineering handbook table or graph. This example illustrates a number of the common problems of scientific and engineering data that we will consider in turn. In Fig. 1, field names are shown to the left of the equal signs, with corresponding field values on the right.

Multivalued Fields—Some fields (such as thickness in Fig. 1) have single values, but others (such as condition in the example) have multiple values. Still others (such as test temperature and impact energy in Fig. 1) are actually vectors of values whose individual data points correspond to one another. The latter type of test measurements can also be represented as pairs of (X,Y) points or, in general, as multidimensional matrices. Representation of such data structures in conventional flat files is awkward at best; it requires breaking up the original hierarchical structure into several different relations and replicating many of the single-valued fields. Rejoining separate relations for input or output of the original table or graph can be quite costly in terms of storage and system efficiency. It also complicates rather than simplifies user understanding of the data.

Footnotes—Footnotes are very important for scientific and engineering data, particularly for evaluated reference data that appears in handbooks. The third impact strength value in Fig. 1 has a footnote flag attached to it (*a) that refers in turn to a separate footnote field. Footnotes also may modify different levels of a given data set. Here, for example, a footnote refers to a single data point. In other cases they may apply to an entire vector (row or column), a particular

material = 4340
condition = quenched and tempered
form = sheet
resistivity = 100 to 150 ohms
thickness = < 0.5 in
test temperature = -40, 75, 200, 300, 400, 500
impact Energy = 5.2, ---, 9.6*a, 10.9, 11.0, 11.3
test Method = Charpy V-Notch
note = *a shattered

FIG. 1—*The simplified hypothetical dataset illustrates various representational problems for materials data.*

observation—for example, the fourth value of each vector—or the entire data set. Users therefore need a general but simple representation mechanism that facilitates "binding" footnote information to the appropriate level.

Null Values, Ranges, and Bounds—Figure 1 also shows some more subtle data problems that present difficulties for both relational and nonrelational data base systems. Note that the second measurement for impact energy in the hypothetical example is missing. Such "null" values must be handled in special ways when calculating summaries (for example, averages), and many data base systems do not provide "built-in" support for calculations involving null values [4].

Ranges and bounds present problems for conventional indexing and query processing. The fields "resistivity" and "thickness" in our example contain numeric values that are a range and a bound, respectively, rather than simple numbers (they also include measurement units, but that issue is addressed separately below). One solution is to index maxima and minima rather than (or in addition to) individual numeric data values.

Representation of Tables and Graphs—Much of the material information users want computerized access to currently exists in published handbooks. Handbook tables and graphs contain a good deal of implicit knowledge, in addition to the underlying data values they represent. Implicit knowledge is imparted by locational features and structural aspects of graphs and tables, the use of printed rules, typeface selection, how tables are nested, the choice of scales in graphs, and so forth. All these devices convey additional information, relationships, emphasis, and constraints regarding the data values themselves. If we simply pull data values from tables and graphs, we lose information contained in those other representational choices made by their designers.

As we put published information into computer-readable form, it is desirable to preserve the original representations for two other reasons as well. First, it facilitates checking against the original published source. Second, it is reassuring to users to see familiar forms of tables and graphs that they are accustomed to in the published handbooks.

On the other hand, computer formats that are convenient for the capture and recreation of published tables and graphs are not well-suited for indexing and retrieval based on standardized names and data values. Storing the original images in raster (bit map) form is cheap to input and easy to display, but it does not support searching.

One compromise solution to this dilemma is to capture data in a structured form that can automatically generate a standard, canonical form to support searching as well as the recreation of facsimiles of the original graphic display. This approach is described elsewhere in this proceedings [5]. It might be even more desirable to have a standard, concise, high-level language to specify reconstruction of original tables and graphs from a single canonical form. Some graphics software languages provide limited capabilities in this regard, but the problem is still largely unsolved.

Names and Measurement Units

Whereas some scientific fields such as chemistry and physics have relatively standard, universally recognized naming systems (for example, chemical formulas, CAS registry numbers, atomic particle names, and so forth), materials science is not so fortunate. As discussed in another paper at this symposium [6], a single material or property may be known by dozens of different names, and a given character string may refer to several different materials or properties. Materials information systems, particularly those that attempt to include different types of data or data from diverse sources, must deal with this diversity of nomenclature.

Multiplicity of names is especially problematic if properties and variables are represented as individual fields. Most data base systems provide rather limited support for field name synonyms and few, if any, permit operations which mix data and metadata (for example, letting certain classes of users other than the data base administrator add new synonyms). Systems that contain diverse data (for example, from handbooks), confront an additional name problem in that each exhibit or data set may contain quite different sets of properties and independent variables. The over-all number of properties and variables may be quite large, while the number applicable to any given data set is relatively small. In such cases, treating each variable or property as a separate field is impractically cumbersome.

One solution to the problems that arise if each property and variable is a separate field is to put the variable names themselves in a general "variable name" field that is bound in turn to a set of one or more values (and possibly footnotes) associated with that variable. Figure 2 shows how the hypothetical data set from Fig. 1 might be reexpressed in this more general type of structure.

Figures 1 and 2 also draw attention to another common problem: scientific data-measurement units. In Fig. 1, measurement units were given as part of the variable value, while in Fig. 2, "units" becomes an optional element (field) associated with each variable structure. In many data bases, measurement units are a relatively simple problem. It is important to specify what they are, but those units remain constant for a given variable across the entire data base.

```
material = 304
variable = condition
   value = quenched & tempered
variable = form
   value = all wrought forms
variable = resistivity
   units = ohms
   minimum = 100
   maximum = 150
variable = thickness
   units = inch
   maximum = 0.5
variable = test temperature
   units = F
   value = -40
   value = 75
   value = 200
   value = 300
   value = 400
   value = 500
variable = Charpy V-Notch
   units = ksi
   value = 5.2
   value = ---
   value = 9.6*a
      note = *a shattered
   value = 10.9
   value = 11.0
   value = 11.3
```

FIG. 2—*An alternate representation of hypothetical data set pictures variable names and units as data.*

Complications arise because:

(1) users may wish to work in terms of different measurement units than data are stored in;
(2) data manipulation may bring together variables with different units; and
(3) data from diverse sources may come in different measurement units.

Forcing data and users to comply with a single measurement unit for each variable is one way to solve the problem. However, users may not agree on what standard measurement units should be used, and it makes control of data integrity more difficult if data are not archived in the same units as their original sources. A more flexible and general solution to the problem is to *archive* data in whatever units are used in the original source, but *index* the data values in terms of specified standard units. The computer system can support automatic conversion between different measurement units within the same unit class (for example, centigrade, Fahrenheit, and Kelvin for temperature). This also permits different users to work in whatever measurement units they wish to employ.

Summary and Abstraction Levels

Another problem frequently encountered in materials data is that material designations, property and variable names, and even data value domains may contain data pertaining to different levels of abstraction. For example, one data set may pertain to 304 steel, while another refers to 304 grade B, and still another to a particular specimen of 304 B from a specified heat (lot or batch) produced by a given manufacturer. The same kind of problem with different levels of specificity arises for properties and variables. For example, the general property "elongation" may be further qualified by measurement type (for example, a 2-in. (50.8-mm) gage length), value type (for example, design), subtype (for example, minimum), and even a particular test method, which itself may be further specified regarding material family, form, and revision date (for example, ASTM Methods of Tension Testing of Metallic Materials [Metric] [E 8] or ASTM Method of Tension Testing Wrought and Cast Aluminum- and Magnesium-Alloy Products [B 557]). Such issues also arise with regard to certain value domains; for example, "form" may have a generic value such as "pipe," which can be more fully qualified by process, application, and size (for example, extruded, hydraulic pipe, with a 3-inch outside diameter). A user whose query is framed on the basis of the more general names (for example, material = 304, or property = elongation, or form = pipe) should get the more specific instances as well.

Just as actual data may exist at different levels of abstraction, so too is there a need to support *summaries* at different levels. Such summaries might include maximum and minimum for numeric values, the set of nominal values for a coded variable, the set of independent variables for a given property, or "existence tables" (as explained in the following section). Different summary levels of interest might include all data bases or specific data bases (for a network or gateway), or summaries for a particular material or for a specified property. These kinds of capabilities are difficult to implement in the context of current commercial data management systems. Object-oriented data base systems that support inheritance and generalization may offer an alternative approach in the future, but multiuser versions of such systems with full data base capabilities are not yet available.

Diverse and Sparse Data

As materials data bases get large and draw from diverse sources, additional complications arise. First, diverse nomenclatures need to be harmonized across the different sources. Different terms for the same thing need to be indexed in common, while a term that means different things for different parts of the data (for example, tensile strength for plastics versus metals) needs to be distinguished as such. Second, data sets that represent different types of measurements for the same property or variable (for example, design values versus typical values) need to be distinguished. Third, any given material may not have measurements for some properties, let alone properties measured at different levels for several different independent variables. The "existence table" of materials by properties by independent variables is usually quite sparse in terms of actual measured values. Data administrators and users both could benefit from being able to access such "existence tables," which can identify data gaps that may need to be filled as well as limiting user menus to items for which data are available.

If a user proceeds through successive menus of materials and properties, it is quite likely that there will not be data for some of the sought properties or property-variable combinations once a material is specified. Rather than seeing a menu of all possible properties, the user would like to see a list that only includes properties data that pertain to the material s/he has previously specified. These lists can be generated dynamically. Similarly, if property is specified first, users would like subsequent menus of materials or independent variables to only include the names of those for which data exists in conjunction with any previously specified criteria.

Distributed Data-Base Issues

In materials information networks and gateway systems, all of the problems already discussed are compounded by multiple data bases, on top of a standard host of distributed data management problems. Heterogeneity of geographic sites, computer hardware, operating systems, data management software, communications mechanisms, and interfaces all represent major challenges to any distributed information system. Most such systems simplify the problems by restricting heterogeneity in one or more domains. For example, most current "gateway" systems provide transparent access across different communications mechanisms, but they only provide "pass through" access to remote systems in a "virtual terminal" mode. That is, they may provide a directory to remote systems and automatic log-on, but once the user is connected, s/he must cope with the idiosyncracies of that remote interface, its data management software, and its operating system.

To provide a uniform interface to multiple remote systems, a gateway must translate user queries into different query languages, interact with different system interaction protocols, translate different output formats, and deal with different types of remote metadata (such as labels, and so forth). Such gateway systems are very expensive to implement and maintain; very few have gone beyond the prototype stage. One notable exception is the Chemical Substances Information Network [7].

Design Approaches for Material Information

Some design approaches to specific problems outlined in the preceding section have already been suggested above. The remainder of this paper focuses on several more general ideas suggested by experience with the MIST project—namely, modular architecture, three-tiered naming, an active thesaurus, class hierarchies, existence table support, and object-oriented representation. The following subsections discuss each of these in turn.

Modular Architecture

Figure 3 shows a schematic diagram of how different aspects of a material information system can be divided into semiautonomous modules. The "horizontal" modules shown here as "rows"—Database Management, Metadata Management, Interfaces, and Analysis—apply to any kind of materials information system. The slices or "columns" in the Database and Metadata Management sections picture "vertical" partitioning of information about different data bases in a multi-data base system. Note that data bases may contain different types of informa-

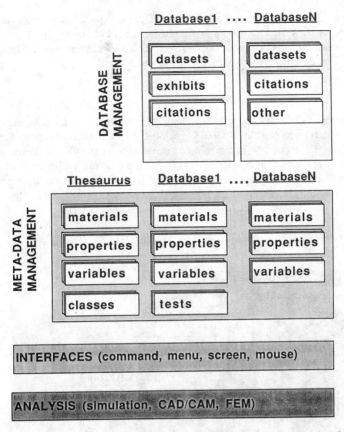

FIG. 3—*Modular architecture partitions implementation by function and data base.*

tion. Database1 has data sets, exhibits (facsimiles of tables or figures from the original source), and citations, while DatabaseN does not include exhibits. Similarly, in the metadata section, Database1 includes information on tests as well as materials, properties, and variables, while DatabaseN does not. The "Global Thesaurus" at the left pictures information that spans and connects the different component data bases. It will be discussed in more detail in Section 3.3 below. The Interface module pictured towards the bottom of Fig. 3 suggests that different types of interfaces—such as command-driven, menu, full screen "fill in the blanks," and mouse-controlled "point and shoot"—can provide access for the more basic levels pictured above them. Finally, different sets of analysis and manipulation functions can also be regarded as a separable component, as shown in the bottom box.

The horizontal lines separating different modules suggest that different components can be partitioned as separate programs, even between different computers. For example, individual data bases might reside on different remote systems, while metadata management and interface components reside on an optical disk connected to a user's workstation, which might also contain different analysis and manipulation programs (for example, spreadsheet, simulation, and finite-element analysis software) specific to each user. In the MIST system, data base and metadata management components currently reside on a single system (the SPIRES data base management system on an IBM mainframe). Those components can be accessed by three different interfaces—the standard SPIRES command interface, a custom line-oriented menu interface developed by the Material Property Data Network [8], and a special Macintosh interface developed by Lawrence Berkeley Laboratory for Sandia National Laboratory [9]. The first two interfaces run on the mainframe, while the third runs on each user's Macintosh, retrieving and caching data and metadata from the central system as necessary.

Three-Tier Architecture

Figure 4 illustrates how one can visualize three components of the modular architecture pictured in Fig. 3 as drawers of a three-tiered file cabinet. In the bottom drawer we have data from Database1 through DatabaseN. In the middle drawer we have data base-specific metadata describing the data in the bottom drawer, such as the names used to identify materials, properties, and variables in each individual data base. Finally, the top drawer contains a thesaurus and indexes that describe the relationships between different data base-specific terms used in the middle drawer.

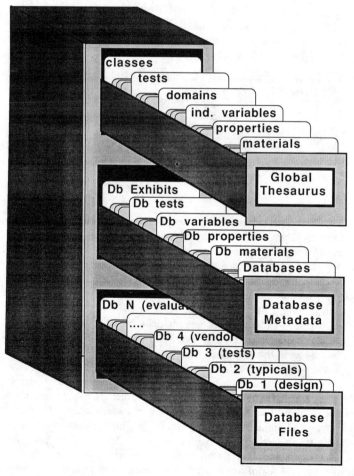

FIG. 4—*A three tier approach to metadata can integrate diverse types of information.*

Figure 5 presents a simplified but more detailed example of how the logical, three-tier namespace design pictured in Fig. 4 is actually implemented in MIST. Data base-specific metadata pictured in the top, right quadrant of this picture describe the materials, properties, and independent variables named in the actual data (for example, AISI 1025 and "Ftu" for the Military Standardization Handbook 5 data; 1025 and "uts" for the Aerospace Structural Metals Handbook data). These data base-specific records include data base name, measurement units, value type (for example, "design" for MH5 and "typical" for ASMH), and any other information specific to each particular version of a particular material or property.

At the left of Fig. 5 are some sample thesaurus entries that show how those records can be used to connect data from different data bases and levels of abstraction. Note how the entry for "ultimate tensile stress," for example, includes aliases (synonyms) of "Ftu" and "uts," which specify that Ftu and uts should be indexed together. Similarly, the entry for G10250 has aliases of "1025" and "AISI 1025," which indicate the identity of those two terms. Finally, note how the G10250 entry has a "broader term" called "carbon steels," which groups specific materials together into material classes that can be used to describe generic qualities as well as to break materials down into manageable size groupings for menus and other such lists. We will return to more detailed consideration of material, property, and variable classes below, following a more detailed discussion of the thesaurus itself.

Implementation of an Active Thesaurus

As Fig. 5 suggests, the thesaurus plays a very central, active role in control and coordination of data and metadata in the MIST project. Figure 6 shows a more detailed example of several thesaurus term records and some of the major kinds of information they contain. Each term is identified by type (for example, material, property/variable, test, measurement unit, and so forth), along with a full description of its meaning, citations to relevant publications (for which

Global Thesaurus Terms

term = ultimate tensile stress
type = property
aliases = uts, Ftu
broader term = tensile properties
definition = stress at maximum load

term = G10250
type = material
aliases = 1025, AISI 1025
broader term = carbon steels
definition = AISI 1025 is a general
 purpose steel for the majority ....

term = carbon steels
type = material class
broader term = steels
narrower terms = G10250, ...
definition = carbon steels are ....

Database Specific Terms

property = MH5!Ftu
measurement units = ksi
value type = design

property = ASMH!uts
measurement units = ksi
value type = typical

material = MH5!AISI 1025

material = ASMH!1025

MH5 Database File

exhibit = MH5!g!2.2.1.0!6/1/83
material = AISI 1025
uts = 55

FIG. 5—*A simplified example of three-tiered naming illustrates the reconciliation of materials and variables from diverse sources.*

```
            term =: yield point
            type =: variable

         term =: 4340
         type =: material

       term =: yield strength
       type =: variable
     aliases =: yield; ys; Fty
related terms =: yield point; upper yield; lower yield
broader term =: tensile properties
  description =: the stress at which a material exhibits a
                 specified limiting deviation from the
                 proportionality of stress to strain. The
                 deviation is expressed in terms of strain
   date added =: 3/20/86
    added by =: J.L. McCarthy
```

FIG. 6—*Term record entries from an active thesaurus illustrate metadata components and hypertext browsing.*

full bibliographic entries are located in another record-type), and so forth. As thesaurus term records are loaded, each narrower, broader, related, and preferred term can automatically generate appropriate reciprocal entries if they do not already exist. Figure 6 also illustrates how the thesaurus can be used for hypertext browsing. Selecting a word or phrase in one record can initiate a search to see what information may be available under that term itself. In this example the user has highlighted the term "yield point" in the record for "yield strength." The system can then automatically bring up a new window to display information from the term entry for "yield point" shown on the "index card" for that record at the back.

An active thesaurus can play many important roles for a materials information system. Some of the major roles played by the MIST thesaurus are pictured in Figure 7. In this exhibit, input thesaurus terms, data base-specific metadata, data records, and user queries are pictured at the left as rounded boxes. Thesaurus records are shown as boxes in the middle column, with material and variable indexes pictured at the bottom right. Major uses of thesaurus information are pictured in ovals, with arrows pointing from the input data to the specific thesaurus items to which they relate.

One of the most basic uses of thesaurus information is to validate and do consistency checks on input to the thesaurus itself. The topmost oval pictures this bootstrap type of operation. In this case the value of the element (field) called "unit class" in the input term record for "ksi" is checked to make sure it contains one of the controlled vocabulary terms permitted for that field. A unit class is a group of measurement units that can be transformed into one another, along with properties and variables that can be measured in terms of those units. For example,

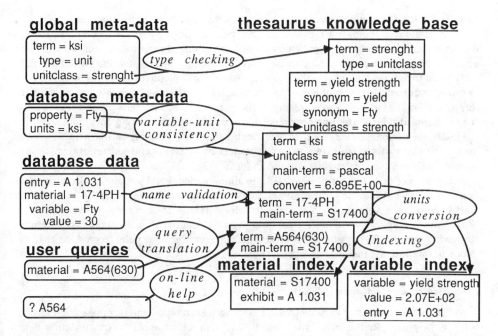

FIG. 7—*Thesaurus knowledge base supports metadata and data validation, indexing, units conversion, and queries.*

Kelvin, centigrade, and Fahrenheit are all members of the "temperature" unit class for variables such as "test temperature" and properties such as "melting point." As the picture implies, unit classes must be defined before the measurement units or variables that are members of that class can be loaded.

The second oval in Fig. 7 pictures how thesaurus information is used to check data base-specific metadata. In this case, the system checks not only the "type" field (not pictured for this record), but also the "property" and "units" fields. Furthermore, it checks to make sure that the property and units identified in the data base-specific variable record are members of the same "unit class."

The third oval from the top shows how names used in individual data bases can be checked against the thesaurus as well as against the data base-specific metadata. Such validation can also include more complex operations, such as checking to see that data values are members of a restricted vocabulary for certain nominal variables, or checking that numbers are within a certain range.

The two ovals at the lower right of Fig. 7 picture how thesaurus information supports indexing and conversion of data values for subsequent retrieval. To permit users to search for information on a given material, property, or independent variable across data bases, MIST creates indexes based on a standard identifier specified in the thesaurus. Whenever data are added, the maximum, minimum, coded, and text values are passed to a central index keyed on the standard identifiers. For example, when the data record for Entry A1.031 shown in Fig. 7 is added, the system adds an entry to the material index that points to that data record. To do this, the material name in the data record (17-4PH) is converted to its standard identifier (S17400) via the thesaurus before passing it to the index. In a similar manner, a variable (here Fty) value (here 30) can be indexed under its standard identifier (in this case "yield strength"). If the variable is numeric (as it is in this case), data base-specific information on measurement units (here ksi) is used in conjunction with conversion factors specified in the thesaurus to convert numeric values to a standard base (here Pascals) before indexing them. This permits archived data to remain in its original form (which may be in any one of several different measurement units) at the same time that a standardized form is automatically used in the overall index.

The next to bottom-most oval in the lower left part of Fig. 7 pictures how the thesaurus is used to translate different variants of names for materials, properties, and variables to the standard forms found in the indexes to do searching. If the query involves a numeric property or variable, the thesaurus also facilitates units conversion, just as it does for indexing data values.

Finally, the bottom left oval pictures how the thesaurus also supports on-line help. Users can access the thesaurus at any time to get explanations of terms and abbreviations, browse broader, narrower, and related terms, and so forth.

A thesaurus of this kind can also be useful in aiding national and international standardization efforts. ASTM Subcommittee E49.03 on Terminology for Computerization of Materials Data may use such a thesaurus to organize its information. The Commission of the European Community is preparing a computerized, multilingual thesaurus for use in its Network Demonstrator Project [10]. Perhaps these parallel efforts will merge at some point to produce an international, multilingual thesaurus for general use.

Class Hierarchies for Materials Information

As data bases grow in size and complexity, users need abstraction and summarization mechanisms to help navigate the information space. One such mechanism is to group similar items into classes, which can be used recursively to provide more and more general items for browsing and searching. As shown briefly above, the MIST thesaurus—with indexes and the assignment of classes to data base specific information—provides just such a mechanism. Each data base-specific material, property, and variable record can be assigned to one or more classes. Names of individual materials, properties and independent variables, as well as other (sub)classes, can be automatically associated with each class via indexing or relational join mechanisms. The resulting groupings can be used to produce lists of classes for menus and so forth.

Figure 8 shows a schematic representation of part of a material class hierarchy from the MIST project. In this instance, each set of classes and materials is ultimately grouped under one of two databases—Military Handbook 5 (MH5) or Aerospace Structural Metals Handbook (ASMH). Certain classes (for example, steel) are further divided into subclasses (for example, carbon, stainless, and high alloy for MH5 versus wrought for ASMH). In addition to providing finer level substantive groupings, subclasses can be used to cut each grouping down to a small enough number to fit comfortably within a single-screen computer menu. Each data base need

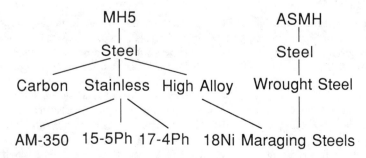

FIG. 8—*Data base specific metadata support material, property, and variable classes.*

not have the same number of levels of classes, nor do the same classes have to appear at the same level in different data bases. Note also that a given material or subclass may appear under different classes in different data bases. For example, 18 Ni Maraging steel is classed under "high alloy steels" in MH5, but under "wrought steel" in ASMH. The relationships are therefore not strictly hierarchical, but rather produce a directed, acyclic graph. For a more detailed discussion of this point, see Ref *11*.

Earlier in this paper, we raised the point that data values can sometimes themselves be classes, particularly for nominal (category) variables. For example, many design values apply to entire classes of forms, such as "all wrought forms." This creates a retrieval problem because simple queries for narrower terms (for example, "sheet") would not retrieve data sets categorized under the broader class. Conversely, queries for a broader class such as "all wrought forms" would not automatically retrieve items listed under narrower categories of that class. One answer to this problem is to have the computer system and its indexing support "automatic term explosion" based on class information in the thesaurus or data base-specific metadata or both. If the user chose to invoke "automatic term explosion," the system would automatically augment each user query to include any broader terms and narrower terms listed for a given data *value*. For example, a query for "form = sheet" would automatically generate an additional search for "all wrought forms" since that would be listed as a broader term for "sheet" in the thesaurus.

Existence Tables

Classes provide a useful mechanism that supports hierarchical menus of materials, properties, and variables, as well as higher-order grouping of such classes. But once users have made an initial selection, a computer retrieval system may not differentiate between subsequent classes that contain data and those that do not. Users become frustrated quickly if the result of successive selections is "no data." Additional mechanisms must be provided to reduce successive lists of materials, properties, and variables on the basis of what items have underlying data that also pertain to previously selected items. Data administrators also need similar mechanisms to summarize where information gaps exist and new data needs to be added.

The MIST project has experimented with several different approaches to this problem, none of them wholly successful to date. The approach that currently seems most promising can be pictured as a three-dimensional matrix like the cube in Fig. 9 for each data base. The three dimensions of the matrix, as shown by the cube, are materials, properties, and independent variables. Since the set of materials, properties, and independent variables are relatively fixed for each data base, each item within each of the three dimensions can be assigned a sequential number. Individual data values in the matrix (cells in the cube) can be represented as simple logical (0/1) values, which indicate the presence or absence of data for that particular combination of material, property, and independent variable. Alternatively, the entries could contain *counts* of data sets containing such information. This type of "existence table" or "existence cube" can be created as data are loaded and even stored at a different location than the data location, along with other types of summary metadata. Since materials data are basically archival, updates will occur rarely, if at all, so the update problems associated with such summary data would be minimal.

Object-Oriented Representation

As discussed above, many aspects of material properties data defy straightforward relational representation. Most of the alternate structures proposed throughout this paper can be more naturally represented in terms of object-oriented, network, or hierarchical data models.

For example, Fig. 10 shows a simple graph, along with a structured representation of the

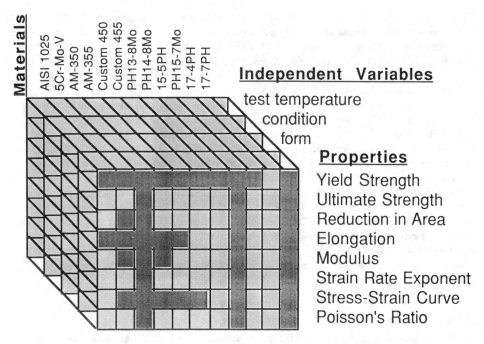

FIG. 9—*"Existence tables" show data availability for materials by property and independent variable.*

Original Exhibit

Fig 3.03112 Stress-strain curves at room and elevated temperatures (23)

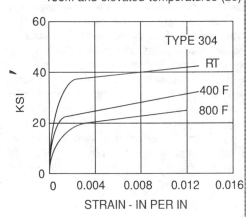

Discrete Components

title = Fig 3.03112 Stress-strain curves at room and elevated temperatures (23)
material = 304
axis = vertical
   label = KSI
   variable = stress *note: this is implied*
   units = ksi
   ticks = 0,20,40,60
axis = horizontal
   label = STRAIN - IN PER IN
   variable = strain
   units = in per in
   ticks = 0,0.004,0.008,0.012,0.016
caption = TYPE 304
curve set = a
   label = 800 F
   xy = 0,0
   xy = 0.002,18
   xy = 0.004,20
   xy = 0.012,24
   curve indep var = temperature
   value = 800

FIG. 10—*Graphic representation has high information density; it can be described in terms of discrete component parts.*

components of that graphical object. Variables and values are identified to facilitate indexing and retrieval, but there is also sufficient information to recreate a reasonably faithful facsimile of the original figure. This type of object-oriented description is relatively easy to understand, while a fully normalized relational version would be difficult to comprehend.

Most of the material properties information systems implemented to date use hierarchical, network, or hybrid rather than relational data base management systems. General Electric's proprietary COMPIS system and the American Welding Institute's Welding Information Network (WIN) both use FOCUS. Chemical Abstracts Service has made the DIPPR data base (created by the American Institute of Chemical Engineers) available on their proprietary, hierarchical Messenger system. General Dynamics' Engineering Materials Property Data Base described elsewhere in this symposium runs on IMS [*12*]. The DOE/NBS MIST prototype is implemented on SPIRES, a hybrid system from Stanford University. There are, of course, notable exceptions that have been implemented on relational data base systems. Dow Chemical's Materials Engineering Data Base, described elsewhere in the symposium, runs on CDC's relational IPF system [*13*]. The National Aeronautics and Space Administration's Materials and Processes Technical Information System (MAPTIS) is implemented on Oracle. Another general approach to the problem of complex, nested data is to model it in a conceptional schema [*14*].

Conclusions

This paper has outlined some of the major challenges that material properties data bases present. It also has described how multiple data bases from diverse sources compound those challenges. It has suggested how modular architecture, three-tiered naming, an active thesaurus, class hierarchies, and existence tables can help overcome many of the problems inherent in the design of materials information systems. Finally, it has suggested that we may need to go beyond the standard relational model to implement some of those solutions.

Acknowledgments

This work was supported in part by the National Bureau of Standards Standard Reference Data Program, the Sandia National Laboratory, and the Office of Energy Research of the U.S. Department of Energy, under Contract No. DE-AC03-76SF00098.

References

[*1*] Northrup, C. J. M., Jr., McCarthy, J. L., Westbrook, J. H., and Grattidge, W., *Materials Information For Science and Technology (MIST): A Prototype Material Properties Data Network System,* J. G. Kaufman, Ed., *Materials Properties Data-Applications and Access,* American Society of Mechanical Engineers, New York, (ASME), 1986.

[*2*] Grattidge, W., Westbrook, J. H., McCarthy, J. L., Northrup, C. J. M., Jr., and Rumble, J., Jr., "Materials Information for Science & Technology (MIST): Project Overview. NBS" Special Publication 726, 1986.

[*3*] Dathe, G., "Peculiarities and Problems of Materials Engineering Data" and Shoshani, A., Olken, F., and Wong, H. K. T., "Characteristics of Scientific Databases" in: *The Role of Data in Scientific Progress,* Elsevier, CODATA, 1985.

[*4*] Codd, E. F., "Missing Information (Applicable and Inapplicable) in Relational Databases," *Sigmod Record,* Vol. 15, No. 4, Dec. 1986, pp. 53-78.

[*5*] Grattidge, W., "Capture of Published Materials Data," see this volume pp. 151-174.

[*6*] Westbrook, J. H., "Designation, Identification, and Characterization of Metals and Alloys," see this volume pp. 23-42.

[*7*] "An Architecture for the Chemical Substances Information Network," Technical Report CCA-83-02, Computer Corporation of America, Cambridge, MA, Jan. 1983.

[*8*] The National Materials Property Data Network, Inc., "Experimental Version: Demonstration Document and User Manual for Pilot Network," July 1987.

[9] Benson, W. H. and McCarthy, J. L., "Designing a Macintosh Interface to a Mainframe Database," *Proceedings of the 22nd Hawaii International Conference on Systems Services* (Jan. 1989), LBL 24176.
[10] Kroeckel, H. and Steven, G., see this volume pp. 63-74.
[11] Dobrucki, M., Kasprzyk, J. M., Krop, E., Liskowacki, J., Rowski, J., and Wroblewski, K., "Structural Materials Corrosion Databank" in Z. S. Hippe and J. E. Dubois, Eds. *Computer Science and Databanks: Papers Presented at the 8th International CODATA Conference,* Jachranka, Poland, 4-7 October 1982, 1984 Polish Academy of Science, pp. 147-155.
[12] Coyle, T. and Little, D., see this volume pp. 200-210.
[13] Petrisko, L., see this volume pp. 229-238.
[14] Griethuysen, J. H., Ed., "Concepts and Terminology for the Conceptual Schema and the Information Base," ISO publication number ISO/TC97/SC5-N695, ISO.

Walter Grattidge[1]

Capture of Published Materials Data

REFERENCE: Grattidge, W., **"Capture of Published Materials Data,"** *Computerization and Networking of Materials Data Bases, ASTM STP 1017*, J. S. Glazman and J. R. Rumble, Jr., Eds., American Society for Testing and Materials, Philadelphia, 1989, pp. 151–174.

ABSTRACT: Computerized data systems involving materials property data are needed to support advanced methods of engineering design and computer supported manufacturing. The challenge is to transform the data values extant in printed handbooks, journals, reviews, or monographs into a coherent, accurate, reliable, and consistent body of computerized data available both for automatic as well as manual access in support of engineering tasks. In print, the data values are presented through four principal formats: Direct; Paragraphs (text including embedded numerics); Tables (text or tabular numerics or both); and Graphs (point sets, curve sets, or range sets). Generic templates have been developed to identify and record the metadata as well as guide the capture of the common and separate features with the associated presentation logics for the different formats. Procedures for data capture for tables and graphs are described.

An extensive metadata schema has been developed covering mechanical properties data included in representative sections of the Aerospace Structural Metals Handbook, and Military Handbook 5. When used in conjunction with data input, the local usage of symbols, property, or material names or designations can be allowed and incorporated and still be consistent with an overall metadata structure.

KEY WORDS: data capture, metadata, graph metadata, table metadata, table structures, re-presentation, mechanical property data, metadata schema, materials property data

Computerized data systems involving materials property data are needed to support advanced methods of engineering design and computer supported manufacturing. While within engineering a broad range of computerized design methodologies and associated application programs have been developed, a comprehensive set of computerized data to automatically interface with and support such design tools has not yet been developed. In most instances, users are required to consult the many printed handbooks and select subsets to manually enter into their programs as required. The challenge is therefore to transform the data values extant in printed handbooks, journals, reviews, or monographs into a coherent, accurate, reliable, and consistent body of computerized data that would be available both for automatic, as well as for manual, access in support of engineering tasks.

The goal, simply stated in this age of work station-based engineering, is the one-stop availability of computerized numeric and factual data for manual and automatic support of computerized engineering and manufacturing activities. However, before such a goal can be achieved, there are a number of other challenges to be faced. These include the following:

- actual conversion of existing printed data to computer-readable form,
- the ability to monitor the converted data for completeness and accuracy,
- organization of the data for generalized search and retrieval,
- the ability to create new, user-controllable formats to present the stored data, and
- the ability to interface selected data with application programs and CAD/CAM.

[1]Principal consultant, Sci-Tech Knowledge Systems Inc., Scotia, NY 12302.

The conversion procedure for transforming technical data from the printed page to an organized body of computerized data is more than producing a machine-readable, character-by-character equivalent for the published data. This paper identifies the considerations and procedures necessary to end up with a useable product.

Significance of Metadata

To fully exploit access to, and manipulation of, numerical data in a computer, a primary requirement is a data base management system for handling and controlling the inputted data. However, before the data base management software can cope with the internal organization of the numerical values contained within a data set, it must be supplied with an extensive set of data descriptors and other associated information that characterizes the individual data values. This necessary "data about the data" is included in the concept of Metadata.

Metadata is the term used to describe all the various features and parameters necessary to ensure that numeric and factual data values that get entered into a data base system do not get lost or become irretrievable [1,2]. By identifying these critical elements and ensuring that the data base system handles the metadata elements appropriately, the integrity of a numeric data base is established and maintained.

All scientific and technical data require associated metadata. A scientific datum cannot exist alone. It exists within a matrix of concepts, parameters, units, terms, and the like that are appropriate to the particular field of science or technology. Consequently, all the relevant elements of this matrix must be contained within the computer data base system if the final stored data values are to be identifiable and extractable.

As we shall also see, the printed forms of data presentation also possess associated metadata features that are essential to record and store if the relationships within the source data are to be retained. As described in this paper, considerable effort has been expended to identify and build a well-structured system of metadata for a wide variety of scientific data presentation formats, including many types of tables and graphs.

Characteristics of Scientific and Technical Data

There are a number of characteristics of scientific and technical data that condition data capture and data storage methods and practices.

Scientific and Technical data (S&T data), consist in general of an entity (a material or a concept); associated variables (dependent, independent, and others, that is, descriptive[2]); values of the variables, which may be expressed in numerical or textual form or may be unexpressed as default values of implicit independent variables; and units, which may be members of several unit classes. S&T data differ from business data in that the characterization of S&T entities, such as materials, usually requires values for several dependent and independent variables. Business data consist of entities with a relatively small number of dependent parameters, each of which has only a small number of associated independent variables and values.

Data capture and organization procedures for S&T data must recognize and accommodate many associated explicit and implicit independent variables and values. It might also be noted that certain variables in S&T data may act as both dependent and independent variables. For example in Materials Properties data: "Hardness" of a material can be a dependent variable, which may vary depending on the independent variable—"heat treatment" or "condition." In other instances, "Hardness," can be an independent variable indicating a specific hardness

---

[2]Descriptive variables are variables that are neither dependent nor independent yet act as descriptors with values that are essential to fully describe the relevant entity. Examples include color, past history, price (if no controlling independent variables are indicated or implied), and so forth.

value used to define an initial or a test condition, where the dependent variable of interest might be "Ultimate Strength" and its variation with temperature. Similarly, "Specification" can act either as a dependent variable when the various forms covered by the specification are given as values of the independent variable "form," or as an independent variable when different specification numbers are associated with a given material name or designation.

A second characteristic of many scientific and technical fields is that several names may be used for the same property, concept, material, and so forth. However, there are also many instances where similar words or phrases do not mean equivalence. This richness of synonyms must be captured and accommodated in the computerized system to merge data from different sources and to accommodate different classes of users. Table 1 provides an example of synonymous terms and similar but not-synonymous terms for the term "Tensile Modulus of Elasticity." A paper by Westbrook at this symposium [3] provides an extensive discussion of the name identification and equivalence problem for material designations.

Similarly, in many scientific and technical fields several systems of measurement units may be used. Although each data set may only use one system of units, computerized systems must be able to accommodate data from different sources. Moreover, users should be able to interact with a computerized system in any one of the applicable unit systems, obtain access to all the data, and display it in any unit system of their choice.

Finally, many scientific and technical fields have developed specialized vocabularies with definitions, symbols, and unit systems that require careful and extensive cross-referencing. This cross-referencing is especially significant when moving from the print form of the data to a computerized format, and must be captured if the full technical relationships are to be preserved.

Data Presentation Formats

One of the most glaring characteristics of printed data sources of technical data is the wide diversity in data presentation formats. Some sources favor tabular presentations, others graphs, and still others a combination of tables and graphs. While the usual convention for graphs is to plot the dependent variable along the y-axis (ordinate), some types of data are conventionally presented with the dependent variable along the x axis. In the case of tables, the choice of the logical form is usually determined by page layout considerations, rather than any regard for technical custom. The challenge for the computer system designer is to include capabilities that can provide access to the full original exhibits regardless of origin, as well as to the basic data values, along with alternate types of presentation.

TABLE 1—*Synonymous terms and similar but not-synonymous terms for the term, "tensile modulus of elasticity."*

TERM: Tensile Modulus of Elasticity	
Synonymous Terms	Not-Synonymous Terms
Elastic modulus	tangent modulus
Young's modulus	secant modulus
Modulus of elasticity (E)	chord modulus
Modulus of extensibility	initial tangent modulus
Stretch modulus	dynamic modulus
Monotonic modulus	kinetic modulus
Tensile modulus	
Static modulus	
Coefficient of elasticity	

Perhaps one of the most important concerns in capturing data from the printed page is that not all the data descriptors are always included within the bounds of an exhibit. Important values of associated independent variables may be found, for example, only in the title, as part of some associated descriptive text, or even buried in a footnote located away from the actual exhibit. All this essential metadata must be identified and captured as part of the exhibit record. Otherwise the computer record will be of little value.

There are many different types of print sources of data, ranging from monographs and journals to handbooks and encyclopedias. Within each of these source types there are many different types of exhibits, or ways of presenting the data. In the case of printed data sources, the eye-brain combination of the scientist or engineer has become highly adept at interpreting how to carry along the pertinent associated information or metadata when extracting a data value. In a computerized system, these critical metadata must be made explicit.

In print, S&T data are presented through four principal formats or types of exhibits:

- direct (in which a property name is given followed by a data value),
- paragraphs (which involves text including embedded numerics),
- tables (text or numerics or both in tabular form), and
- graphs (point sets, curve sets, or range sets).

There are other types of formats that are less specific in terms of ease of interpretation of the inherent data values. These include: Diagrams; Images such as micrographs, X-ray photographs, and so forth; and Equations. This paper focuses on the first four types of formats.

Exhibit Metadata

As part of our company's work in this field, generic templates have been developed for describing the generalized features of graphs [4] and of tables [5]. These templates constitute sets of metadata features associated with the various formats. No standards or proposed standards for exhibit metadata have been presented yet, though that would be desirable for wider acceptance of the methodology. In addition to ensuring accurate storage of the parametric relationships within the data, such templates can also be used in data entry to guide the capture of the parameters and variable values, and serve as protocols to display different formats.

Figure 1 shows examples of four presentation formats used in printed data handbooks: Direct, Paragraph, Tabular, and Graphical.

Direct

In the Direct Type, the format is usually

$$\text{Name of Property : Material} = \text{Value : Unit}$$

or

$$\text{Name of Property : (Independent Variable Name : Value) : Material} = \text{Value : Unit}$$

The electronic capture of such direct format data entries is relatively straightforward, providing that (1) the names of variables, units, and so forth are acceptable as terms or synonyms in the data system and (2) the print source of the data values is identified.

Paragraph/Text

The paragraph form of data presentation includes numerical or textual data values surrounded by text, the latter often describing the test environment or giving the names and values

Handbooks: Direct, Paragraph, Tabular and Graphical

DIRECT:

M.P. Lead = 327.502 C

From: CRC HDBK CHEM & PHYS 55th ED.

PARAGRAPH:

3.061   Poisson's ratio: 0.291 at room temperature for all heat-treated conditions (45).

From: ASMH 2-1501   P3.061   (17-4PH)

TABULAR:

Alloy	Fe-0.4C-1.8Ni-0.8Cr-0.25Mo				
Form	1-1/2-inch Plate				
Condition	1550 F 1 hr. OQ + Temper 3 hr. AC				
Temper, F	Orientation	F_{ty}, ksi	F_{tu}, ksi	e (4D), percent	R.A. percent
300	L	227.3	340.0	10.0	37.4
		228.0	307.6	11.8	41.3
375	L	226.6	318.0	13.7	38.1
	T	216.4	283.1	13.0	45.2
450	L	236.5	295.7	12.7	43.5
	T	228.5	273.5	14.0	44.3
500	L	241.2	286.4	11.2	44.0
Condition	1550 F 1 hr. Salt Q at 450 F 5 Min. AC + Temper 3 hr. AC				
375	L	229.0	325.1	12.3	37.8
450	L	239.9	309.2	10.0	36.1

TABLE 3.0219.   EFFECTS OF VARIOUS TEMPERING TEMPERATURES ON TENSILE PROPERTIES OF ELECTROSLAG REMELTED PLATE AFTER CONVENTIONAL OIL-QUENCHING AND AFTER MARTEMPERING (27)

From: ASMH: 1-1206 p 18 (4340)

GRAPHICAL:

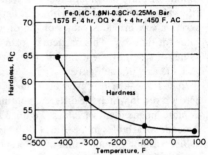

FIGURE 1.064. EFFECT OF TEMPERATURES FROM 80 F TO -423 F ON HARDNESS OF BAR HEAT TREATED TO A HARDNESS LEVEL OF RC 51 (31)

From: ASMH 1-1206 G1.064 (4340)

FIG. 1—*Examples of four presentation formats used in printed data handbooks: Direct, Paragraph, Tabular, and Graphical.*

of the associated independent variables. These variable values may be qualitative as well as quantitative. Taken by itself, or out of context, a paragraph may not specifically include all the required metadata, as in the example, Fig. 1, where the designation of the material (17-4PH) is not part of the paragraph text.

Tabular/Tables

A table is an array made up of character strings, symbols, and numerical values whose purpose is to succintly provide the reader with specific values of one or more dependent variables (properties) of an entity (material) for stated values of associated independent variables. The significance of the layout of tabular data in conveying information regarding the relationships between variables and data values has been discussed by Dolby [6] and Clark [7]. A number of different forms of the structure of the tabular arrays are possible, resulting in a number of types of "table logics." These table logics define just how the numerical values given in the different cells are tied to the table variables (principally the dependent variable). Attention to these logics is necessary if the relationships between variables implicit in a particular table structure are to be captured and properly represented in a computerized record. These structures have been categorized as column-table, row-table, combined row/column-table, column-list, row-list, "mathematical" type tables, special format tables and other table structures.

The first three designate specific types of logical relationships for the dependent variable entries—column-wise, row-wise, or whole-table-wise, respectively. The next two (column/row-lists) cover tabular formats for which there is solely a one-dimensional connection for the column or row contents, rather than a two-dimensional one. The "mathematical" type cover tables that involve an associated access algorithm or access procedure to retrieve a value from the table. Special format tables include those that are "folded" or "stacked" with respect to spatial repetitions of variables and so forth. The "other" category designates other types of table structures not included in the above, such as logic type tables.

Column-Table—In this type of format, the values of the dependent variable are located as entries in a column, each row stub (or header) of which defines the specific value of an associated independent variable. Figure 2 is an example of a column table [8]. In this table, the first column displays the independent variable temperature, with the subsequent rows showing specific values of this variable. Different dependent variables occupy the several columns. Reading across a row provides values of the different dependent variables for the stated temperature value for that row. This particular form of table is comparable to a "stacked graph," in which values of several dependent variables are plotted separately but with a common independent variable axis in one figure.

Row-Table—In this type of format, the several values of a dependent variable are given across a row in which the different columns correspond to various combinations of values of one or more independent variables. Figure 3 shows an example of a row table [9]. Inclusion of other dependent variables in the same exhibit (given on separate rows) also implies the simultaneity of values within a given column of different dependent variables for the stated values of the independent variable(s).

Combined Row/Column-Table—In this type of format, all the values given within the table relate to a single dependent variable, with different values of the independent variables (or entities) given in both columns and rows. Figure 4 shows a combined table [10]. Note in this example that the name and units of the dependent variable (Allowable Unit Stresses in Bending, pounds per square inch) do not appear within the table itself, but only as part of the table title and legend.

Table-Lists (Column-List, Row-List)—In these types of formats, the data elements of a row or column are interpreted as having a logical connection only up or down the respective column or along the respective row. Figure 5 shows an example of a column-list [11], and Fig. 6, a row-

Table 2. Properties of Air at Standard Atmospheric Pressure.[a]

Temp (°F), t	Density (lbm/cu ft), $\rho \times 10^2$	Specific Heat (Btu/lbm-°F), $C_p \times 10$	Viscosity (lbm/sec-ft), $\mu \times 10^5$	Kinematic Viscosity (ft²/sec), $\nu \times 10^3$	Thermal Conductivity (Btu/hr-ft-°F), $K \times 10^2$	Thermal Diffusivity (ft²/hr), a	Prandtl Number, P_r
−280	22.48	2.452	0.4653	0.020700	0.5342	0.09691	0.770
−200	15.64	2.416	0.6659	0.043856	0.7648	0.20863	0.755
−100	11.04	2.403	0.8930	0.080620	1.0450	0.39390	0.739
0	8.66	2.401	1.0926	0.10960	1.3124	0.54874	0.720
100	7.10	2.404	1.2750	0.18102	1.5647	0.92477	0.706
200	5.99	2.414	1.4413	0.24213	1.8047	1.25633	0.694
300	5.23	2.429	1.5951	0.29293	2.0320	1.53656	0.686
400	4.62	2.450	1.7390	0.36471	2.2481	1.92489	0.681
500	4.14	2.474	1.8743	0.45420	2.4570	2.40600	0.680
600	3.75	2.512	2.0027	0.53587	2.6536	2.82656	0.680
700	3.42	2.538	2.1231	0.62122	2.8431	3.27811	0.682
800	3.14	2.568	2.2390	0.71310	3.0220	3.74800	0.684
900	2.92	2.596	2.3498	0.80562	3.2003	4.22567	
1000	2.71	2.628	2.4569	0.90602	3.3710		
1100	2.54	2.659	2.5600	1.0075			
1200	2.39	2.690	2.6569				
1300	2.25	2.717					
1400	2.13						

FIG. 2—*Example of column-table grouping.*

TABLE 2.6.9.0(i). *Design and Physical Properties of 17-4PH Stainless Steel Investment Casting*

Specification	AMS 5344	AMS 5343	AMS 5342
Form	Investment casting		
Condition	a	H1000[b]	H1100[c]
Thickness, in.	...	...	...
Basis	S	S	S
Mechanical properties:			
F_{tu}, ksi	180	150	130
F_{ty}, ksi	160	130	120
F_{cy}, ksi	...	132	...
F_{su}, ksi	...	98	
F_{bru}[d] ksi: (e/D=1.5)			

FIG. 3—*Example of row-table grouping.*

Table 4. Allowable Unit Stresses for Timber in Bending [a]

Recommended by the Forest Products Laboratory, Forest Service, U. S. Dept. of Agriculture.[†] All values are in pounds per square inch.

Species	Continuously Dry		Occasionally Wet but Quickly Dried				More or Less Continuously Damp or Wet			
	All Thicknesses		4 in. and Thinner		5 in. and Thicker		4 in. and Thinner		5 in. and Thicker	
	Select	Common	Select	Common	Select	Common	Select	Common	Select	Common
Ash, black	1000	800	860	680	900	720	710	600	800	640
" commercial white	1400	1120	1070	910	1200	960	890	760	1000	800
Aspen and large tooth aspen	800	640	580	490	650	520	440	370	500	400
Basswood	800	640	580	490	650	520	440	370	500	400
Beech	1500	1200	1150	980	1300	1040				

FIG. 4—*Example of combined row/column-table grouping.*

list [12]. Such column-list or row-list formats are usually used when presenting multiple values associated with a column or row variable. However, with this type of presentation format, there sometimes can be ambiguity as to whether relationships between individual values in adjacent rows and columns are to be inferred.

Mathematical Tables—For most types of tables, the values of the variables can be easily located within a column or a row. However, for certain forms of tables, the specification of a procedure, or access algorithm, is necessary (or is implied) to locate a table value. In such instances, the computerized record must include the access algorithm as part of the associated metadata. One example of such a table is presented in Fig. 7 [13].

Table 1. Applications for Manufactured Carbon and Graphite[a]

Aerospace	Metallurgical
nozzles	electric furnace electrodes for
nose cones	the production of iron and
motor cases	steel, ferroalloys, and
leading edges	nonferrous metals
control vanes	furnace linings for blast
blast tubes	furnaces, ferroalloy
exit cones	furnaces, and cupolas
thermal insulation	aluminum pot liners and
Chemical	extrusion tables
heat exchangers and centrifugal	run-out troughs
pumps	for molten iron
electrolytic anodes for the	from blast furnaces
production of chlorine,	and cupolas
aluminum, and other	metal fluxing and inoculation
electrochemical products	tubes for aluminum and ferrous furnaces
electric furnace electrodes for	ingot molds for steel, iron,
making elemental phosphorus	copper, and brass
activated carbon	extrusion dies for copper and
porous carbon and graphite	aluminum
reaction towers and accessories	Nuclear
Electrical	moderators
brushes for electrical motors and	reflectors
generators	thermal columns
anodes, grids, and baffles for mercury	shields
arc power rectifiers	control rods
electronic tube anodes and parts	fuel elements
telephone equipment products	Other
rheostat disks and plates	motion picture projector carbons
welding and gouging carbons	turbine and compressor packing and seal rings
electrodes in fuel cells and batteries	spectroscopic electrodes and powders for
contacts for circuit breakers and	spectrographic analyses
relays	structural members in applications requiring high
electric discharge machining	strength-to-weight ratios

FIG. 5—*Example of table-list: column-list (also folded).*

Material = UNS N06600, the Ni-Cr-base alloy, Inconel 600

Organization	Specification Numbers
AMS	5540, 5580, 5665, 5687, 7232
ASME	SB163, SB166, SB167, SB168, SB564
ASTM	B163, B166, B167, B168, B366, B516, B517, B564
FED	QQ-W-390
MIL	MIL-R-5031 (Cl 8), MIL-T-23227, MIL-N-23228, MIL-N-23229

FIG. 6—*Example of table-list (row-list).*

In this particular table, all the values given in the table are in thermocouple emfs, so this could be designated a "combined" table according to the earlier classification. The row headers to the table represent temperature (in steps of 100°C), whereas the column headers are 0, 20, 40, 60, and 80°C respectively. The column headers represent incremental values of temperature (in intervals of 20°C), between the first column (headed "0 degrees") of one row (say 100°C), and the first column ("0 degrees") of the next row (200°C). Note however, that intervals of 20°C are given as separate increments, with no indication as to how to interpolate for values within any 20°C interval.

22.11. TABLE: emf FOR PLATINUM VERSUS PLATINUM-RHODIUM THERMOCOUPLES (ABSOLUTE mV)

Reference junction at 0°C. Based upon the original table in American Institute of Physics, *Temperature, Its Measurement and Control in Science and Industry* (Reinhold, New York, 1941). Values of the emf have been adjusted to correspond to temperatures expressed on the International Practical Scale of 1968.

Temp.	Platinum–10% rhodium vs platinum				
(°C)	0	20	40	60	80
0	0.000	0.113	0.235	0.385	0.502
100	0.645	0.795	0.950	1.109	1.273
200	1.440	1.611	1.785	1.962	2.141
300	2.323	2.506	2.692	2.880	3.069
400	3.260	3.452	3.645	3.840	4.036
500	4.234	4.432	4.632	4.832	5.034
600	5.237	5.442	5.648	5.855	6.064
700	6.274	6.486	6.699	6.913	7.128
800	7.345	7.563	7.782	8.003	8.225
900	8.448	8.673	8.899	9.126	9.355
1000	9.585	9.816	10.048	10.282	10.517
1100	10.754	10.991	11.229	11.467	11.707
1300	13.155	13.397	13.640	13.883	14.125
1500	15.576	15.817	16.057	16.296	16.534
1700	17.942	18.170	18.394	18.612	

FIG. 7—*Example of mathematical table (also combined).*

The access algorithm[3] that must be made a part of the electronic record consists in this case of something equivalent to the following: "To find the emf associated with a given temperature, divide the given temperature by 100 and the integer result sets the row to be accessed. Divide the remainder by 20 and the integer result is then matched with the column header. The required value of emf falls between the values given by the cell, which is the intersection of the selected row and the selected column and its immediate neighbor to the right." Note that no further details are given in the table for interpolating within the 20°C column header intervals.

Other "mathematical" type tables include the traditional logarithm or trigonometric tables, where in a similar manner to the thermocouple emf's example both a row and a column header value must be selected as part of the table access procedure. For additional accuracy in such tables, a further interpolation subtable is often provided.

Another type of "mathematical" table is represented by the Spreadsheet or SS-table. In this type of table, each cell or square may have a specific algorithm associated with it that defines the value within that particular cell in terms of values (or contents) in other cells of the table or matrix. Thus the metadata description needs to include the access algorithms for every cell, where appropriate.

A third type of "mathematical" table is illustrated by an input-output table in economics. Here there are implicit relationships between the totals for the rows and the totals for the columns. For example, if percentages are involved, then the ways such percentages are to be computed must be recognized (either as a percentage of the row total or the column total). Capturing such a table must include access procedures to extract appropriate subsets.

Special Format Tables

Folded tables—A folded table is a divided table whose original lower part is located side by side with its original upper part to save space. As an example, if one has a folded, four-column table, columns 5 through 8 repeat the column headings in columns 1-4. Figure 8 shows such a

---

[3]In practice, the emf is measured and a table such as this consulted to find the temperature required to produce that emf value. In this second case, the access algorithm involves finding the cell contents that are nearest (both above and below) the given emf values and then ascertaining the temperature (to the nearest 100°C) from the row and the additional 20°C intervals from the column.

22.08. TABLE: THERMAL CONDUCTIVITY, SPECIFIC HEAT, AND VISCOSITY OF ELEMENTS WHICH ARE FLUID AT NTP, AT 300 K [a]

Element	λ (W m^{-1} K^{-1})	c_p (kJ kg^{-1} K^{-1})	η (10^{-4} Pa s)	Element	λ (W m^{-1} K^{-1})	c_p (kJ kg^{-1} K^{-1})	η (10^{-4} Pa s)
Argon	0.0179	0.521	0.229	Iodine	...	0.145	...
Bromine	0.0047	0.226	0.155	Krypton	0.0095	0.249	0.256
Chlorine	0.0089	0.479	0.136	Neon	0.0491	1.030	0.318
Deuterium	0.1397	7.248	0.126	Nitrogen	0.0258	1.042	0.180
Fluorine	0.0256	0.827	0.227	Oxygen	0.0263	0.920	0.207
Helium-3	0.1781	5.191	0.172	Ozone	...	...	0.134
Helium-4	0.155	5.193	0.199	Radon	0.0036	...	...
Hydrogen, normal	0.183	14.31	0.090	Tritium	0.1083	4.837	0.155
Hydrogen, ortho	0.181	14.10	0.081	Xenon	0.0056	0.160	0.234
Hydrogen, para	0.188	14.85	...				

[a] *McGraw-Hill/CINDAS Data Series on Material Properties* (McGraw-Hill, New York, 1980), Vol. III-2.

FIG. 8—*Example of folded column table.*

folded column table [14]. Thus, in accessing a folded table, the computer program must be instructed to seek the values for a given variable in more than one row or column.

Similarly, with table-lists, the list is often presented as a folded table, as shown in Fig. 5. For accurate data capture, the metadata must contain the identification of the repeat columns.

Stacked tables—In a stacked table, headers and values of independent variables are only listed in one row or column, or one set of rows or columns, but are associated with dependent variables appearing in more than one row or column or one set of rows or columns. One example of a simple stacked table was illustrated in Figure 2—a stacked column table, in which temperature (independent variable) values were given in the first column, but applied to several dependent variables given in the other columns. Another example is given in Figure 9 [15], in which the first four columns include values of several independent variables that are common to two sets of dependent variables (along with additional local independent variables) given in columns 5 to 9 and 10 to 14, respectively. The metadata for this table must therefore indicate that columns 1 to 5 represent common variables with values applicable across the two sets of dependent variables. Additional column interrelationships (such as between columns 7 and 9 with respect to column 8, and columns 12 and 14 with respect to column 13) also need to be captured.

Other Table Structures

Logic tables or existence tables—There exists a type of table in which the cell contents represent logical operators rather than the conventional cell values. A simple form of such tables is an

TABLE 5.1.2.1.1. *Typical Values of Room Temperature Plane-Strain Fracture Toughness of Titanium Alloys* [a]

Alloy	Product	Heat Treat Condition	TYS Range, ksi	Product Thickness Range, inch	L-T [b]					T-L [b]				
					No. of Lots/ Specimens	Specimen Thickness Range, inch	K_{Ic}, ksi-in.$^{1/2}$ Test Data Minimum	K_{Ic}, ksi-in.$^{1/2}$ Test Data Average	K_{Ic}, ksi-in.$^{1/2}$ Test Data Maximum	No. of Lots/ Specimens	Specimen Thickness Range, inch	K_{Ic}, ksi-in.$^{1/2}$ Test Data Minimum	K_{Ic}, ksi-in.$^{1/2}$ Test Data Average	K_{Ic}, ksi-in.$^{1/2}$ Test Data Maximum
Ti-6Al-4V	Bar	ANN	120-140	—	8/18	0.5-1.0	38	59	77	9/27	0.5-1.25	33	56	81
Ti-6Al-6V-2Sn	Bar	ANN	149	1.75	1/3	1.0	57	59	62	—	—	—	—	—
Ti-6Al-6V-2Sn	Bar	STA	184	1.75	1/3	1.0	30	31	32	—	—	—	—	—

[a] These values are for information only.
[b] Refer to Figure 1.4.12.3 for definition of symbols.

FIG. 9—*Example of stacked column table.*

existence table, given in Figure 10 [*16*]. In this table, the cell values consist of either an X or a blank. In those cells in which an X occurs, the subject area identified by the row header is contained in the source indicated by the corresponding column header.

More complicated forms of logic tables occur as truth tables, in which the cell values represent one of several logical states relating row and column headers, or as conversion tables, in which the cell contents are used in mathematical operations to convert values in one system of units (row headers) to corresponding values in another system denoted by the column header.

Symbols

There are a class of tables in which the cell values are not given as numerics or references to footnotes, but are given in terms of a set of symbols. Figure 11 [*17*] illustrates such a table. In this table there are four operator-type symbols distributed among the 288 cells. Computer access to the individual cell values in such a table must provide as output the appropriate interpretations and procedures for the various symbol values.

Metadata for tables—Table 2 provides a summary of the types of table logic and format structures identified. Table 3 categorizes the features of tables that must be specified if the full logic, captions, and labels associated with various table formats are to be transferred to a fully accessible computerized record. The principal metadata categories include data about the exhibit as a whole; items relating to the table structure; items relating to the columns/rows; notes and footnotes, references or sources of the data values by table, row, column, or cell; symbols and their representation; procedures required for any associated access algorithm; and any associated equation(s).

Graphical formats—A graphical exhibit is generally of the form of a continuous line(s) or set of points expressing the relationship between two variables (axes). Even if the form of the relationship can be represented by a mathematical equation, a printed graphical exhibit over either

EXISTENCE TABLE FOR MATERIAL-PROPERTIES MATRIX IN DATA SOURCES
6.3: MATERIAL - 4340

PROPERTY	ASMH	SAH	MIL-5	SAE/J1099	DAM-T
Mechanical Properties					
a. Tensile Properties					
Tensile Yield Strength	X	X	X	X	X
Tensile Ultimate Strength	X	X	X	X	X
Tensile Reduction in Area	X	X		X	
Tensile Elongation	X	X	X	X	
Tensile Modulus	X	X	X	X	
Strain Rate Exponent		X			
Tensile Stress-Strain Curve	X	X	X		
Poisson's Ratio	X	X	X		
b. Torsion Properties					
Yield Strength in Shear (Torsion)		X			
Ultimate Strength (Torsion)	X	X	X		
Torsion Modulus	X	X	X		
c. Compression Properties					
Compressive Yield Strength	X	X	X		
Compressive Modulus	X	X	X		
Compression Stress-Strain Curve	X	X	X		
Bulk Modulus					

FIG. 10—*Example of logic (existence) table.*

Table 16. Influence of Alloying Elements upon the Properties of Steel
Elements Taken Individually in Considering Their Influence*

	C Carbon	Mn Manganese	Si Silicon	Al Aluminum	Ni Nickel	Cr Chromium	Mo Molybdenum	V Vanadium	W Tungsten	Co Cobalt	Cu Copper	S Sulfur	P Phosphorus	Ti Titanium	Ta Tantalum	Nb Niobium (Cb)
Yield strength	↗	↗	↗		↗	↗	↗	↙	↗							
Tensile strength	⇗	⇗	↗		↗	↗	↗	↗	↗	↗	↗			↗	↗	↗ ↗
Elongation	↙	↙	↙		↙	↙	↙		↙	↙	↙		↙			
Tensile strength at elevated temperature	↗					↗	⇗	↗	↗	↗			↗			
Creep strength	↗	↗	↗	↙	↗	↗	↗	↗								
Fatigue strength	↗				↗	↗	↗	↗	↗	↗						
Ac1 point		↙	↗	↗	↙	↗	↗	↗	↗							
Ac3 point		↙	↗	↗	↙	↗	↗	↗	↗	↙	↙			↗	↗	↗ ↗
Austenite field		↗	↙	↙	↗	↙	↙	↙	↙	↗	↗			↙	↙	↙
Grain growth	↗	↗	↗	↙	↙	↗		↙		↙				↗	↙	
Susceptibility to overheating		↗	↙		↙		↙	↙	↙				↗			
Oxidation resistance		↗	↗		↗			↗								
Red shortness												↗				
Critical cooling rate		↙	↙		↙	↙	↙	↙	↙	↗						
Hardenability	↗	↗	↗		↗	↗	↗	↗	↗							
Hardness	↗	↗	↗		↗	↗	↗	↗	↗	↗				↗	↗	↗
Tempering stability	↙		↗		↗	↗	↗	↗	↗						↗	↗ ↗
Carbide formation	↗					↗	↗	↗	↗					↗	↗	↗

* Influence of element is ↗ increased, ⇗ greatly increased, ↙ decreased, ⇙ greatly decreased.

FIG. 11—*Example of symbol (logic) table.*

the printed equation or a table of values enables the reader to spot trends or anomalies in the data values, or gain insight into the mathematical form of the relationship between the variables.

There are many different forms of graphical presentation used in science and technology. Each field has its preferred set. The general metadata characteristics for a graph are given in Table 4. The principal categories for metadata are those items that relate to the exhibit as a whole; relate to the graph axes; or relate to the curves included in the exhibit, to point sets, to notes/references/symbols, and to access algorithms. For graph axis scales there are, for example, linear, logarithmic, probability, reciprocal, and other forms. For curves and points there are many identification options, such as solid or dotted lines, or circles or triangles, respectively.

TABLE 2—*Summary of types of table logic & format structures.*

Table	Description
Column table [table-col]	all values of a dependent variable are in the same column but occupy different rows depending on different values of one or more independent variables
Row table [table-row]	all values of a dependent variable are in the same row but occupy different columns depending on different values of one or more independent variables
Combined table [table-cbn]	all values in the table for all rows and columns are values of a single dependent variable with the rows and columns corresponding to various combinations of values of the independent variables or materials or both
List-table-row [list-row]	all values in the table (row) have only a one-dimensional relationship to the row variable, for example, multiple values; note there is no significance to any alignment in columns
List-table-column [list-col]	all values in the table (column) have only a one-dimensional relationship to the column variable, for example, multiple values. Note there is no significance to any alignment in rows
Math-table [table-math]	in addition to a logic structure, an access algorithm must be given to direct the way the table values are to be accessed
Special-format table	define the nature of the format, say folded or stacked, including the access and presentation algorithms
Logic-table	each cell value is an operator linking the row and column entities or variables

Metadata Associated with Materials Property Data

As part of the MIST demonstration data base [18,19], an extensive metadata schema has been developed covering the mechanical properties data included in representative sections of the Aerospace Structural Metals Handbook [20], and Military Handbook 5 [21]. The resulting metadata, covering over 100 materials (UNS number, associated designations or aliases, and material classes); 60 property variables (each with associated synonyms and property classes); 100 independent variables; 140 units and unit classes; as well as 50 general terms, together support a comprehensive search and retrieval system and also provide the basis for a comprehensive validation system for use in data input. The overall metadata schema accommodates the local usage of symbols and property or material names or designations as long as they are allowed terms or synonyms. In this way the metadata schema is able to orchestrate a distributed system of separately organized materials properties data bases.

Data Capture

Data capture is the process of selecting and entering materials property data into a data base. It involves a series of steps that consist of meeting general considerations on adding data to an existing computerized data base; identifying the data to be entered; editing the original source data and associated information; keying the data to specific input formats; updating the data base; and verifying that the data have been correctly entered.

On Adding Computerized Data to an Existing Data Base System

In most cases, a new subset of data will be added to an existing system of data, or more generally, to a system of distributed data bases. To maintain the integrity of the overall system,

TABLE 3—*Metadata categories for a table.*

Exhibit as a whole

- source (bibliographic citation)
- ID (for example, Table 2.013)
- Number of Tables (some exhibits consist of more than one table with common headers)
- table logic group (that is, table-col, table-row, table-cbn, and so forth)
- title
- entity

Items relating to table structure

- number of columns
- number of rows
- column/row variable type (dependent, independent, descriptive)
- column/row labels

Items relating to columns/rows

- identification
- variables (dependent, independent)
- descriptors (test, statistical, and so forth)
- units
- label(s) (headers/stubs)
- format (column separator format)
- data values
- internal relationships (internal hierarchies)

Notes/footnotes/references

- identification
- text/citation
- location

Symbols

- identification

Access algorithm

- procedure

Associated equation(s)

- equation
- substitution procedure(s)

---

the metadata and data values for the new data must satisfy certain criteria and conditions in advance. Thus, before actual data capture is undertaken, it is appropriate to consider and respond to these constraints and requirements.

The first condition involves the metadata compatibility constraints. The term names used with the new data must be found acceptable (either as the same, equivalent, or synonymous to terms already in the data base system or as new terms that can be added within the existing schema). A similar condition holds for the units to be used and associated unit conversions.

Beyond the consistency of metadata, there are constraints regarding the actual characterization of the data. The overall data base system may call for specific review procedures to be completed before the data will be accepted. This may require specific annotations as to the reliability of the data values. Specific designation of the data values as to whether they represent measured, estimated, or derived values may be required, and if the latter, specification of the derivation procedure may be called for.

TABLE 4—*Metadata categories for a graph.*

Exhibit as a whole

- source (bibliographic citation)
- ID (for example, Fig. 2.013)
- Number of graphs (some exhibits consist of more than one graph with common axis)
- Number of quadrants (that is, Do x- or y-axis or both show negative as well as positive values?)
- legend
- inset caption (if any)

Items relating to axes

- scale type (linear, logarithmic, probability, reciprocal, and so forth)
- variable type (dependent, independent)
- location (left, right, bottom, top, rotated, and so forth)
- scale labels
- units
- axis length
- axis length unit
- max and min values
- axis breaks (number, location)
- grid/tic lines (number, location)
- grid/tic values
- grid labels

Items relating to curves

- number of curve sets/curves
- analytical form (linear, curvilinear, parabolic, hyperbolic, and so forth)
- line type (solid, dotted, and so forth)
- breaks (number, location)
- curve labels
- digitization (data tuples)

Items relating to points

- number of point sets/points
- point symbol (circle, square, triangle, error bars, and so forth)
- point labels
- digitization (data tuples)

Notes/references/symbols

- identification
- text/citation
- location

Access algorithm

- access procedure
- interpolation/extrapolation

Regarding the form of the data records, the overall system may have specific rules for designating the materials covered by a given data set. The UNS number code may be desirable for metals and alloys, though not all metals and alloys, particularly new materials, have yet been assigned an official UNS number. Similarly for chemical compounds, the ACS Registry Number provides a similar unique designation capability (within limits).

Finally, the overall system may require specific ways of identifying the various "sources" involved explicitly or implicitly with the new data set. In addition to the unique identifiers for the data set within the data base system, there are sources for test methods, sources for specimens, and sources for the data values themselves (organizations or persons performing the test) or references in which the data was reported.

Data Selection and Identification

The criteria for the selection of the materials and properties that have been used to-date include the following:

(1) identifying and setting priorities for the materials to be included, as well as associated properties based on pertinent application areas,

(2) identifying the property data available and their occurrence in the various data sources or handbooks,

(3) selecting the property data to be captured by identifying from which source the data is to be obtained and in what form it is presented (paragraph, table, or graph), and

(4) identifying the bibliographic and secondary references from which the data were obtained.

Source Editing and Data Transcription

This is the process of preparing the data sources for the data input stage. The procedures involved are dependent on the type of exhibit being processed. The amount of editing depends on the level of technical sophistication of the data entry personnel as well as the extent to which the data entry system provides guidance regarding the interpretation of various exhibit features. Paragraphs require essentially no editing other than identifying the specific variables and values that are to be recorded along with the text. Tables need to be classified as to table logic type; numbers of columns and rows; and the identification of materials as well as dependent and independent variables. Complex tables may require additional editorial attention.

Graphs require several types of editing activities, including:

- Identifying the dependent and independent variables and their related graph axes.
- Identifying the various point and line sets and the associated symbols to be used. If an electronic tablet digitizer is to be used, guidance given on the number of points be taken on the digitizer to obtain an adequate representation of the various portions of the curves.
- Identifying the associated graphic and other metadata needed. As noted earlier, significant relevant parametric data often appear as part of the associated descriptive text, as well as in titles or legends, captions or keys, or footnotes or references.

Data Entry

Data entry involves the keying or capturing (OCR or audio) of data to end up with machine-readable data in a predesigned input format, the exact format being dependent on the type of exhibit and on the record type. In our system, data entry is performed using an interactive type of input program that guides the operator through the order and the content formats of the successive fields [22]. When coupled with on-line validation checks, such a procedure minimizes data entry errors, particularly those involving the consistency of variable and unit names and allowed data values.

Paragraphs/References—The data for paragraphs and references are extracted directly from the source provided.

Tables—Tabular data is best entered to a controlled protocol so that the table logic, with the unique column and row structure, is preserved and a subsequent computer program can identify the individual data tuples to be extracted and stored in the data base.

Figures/Graphs—The treatment of graphs is analogous to that of tabular data except that the axis-based structure of the figure or graph is captured rather than the column and row structure of a table. Input of the digitized data from an electronic tablet is straightforward once the structure of the graph is established.

Data Capture Procedure Controls

In Materials Information for Science and Technology (MIST), special controls are exercised during the loading of the data into the data system to ensure that the captured data are consistent with the system metadata schema; that the type of data records used are correct; and that the records accurately reflect the original data.

Data Validation

Data validation is performed on all data entered into the MIST data base, either during data entry by testing the input against validation tables, or after data entry by obtaining a re-presentation of the data in a format appropriate for the type of data being validated, (that is, table, graph, paragraph, and so forth), and then comparing it with the original data source. Checking for correctness of terms and other factors can be performed during data entry using built-in file definition and thesaurus validation procedures within the system. Additional system-level validation checks are also made.

Re-Presentation of Captured Exhibits

A convenient inspection method for ensuring the accuracy of the data captured is to re-present the original exhibit from the data record. Figure 12 shows a table exhibit (table-row) taken from MIL-HDBK-5, Chapter 2, Table 2.2.1.0 (b). Figure 13 shows the data record (slightly abridged) for the table in Fig. 12. (The abridgements are due to dropping portions of the text in the long character fields.) Figure 14 shows the re-presentation of the captured data record for the table in Fig. 12. Similarly, Fig. 15 shows a graph exhibit taken from ASMH, Section 2-1307, Fig. 2.015. Figure 16 shows the data record for the graph in Figure 15, and Fig. 17 shows the re-presentation of the captured data record for the graph in Fig. 16.

User Entered Data

In an ongoing system, particularly one that links separate data bases within a distributed data base system, users will be provided the option of entering their own data, either to be retained for their own exclusive use, or else, following some review and acceptance procedure, relayed into a central registered file for general access. If the data are to be used privately, then the individual user will be responsible for entering and maintaining the data and associated metadata. If the data are to be available for general use, then a central control will probably accept, process, and validate such data. In either case, the procedures outlined above will need to be followed to ensure the consistency and integrity of such data.

Data Organization

In the absence of a Neutral Materials Property Data Exchange Format, the data capture process needs to be undertaken with some prior knowledge of the data organization of the data base management system being used. However, this knowledge need only consist of the various record types and the constituent data elements or fields, including any validation controls being

MIL-HDBK-5D

1 June 1983

TABLE 2.2.1.0(b). *Design Mechanical and Physical Properties of AISI 1025 Carbon Steel*

Specification	MIL-S-7952, 1025	MIL-T-5066	MIL-S-7097, Comp. 3
Form	Sheet and strip	Tubing	Bars
Condition	Cold rolled	Normalized	All
Thickness, in.	...	...	...
Basis	S	S	S[a]
Mechanical properties:			
F_{tu}, ksi:			
L	55	55	55
LT	55	55	55
ST	...	...	55
F_{ty}, ksi:			
L	36	36	36
LT	36	36	36
ST	...	...	36
F_{cy}, ksi:			
L	36	36	36
LT	36	36	36
ST	...	...	36
F_{su}, ksi	35	35	35
F_{bru}, ksi:			
(e/D = 1.5)	...	...	...
(e/D = 2.0)	90	90	90
F_{bry}, ksi:			
(e/D = 1.5)	...	...	...
(e/D = 2.0)	...	...	...
e, percent:			
L	...	b	ab
LT	b	...	...
ST	...	...	...
E, 10^3 ksi		29.0	
E_c, 10^3 ksi		29.0	
G, 10^3 ksi		11.0	
μ		0.32	
Physical properties:			
ω, lb/in.^{3}		0.284	
C, Btu/(lb)(F)		0.116 (122 to 212 F)	
K, Btu/[(hr)(ft^2)(F)/ft]		30.0 (at 32 F)	
α, 10^{-6} in./in./F		See Figure 2.2.1.0.	

[a] Grain direction not specified.
[b] See applicable specification for variation in minimum elongation with ultimate strength.

FIG. 12—A table exhibit (table-row) taken from MIL-HDBK-5, Chapter 2, Table 2.2.1.0 (b).

```
RECO2 KEY = mhS12tt!2.2.1.0(b)!6/01/83;        ROW = 6;                            ROW = 26;
SOURCE_FORMAT = table;                          FORMAT = 1|c|c|c;                   FORMAT = 1|d|d|d;
ERRORS = none;                                  ENTRY = Hechanical properties!!!    ENTRY = e, percent:!!!;
AUDIT TRAIL = Mon 05/04/87 AT 14:21 BY FH9I     ROW = 7;                            ROW = 27;
MODIFIER;                                       FORMAT = 1|c|c|c;                   FORMAT = 1|d|d|d;
MOD DESCRIPTION;                                ENTRY = Ftu, ksi!!!;                ENTRY = L!...!(n b)!(n e)(n b);
COMMENTS = TAD0001;                             ROW = 8;                            RCELLVAR = e!!;
SECTION = 2;                                    FORMAT = 1|d|d|d;                   ROW = 28;
SOURCE PART = 2.2.1.0(b);                       ENTRY = L!55!55!55;                 FORMAT = 1|d|d|d;
REVISION DATE = 06/01/83;                       RCELLVAR = ftu!!;                   ENTRY = LT!(n b)!...!...;
PAGE NUMBER = 2-7;                              ROW = 9;                            RCELLVAR = e!!;
TITLE = TABLE 2.2.1.0(b). Design Mechanical and FORMAT = 1|d|d|d;                   ROW = 29;
025 Carbon Steel;                               ENTRY = LT!55!55!55;                FORMAT = 1|d|d|d;
TABLE ID = 1;                                   RCELLVAR = ftu!!;                   ENTRY = ST!...!...!...;
  NROWS = 43;                                   ROW = 10;                           RULE = under;
  NCOLS = 4;                                    FORMAT = 1|d|d|d;                   RCELLVAR = e!st;
  GROUPING = table-row;                         ENTRY = ST!...!...!55;              ROW = 30;
  WHOLE TABLE INFO;                             RCELLVAR = ftu!st;                  FORMAT = 1|n s s;
    TABLE MATERIAL = AISI 1025;                 ROW = 11;                           ENTRY = E, 10(e 3) ksi!29.0;
    TINDVAR = basis;                            FORMAT = 1|d|d|d;                   RCELLVAR = modulus of elasticity - tension;
      VALUE = S;                                ENTRY = Fty, ksi!!!;                ROW = 31;
    COLUMN = 2;                                 ROW = 12;                           FORMAT = 1|n s s;
      CINDVAR = specification!mil;              FORMAT = 1|d|d|d;                   ENTRY = E(s c), 10(e 3) ksi!29.0;
        VALUE = MIL-S-7952, 1025;               ENTRY = L!36!36!36;                 RCELLVAR = modulus of elasticity - compression;
      CINDVAR = form;                           RCELLVAR = fty!!;                   ROW = 32;
        VALUE = Sheet;                          ROW = 13;                           FORMAT = 1|n s s;
      CINDVAR = form;                           FORMAT = 1|d|d|d;                   ENTRY = G, 10(e 3) ksi!11.0;
        VALUE = Strip;                          ENTRY = LT!36!36!36;                RCELLVAR = modulus of rigidity!10(e 3) ksi;
      CINDVAR = cond;                           RCELLVAR = fty!!;                   ROW = 33;
        VALUE = Cold rolled;                    ROW = 14;                           FORMAT = 1|n s s;
      CINDVAR = thickness!in;                   FORMAT = 1|d|d|d;                   ENTRY = (g mu)!0.32;
        VALUE = ...;                            ENTRY = ST!...!...!36;              RULE = under;
    COLUMN = 3;                                 RCELLVAR = fty!st;                  RCELLVAR = pr-nu;
      CINDVAR = specification!mil;              ROW = 15;                           ROW = 34;
        VALUE = MIL-T-5066;                     FORMAT = 1|d|d|d;                   FORMAT = 1|c s s;
      CINDVAR = form;                           ENTRY = Fcy, ksi!!!;                ENTRY = Physical properties!;
        VALUE = Tubing;                         ROW = 16;                           ROW = 35;
      CINDVAR = cond;                           FORMAT = 1|d|d|d;                   FORMAT = 1|n s s;
        VALUE = Normalized;                     ENTRY = L!36!36!36;                 ENTRY = (g omega), lb/in.(e 3)!0.284;
      CINDVAR = thickness!in;                   RCELLVAR = fcy!!;                   RCELLVAR = density;
        VALUE = ...;                            ROW = 17;                           ROW = 36;
    COLUMN = 4;                                 FORMAT = 1|d|d|d;                   FORMAT = 1|n s s;
      CINDVAR = specification!mil;              ENTRY = LT!36!36!36;                ENTRY = C, Btu/(lb)(F)!0.116(n c);
        VALUE = MIL-S-7097, Comp. 3;            RCELLVAR = fcy!!;                   RCELLVAR = specific heat;
      CINDVAR = form;                           ROW = 18;                           ROW = 37;
        VALUE = Bars;                           FORMAT = 1|d|d|d;                   FORMAT = 1|n s s;
      CINDVAR = cond;                           ENTRY = ST!...!...!36;              ENTRY = K, Btu/(hr)(ft(e 2))(F)/ft!30.0(n d);
        VALUE = All;                            RCELLVAR = fcy!st;                  RCELLVAR = thermal conductivity;
      CINDVAR = thickness!in;                   ROW = 19;                           ROW = 38;
        VALUE = ...;                            FORMAT = 1|d|d|d;                   FORMAT = 1|n s s;
  ROW = 0;                                      ENTRY = Fsu, ksi!35!35!35;          ENTRY = (g alpha), 10(e -6)in./in./F!(n e);
    RULE = under;                               RCELLVAR = fsu;                     RULE = under;
  ROW = 1;                                      ROW = 20;                           RCELLVAR = thermal coefficient of expansion;
    FORMAT = 1|c|c|c;                           FORMAT = 1|d|d|d;                   ROW = 39;
    ENTRY = Specification!MIL-S-7952, 1025!MIL- ENTRY = Fbru, ksi.!!!;              FORMAT = 1 s s s;
    RULE = under;                               ROW = 21;                           ENTRY = (n a) Grain direction not specified.;
  ROW = 2;                                      FORMAT = 1|d|d|d;                   ROW = 40;
    FORMAT = 1|c|c|c;                           ENTRY = (e/D=1.5)!...!...!...;      FORMAT = 1 s s s;
    ENTRY = Form!Sheet and strip!Tubing!Bars;   RCELLVAR = fbru!e/d=1.5;            ENTRY = (n b) See applicable specification for variation in minimum elong
    RULE = under;                               ROW = 22;                           ation with ultimate strength.;
  ROW = 3;                                      FORMAT = 1|d|d|d;                   ROW = 41;
    FORMAT = 1|c|c|c;                           ENTRY = (e/D=2.0)!90!90!90;         FORMAT = 1 s s s;
    ENTRY = Condition!Cold rolled! Normalized!A RCELLVAR = fbru!e/d=2.0;            ENTRY = (n c) (122 to 212F);
    RULE = under;                               ROW = 23;                           ROW = 42;
  ROW = 4;                                      FORMAT = 1|d|d|d;                   FORMAT = 1 s s s;
    FORMAT = 1|c|c|c;                           ENTRY = Fbcy, ksi!!!;               ENTRY = (n d) (at 32 F);
    ENTRY = Thickness, in.!...!...!...;         ROW = 24;                           ROW = 43;
    RULE = under;                               FORMAT = 1|d|d|d;                   FORMAT = 1 s s s;
  ROW = 5;                                      ENTRY = (e/D=1.5)!...!...!...;      ENTRY = (n e) See Figure 2.2.1.0.;
    FORMAT = 1|c|c|c;                           RCELLVAR = fbcy!e/d=1.5;
    ENTRY = Basis!S!S!S(n a);                   ROW = 25;
    RULE = under;                               FORMAT = 1|d|d|d;
                                                ENTRY = (e/D=2.0)!...!...!...;
                                                RCELLVAR = fbrv!d=2.0;
```

FIG. 13—*The data record (abridged) for the table in Fig. 12. (The abridgements are due to dropping portions of the text in the character fields).*

imposed at the system level. An overview of the three-tier approach of a global thesaurus, data base metadata, and data base files included in the MIST Demonstration Database is given in McCarthy [23].

A significant transition (involving a breaking down of the captured metadata and data values) is required between capturing the original exhibit structure and data values in machine readable form and having identifiable elemental materials property data values (with all the associated independent variables and their values) organized and stored in the data base management system. As has been emphasized, only if the appropriate exhibit structure and associated metadata categories have been assigned and captured can such a disassembly operation be performed correctly.

Summary

This paper has attempted to show that the capture of materials property data from print sources and their transfer to computer data management systems is complex for a number of reasons:

1. An extensive system of metadata is needed for the materials properties field; otherwise, captured data ends up being irretrievable.
2. The logical structures inherent in tabular and graphical formats used in print contain

Specification	MIL-S-7952, 1025	MIL-T-5066	MIL-S-7097, Comp.3
Form	Sheet and strip	Tubing	Bars
Condition	Cold rolled	Normalized	All
Thickness, in.	...	...	...
Basis	S	S	S(n a)

Mechanical properties
Ftu, ksi
  L
  LT
  ST
Fty, ksi:
  L
  LT
  ST
Fcy, ksi:
  L
  LT
  ST
Fsu, ksi:
Fbru, ksi.
  (e/D=1.5)
  (e/D=2.0)
Fbry, ksi:
  (e/D=1.5)
  (e/D=2.0)
e, percent:
  L
  LT
  ST
	
	55
	55
	...
	
	36
	36
	...
	
	36
	36
	...
	35
	
	...
	90
	
	...
	...
	
	...
	(n b)
	...
		
		55
		55
		...
		
		36
		36
		...
		
		36
		36
		...
		35
		
		...
		90
		
		...
		...
		
		(n b)
		...
		...
			
			55
			55
			55
			
			36
			36
			36
			
			36
			36
			36
			35
			
			...
			90
			
			...
			...
			
			(n a)(n b)
			...
			...

E, 10(e 3) ksi E(s c), 10(e 3) ksi G, 10(e 3) ksi (g mu)		29.0 29.0 11.0 0.32	
Physical properties (g omega), lb/in.(e 3) C. Btu/(lb)(F) K, Btu/[(hr)(ft(e 2))(F)/ft] (g alpha), 10(e -6)in./in./F		0.284 0.116(n c) 30.0(n d) (n e)	

(n a) Grain direction not specified.
(n b) See applicable specification for variation in minimum elongation with ultimate strength.
(n c) (122 to 212F)
(n d) (at 32 F)
(n e) See Figure 2.2.1.0.

TITLE : TABLE 2.2.1.0(b). Design Mechanical and Physical Properties of AISI 1025 Carbon Steel

FIG. 14—*The re-presentation of the captured data record for the table in Fig. 12.*

GRATTIDGE ON PUBLISHED DATA CAPTURE   171

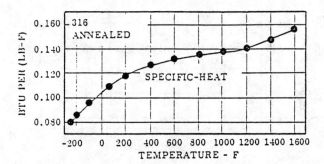

FIG. 15—*A graph exhibit taken from ASMH, section 2-1307, Fig. 2.015.*

FIG. 16—*The data record for the graph in Fig. 15.*

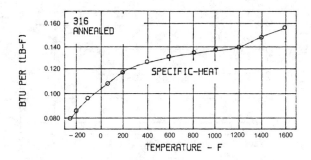

FIG. 17—*The re-presentation of the captured data record for the graph in Fig. 15.*

sophisticated internal relationships between the variables and data values. These relationships present significant challenges when isolating the basic elements of the individual data values.

3. Through analysis of the logical structures of print exhibits and categorization of the associated metadata, data capture procedures can be devised that result in accurate data base entries for the building of data base management systems of materials properties data.

Continuing Challenges to the Capture of Technical Data from Print

A number of continuing challenges exist and must be faced if the progress made to date is to continue.

The first challenge is to extend the capture capability to all forms of materials property data from printed exhibits. This paper has dealt with the capture of data from direct, paragraph, table, and graphic forms of exhibits. Still remaining are diagrams, symbols, equations, and specialized forms of graphs. Among the latter are noncartesian graph types (polar, curvilinear, triangular coordinate, and so forth), and 3-D plots.

There is the challenge to move from the R&D phase into a production phase for the capture of materials data. At this stage, major segments of printed materials data could be captured at the production level. A production level system integrates the various capture methods and procedures; validates the included metadata and data; and provides re-presentation capabilities to allow confirmation of the accuracy of the captured data.

A third challenge is to rationalize the metadata needed in the capture of materials property data through the development of consensus standards for term names, designations, definitions, hierarchical classes, and so forth. This is necessary not only in order that data system design can be rationalized, but also that multiple parallel access routes to the data can be established. This challenge impinges directly on the scope of work of ASTM Committee E-49 on the Computerization of Material Property Data as well as on similar work underway in Europe and in other parts of the world.

The fourth challenge is to improve the automation of the capture process by the extension and improvement of available techniques such as OCR, voice, and processed raster scanning applied to the capture of data from print sources.

A fifth challenge is to quantify the efficacy of this metadata template method with respect to time (cost), error rate, and completeness of information capture. To date no specific comparison tests have been conducted on alternative data capture methods, though informal evaluations indicate that validation checks applied during the data entry procedure result in fewer errors (and therefore less costly corrections) at the subsequent data base loading stage.

A final challenge is to extend the reconnaissance of the applicability of graphic and tabular structure types to other disciplinary fields of science and technology, for example, the geosciences, chemistry, physics, the life sciences, and so forth. The identification and organization of the relevant metadata would also be required. All S&T fields use tables and graphs as presentation formats and have significant portions of their "working data" available only in printed versions. We intend to continue to develop these methods to capture archival print data and transform them into accurate, computer-usable form.

Acknowledgments

The work on the metadata descriptions of graphs has been performed under NSF (SBIR) Program ISI-85-60304; that for tables, under New York State Program #SBIR (86)—79. Specific data capture activities have been supported under NBS 60NANB6D0661. Acknowledgment is made to the seminal ideas of Dr. J. L. McCarthy for the metadata descriptions and their representation in the MIST Data Base Project. The support and guidance of J. Rumble, NBS; C. Northrup, Sandia Laboratories; and J. C. Kaufman, NMPDN under the MIST development program is gratefully acknowledged. Other major contributors to this program of work have included J. H. Westbrook, Jean Murray, Jean Fecteau, and Bruce Lund.

References

[1] McCarthy, J. L., "Metadata Management for Large Statistical Databases," *Proceedings of the 8th International Conference on Very Large Data Bases,* Mexico City, 1982, pp. 234-244.
[2] Shoshani, A., Oklen, F., and Wong, H. K. T., "Data Management Perspective Of Scientific Data," in *The Role of Data in Scientific Progress,* P. S. Glaeser, Ed., Elsevier Science Publishers B. V., North-Holland, Amsterdam, CODATA, 1985.
[3] Westbrook, J. H., this volume pp. 23-42.
[4] Westbrook, J. H., and Grattidge, W., "Metadata for the Representation of Graphical Information in Computer Systems," Final Report, NSF SBIR Grant No. ISI-85-60304, Sci-Tech Knowledge Systems Inc., Scotia, NY, July 1986.
[5] Grattidge, W. and Westbrook, J. H., "Metadata for the Representation of Tabular Data in Computer Systems," Final Report, New York State Science and Technology Foundation, Program #SBIR (86)—79, Sci-Tech Knowledge Systems Inc., Scotia, NY, Nov. 1987.
[6] Dolby, J. L., "A Theory of Data: Implications for Information Retrieval," *Proceedings of the International Conference on Data Engineering,* IEEE Computer Society, Los Angeles, 1984, pp. 111-117.
[7] Clark, N., "Tables as a Medium of Communication," presented at Online '84, San Francisco, 29 October 1984.
[8] Eshbach, O. H., *Handbook of Engineering Fundamentals,* Third Edition, Wiley & Sons, NY, 1975, Section 7, p. 650.
[9] Eshbach, O. H., *Military Standardization Handbook,* Metallic Materials and Elements for Aerospace Vehicle Structures, Military Handbook 5, U.S. Department of Defense, Chapter 2, p. 2-167, Revision dated June 1983.
[10] Eshbach, O. H., *Handbook of Engineering Fundamentals,* Third Edition, Wiley & Sons, NY, 1975, Section 5, p. 541.
[11] Kirk-Othmer, *Concise Encyclopedia of Chemical Technology,* John Wiley and Sons, NY, 1985, Section: Carbon, p. 204.
[12] Westbrook, J. H., this volume pp. 23-42.
[13] Anderson, H. L., Ed., *Physics Vade-Mecum,* American Institute of Physics, 1981, Section 22, Page 325.
[14] Anderson, H. L., Ed., *Physics Vade-Mecum,* American Institute of Physics, 1981, Section 22, Page 323.
[15] Anderson, H. L., *Military Standardization Handbook,* Metallic Materials and Elements for Aerospace Vehicle Structures, Military Handbook 5, U.S. Department of Defense, Chapter 5, p. 5-3, Revision dated June 1983.
[16] Grattidge, W., Westbrook, J. H., McCarthy, J. L., Northrup, C. J. M., Jr., and Rumble, J., Jr., "Materials Information for Science & Technology (MIST): Project Overview." NBS Special Publication 726, National Bureau of Standards, Washington, DC, Appendix G, p. 82, 1986.

[17] Eshbach, O. H., *Handbook of Engineering Fundamentals,* Third Edition, Wiley & Sons, NY, 1975, Section 16, p. 1353.
[18] Northrup, C. J. M., Jr., McCarthy, J. L., Westbrook, J. H., and Grattidge, W., "Materials Information For Science and Technology (MIST): A Prototype Material Properties Data Network System" in: *Materials Properties Data—Applications and Access,* ASME, PV&P and Computer Engineering Division, 1986.
[19] Grattidge, W., Westbrook, J. H., McCarthy, J. L., Northrup, C. J. M., Jr., and Rumble, J., Jr. "Materials Information for Science & Technology (MIST): Project Overview." NBS Special Publication 726, National Bureau of Standards, Washington, DC, 1986.
[20] Grattidge, W., Westbrook, J. H., McCarthy, J. L., Northrup, C. J. M., Jr., and Rumble, J. Jr., *Aerospace Structural Metals Handbook,* MCIC, Battelle-Columbus Laboratories, Columbus, OH, 1986.
[21] Grattidge, W., Westbrook, J. H., McCarthy, J. L., Northrup, C. J. M., Jr., and Rumble, J. Jr., *Military Standardization Handbook,* Metallic Materials and Elements for Aerospace Vehicle Structures, Military Handbook 5, U.S. Department of Defense, Revision dated June 1983.
[22] Grattidge, W., Westbrook, J. H., Brown, C., and Novinger, W. B., "A Versatile Data Capture System for Archival Graphics and Text," in: *Computer Handling and Dissemination of Data,* P. S. Glaeser, Ed., Elsevier Science Publishers B. V., North-Holland, Amsterdam, CODATA, 1987.
[23] McCarthy, J. L., this volume pp. 135–150.

Shuichi Iwata[1]

Expert Systems Interfaces for Materials Data Bases

REFERENCE: Iwata, S., **"Expert Systems Interfaces for Materials Data Bases,"** *Computerization and Networking of Materials Data Bases, ASTM STP 1017*, J. S. Glazman and J. R. Rumble, Jr., Eds., American Society for Testing and Materials, Philadelphia, 1989, pp. 175–184.

ABSTRACT: The future course of more intelligent systems as interfaces for materials data bases is reviewed in comparison with an ideal information service supplied by human experts. The ability to understand a wide range of materials information is discussed in connection with a variety of services, ranging from a retrospective search of the original measured information and continuing through value-added procedures of grouping and abstraction. Linking computerized materials data bases and information systems poses problems in handling and organizing incomplete knowledge, which can be solved by valued-added procedures. An idea "simulator" that poses and answers a wide variety of questions to substitute for experiments is proposed. This simulator integrates the relevant data on the materials.

KEY WORDS: expert system, interface, knowledge base, distributed data bases, semantic capacity, simulator

A large gap exists between the present capabilities of materials data systems and users' expectations of them. This is also true for the available expert systems applications. For example, materials could be tailored in accordance with a given requirement, but presently, results retrieved from a data base are nothing more than a set of discrete data. Users want more intelligent systems. To illustrate these gaps and to identify ways to break through existing barriers, an intelligent information service based on users' needs is discussed. A review of the limited artificial intelligence (AI) applications in the materials field will not be given in this paper.

An ideal information service is that provided by human experts. Usually materials users are interested in the performance of the materials. Materials investigators consider structure property correlations; others, such as fabricators, want to organize processing. Each has his own viewpoint or data model to represent the subject. The first query from a customer is usually vague and ambiguous. Moreover, knowledge representations of their understanding of materials reflect different focal points. Human experts must first know the essence of a customer's requirements through interactive communication.

Even in the simple retrieval of a set of specific factual data, some type of map showing its position in reference to the entire materials world is required to make the information system "user friendly." Almost all other important materials problems, such as materials selection, materials design, and materials evaluation, are more sophisticated and dynamic problems, and the ability to carry out other tasks beside tracing relations in a thesaurus is required. A total view of each application, integrating all relevant data models is needed. Various processing techniques must be carried out, namely, relating, conversion/normalization, and structuring.

---

[1]Associate professor, Department of Nuclear Engineering, Faculty of Engineering, University of Tokyo, 7-3-1 Hongo, Bunkyo-ku, Tokyo, Japan.

No one knows everything perfectly. No perfect data base exists, and each information system is incomplete. Therefore, the ambiguous parts must be made clearer and more concrete and such parts must be complemented by higher-level knowledge to reach a given objective. The ideal information system requires an iterative question-and-answer (QA) procedure. Such a system adds complementary information step by step until a given objective is satisfied. Various kinds of intelligence levels for materials data systems are considered in the next section.

Practical results from artificial intelligence (AI) applications have been obtained in analysis and diagnostics, in which a specific solution can be given for a closed subproblem. In such established fields, a well-defined knowledge data set can be cost-effectively reused. However, almost all problems in materials-related fields are ill-structured and open-ended. So a serious reevaluation of available AI applications is necessary before proceeding further.

In this paper, the need for materials data systems is first summarized in accordance with the previous discussion. Second, discrepancies between ideal and feasible information services are discussed to uncover future fruitful lines of development. Finally, a method of integrating relevant information is proposed.

An Ideal Information Service

After successes with large integrated data bases of simple data and data structures (for example, bibliographic data bases), we are facing the problem of managing sophisticated data on subjects such as materials properties. Complexities in the data and consequent difficulties in evaluation suggest that distributed data systems based on different data models could be an initial requirement of materials information services. Three intelligence levels are considered: (1) transparent computer networks between a user and each data base, (2) intelligent gateways between each user and the relevant data bases, and (3) expert systems interfaces adding complementary information to retrieved data.

A user needs to know how to access each data base in the first instance. In the second, the gateway knows how to translate a user's queries into those suitable for the relevant data bases, for use by computer networks, and for reporting results meeting the requirements. The user need only know how to manipulate his or her own terminal for data retrieval. In last case, a system must have a comprehensive data model and knowledge about materials, which is partly realized by a group of human experts.

Thus, the most effective way to get the most suitable information is still to consult the best (most informed) person. One implementation of this concept is a so-called "telephone system" of experts. This kind of system works very well as long as real experts can be identified. Substitution of such "telephone system" by a computerized system has been attempted to date only for various well-defined procedures, such as statistical analysis, information retrieval, reporting, and so on, but many ill-structured subjects still need a solution.

One major difference between a computerized system and this "telephone system" is that data models of human experts and computerized systems differ with respect to their flexibility, modifiability, extendability, or model building capability, their descriptive capability and expressiveness for various types of data, and their ability to structure representations at various levels of abstraction. Human experts know or abstract the general features of each subject, as well as its details. Materials information is categorized (with the details abstracted) with respect to fabrication method, shape, size, microstructure, and properties. General features are extracted from each category to be incorporated into a map and used as higher-level knowledge for more specific information.

Because of the lack of information, attributed to the economics of data base building as well as limitations on the measurement of materials, almost all implemented data models are incomplete, and only an approximation of part of the huge materials world. A flexible human expert or system can examine the content and make a good estimate of the relevant incomplete information. Sometimes the data models can be improved by adding new vocabulary terms and altering representational structures on the basis of newer information.

Computerized and telephone systems also differ in their processing of data. One is parallel, the other sequential. One offers the dynamic modification of models, while the other employs the static execution of predetermined procedures. Finally, one has strong abductive and inductive capabilities, while the other features completeness of deductive execution.

A human expert can make a comprehensive data model for a given objective step by step, posing questions of users, systems, and materials and learning their answers. Different levels of questioning can be assumed, for example, to translate users' requirements from global functions at the system level into materials performance levels, to retrieve fundamental or general data needed to roughly estimate a material's use in a product, or to retrieve relevant information for more precise estimation. At each step, information that is useful in getting more comprehensive explanations is taken into account and used to create more specific queries for a better understanding of the materials. An essential difference here is that a human expert examines all relevant information simultaneously to create well-evaluated, comprehensive descriptions of a given subject.

The appropriate set of suitable interactive queries for an information source has been learned only through trial and error for each individual, and paradigms for reaching better solutions could be selected from among the attempts. Essentially, the information sources are the materials themselves. Queries and answers correspond to tests and materials data, respectively. Descriptions of such question-and-answer procedures, in other words, "discussions with the nature," become reports, papers, models, data bases, and knowledge bases. They are distributed in various forms, including computer data.

Distributed data systems must be able to integrate relevant information retrieved from different types of materials data systems. It is also very important for a data system developer or vendor to prepare an interface to guide use of his or her system for such types of access with several other systems.

Materials selection should be one of the most fruitful applications of materials data systems. We refer to available examples of mistakes in selecting materials for various products as well as other stories of the successful use of materials. We sometimes go to great depths to obtain a comprehensive view, but it is often necessary to stop when a practical compromise is reached. The deeper we go, the less random the results, but the level sought depends on the sensitivity of variables in each area.

After much experience, the path the search must follow is partly predetermined, and other aspects are modified step by step as users learn more about the material or application. This suggests that we have a set of predetermined patterns of queries for rough and global searches and a set of abilities to modify and dynamically specify queries to reach the final decision. Such patterns are multilayered in accordance with the structure of data and knowledge, in which comparisons of all possible methods of meeting users' needs are carried out in parallel as well as sequentially and a decision is made based on a total evaluation.

The essential gaps between such an ideal information service and a computerized service arise from the differences in flexibility in dealing with a wide range of semantic capacities, learning capability, and the ability to add value to the data retrieved.

Semantic Capacities

Because of the nearly infinite features of materials information, almost all materials descriptions are an approximation of the real world. The spectrum of such approximation spreads from microscopic to macroscopic, from image to pattern, from static to dynamic, from deterministic to probabilistic, from instantaneous to long-term, from phenomenological and empirical to theoretical, and from numerical to symbolic. Therefore, the semantic capacity of materials information is extremely large, and an expressive set of representations is required for the correct positioning (categorization) of each data set and their elaborate relations to eliminate combinatorial searching explosions.

To describe such complex facts, rough classifications are introduced based on the requirements of design, fabrication, microstructure, property, performance, and test method, following typical ways of grouping materials information. It is necessary to classify each of these elements, their collection as sets, and the relation defining the sets.

A relation is comprised of a set of descriptions, for example, disjoint, equivalent, hierarchical, member of, and independent variable. This information is defined in the system dictionary. Dictionaries support many uses, such as those listed in Table 1. However, there is no definitive standard for descriptions of materials information, so data are compiled into arbitrary forms, as was pointed out before.

Although many iterative queries are usually required to reach a comprehensive integration of the retrieved data, abstraction processes for materials information can be defined explicitly in principle. Such processes add value, and they should be carefully examined during the evaluation of each data set. General data such as those listed in materials handbooks are usually derived from a large number of measurements and can be used for categorizing specific data and rough materials selection. For a design application, additional statistical information is required to evaluate reliability. This illustrates the need for semantic traceability of each set of information through even a long life cycle of value-added materials information. Several examples of this sort of life cycle are listed in Figs. 1 and 2.

The final goal is establishing a type of retrospective capability, that would allow you to locate the original information alongside the results of value-added procedures of grouping, abstraction, and analysis. Almost all studies of materials performances refer to macro- and microstructural aspects, which are described by shape, form, geometry, and interrelations among components. General predicates for descriptions of shape and geometry could be listed as in Table 2.

These predicates have logical relations. For example, "tube" has descriptions of inner and outer diameters and length as predicates. Sometimes thickness is defined instead of inner diameter. Other important predicates concern statistics on components and correlations among them, for example, "in," "out," "attractive," "repulsive," "in the vicinity," "pinned," "coherent."

Virtual reconstruction of structures from such descriptions of microstructures guided by declarations of constraints (for example, space filling, symmetry, and free energy) also helps users

TABLE 1—*Uses of dictionaries in data systems.*

table look-ups
group names
element names
query names
hierarchy by naming
synonymous terms or implication
disjoint or "not synonymous"
definition
conversion factors between synonyms and relevant properties
variable relations in independent variables
dependent variables
formulae
grouped categories
naming criteria

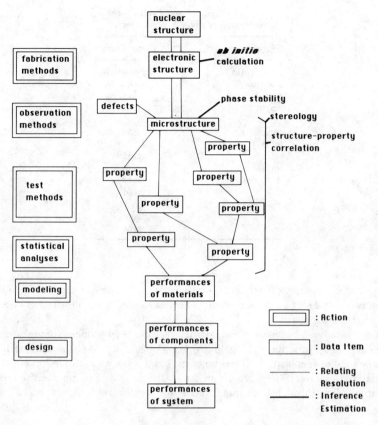

FIG. 1—*The value-added life cycle of materials information: the correspondence of actions and data on materials.*

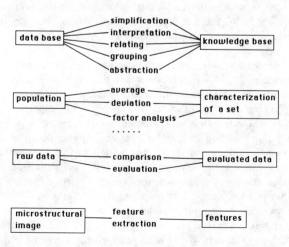

FIG. 2—*The processing of adding values.*

TABLE 2—*General predicates for the description of shape and geometry.*

Geometry	Shape/Form
aspect ratio	complex solid
diameter	cone
depth	cup
length	dish
	flanged
perimeter	grid
pitch	hollow
volume	manifold
surface area	plate
thickness	ribbed
	rod
width	sheet
	tube
	wire

to develop views on materials. Eventually, multimedia data systems are expected to set up an infrastructure for this sort of human interface, for example, in the illustration of analogous images of microstructures.

All qualitative reasonings based on microstructures have been deduced through quantitative evaluations of relevant experiments. Qualitative knowledge based on phenomenological causalities has been used to demonstrate AI technologies, but it is not yet useful for practical applications for materials. Universal facts are the processes that generate a comprehensive explanation for each application through both quantitative and qualitative analyses in each semantic capacity.

Learning by Example

Conventional formats used for industrial standards have a long history of compilation and natural selection. The designer normally uses examples of past designs with good or bad experiences or both that refer to well-established industrial standards, then evaluates, modifies, or improves them according to his sense of judgment. Therefore, a history of design procedures could be summarized as a network, in which nodes and branches correspond to designs and their improvements, respectively.

Practical views of materials selection reflect this confidence. That is, material performance is defined only in the framework of a specific design in the context of such a network. This is especially true for structural materials. Although there is a naive expectation that a "virtual" flat table formed by joining the relevant tables containing all the important variables as fields can be used for materials selection, it is very difficult for general users to prepare such a virtual flat table. For example, in selecting an alloy for a certain high-temperature structural application, mechanical properties and corrosion resistances might be the main concern. But in selecting ceramics, it is very important to know the probability density function of the mechanical properties and a design criteria of the system, for example, brittle materials design. This suggests that it is possible, but not practical, to prepare a virtual flat table for materials selections. Moreover, because materials performance is determined after quantitative evaluation of real service conditions, dynamic and asymptotic approaches to materials data along the design procedure are required. Many standards views of stereotyped situations, or templates, for understanding and using materials could be employed in the design procedure. Grouping with reference to a specific criterion is helpful for general users. From a designer's viewpoint, the overall objectives at the system level are resolved into objectives at the materials level with the available descriptions or precise constraints. The next step is identifying or estimating service conditions by setting up a minimum configuration of the selected set of materials, by selecting similar examples from available designs, or by doing both.

The final stage of materials evaluation might be to generate a set of "simulators" to examine materials performances under the potential design conditions instead of performing experiments or benchmark tests. There are various kinds of models, from atomistic to continuous, for the simulation of microstructural changes (Fig. 3). Such models should be selected from well-described design procedures in accordance with those procedures. Rough or generic models are usually used in the first-order approximation to identify essential design points. These models are generally universal. More specific models are used to reach an optimum solution.

Similar aspects are also found in the selection of test methods and properties. Test methods span a wide spectrum, from general tests that outline materials' behavior to specific tests under operating conditions. A set of definitions for correct application are needed. The important aspects to be described precisely are as follows:

1. test (test method, condition)

    a. internal model

    (1) grouping of test methods
    (2) dominating mechanisms

    b. external model

    (1) naming of tests on the basis of each mechanism
    (2) test objectives
    (3) correlations between tests
    (4) independent variables
    (5) dependent variables
    (6) standards
    (7) operator, laboratory, data

    c. constraints

    (1) application
    (2) test methods

These precise descriptions supply good examples for learning to refine the proposed simulator. Replacements of conventional tests by simulators and materials data bases are carried out through serious evaluation of their tautologies, time and space resolution, and total efficiency in the context of each design procedure.

Query Writing and Translation

Dialogue by natural languages is theoretically the most natural man-machine interface. The best language for one specific field is the one experts use every day. All values and relations are included there. However, immense software problems exist, such as difficulties in the interpretation of input owing to the ambiguity of language. Limited input of natural language, such as when the user employs words he is familiar with, is one of the most suitable methods of dialogue. But some users tend to overestimate the intelligence of the machine and overstep the tight restrictions on input wording. Abstract views are necessary for general users to reach specific information. Question-and-answer dialogues in which the computer asks the operator a series of questions on such abstract views are friendly, but of limited flexibility.

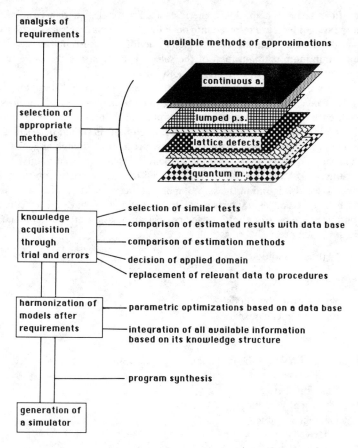

FIG. 3—*Information flows to generate a "simulator."*

Programming language dialogue is concise, precise, powerful, and flexible, but inappropriate for the vast number of users who have not learned to program and do not want to if possible. Therefore, providing the best service for users involves speaking with them in their own language or universal ad hoc language for materials.

In general, languages used in materials fields can be divided into two categories: those for materials users and those for materials investigators and producers. Users are interested in materials performance in a component, whereas the others want to describe materials data by fabrication methods and properties using microstructural information as "keys" to connect them. In selecting a material, it is necessary to have two comprehensive descriptions for these two groups.

The key is to relate all the information by descriptions of microstructures. Icon models covering macro- to microscopic aspects of materials could be used as a universal mode of communication among members of the two groups. Several examples of icons that are suitable for this sort of approach are listed in Table 3.

For time-dependent properties such as creep, fatigue, environmental effects, and irradiation effects, it is user-friendly enough for materials users to prepare icons to define several conditions, for example, loading conditions, heat transfer. For materials investigators, however, mi-

TABLE 3—*Examples of more friendly interfaces based on human interaction to get an information set.*

Items	Conventional Approach in Materials Data System	More Friendly Interfaces
Designation of fabrication methods	identification of a fabrication method and its parameters	selection of an icon corresponding to a method and of parameters in accordance with a process sheet; for a casting, it is better to show the shape and size of the product with the procedure, and modify parameters using it.
Microstructural information	pattern name parametric descriptions image data base	Retrospective search from the pattern to the original image. More direct representation of observed results with more stereological dimensional imaging, by using zoom-up, defocus, etc. Constraints of each method can be illustrated (e.g., surface effect).
Test results	Specification of independent variables and their values	Making test order sheets instead of query writing. Operating a measuring instrument on the screen and getting information, as is done through experimentation. Retrospective search from a set of experimental values to the dynamic and analogous behavior of materials using a simulator.

crostructural changes need to be taken into account to explain property changes explicitly, and icons of various microscopies, probe beams, and so forth are needed to get microstructural information. Many competitive processes are assumed in this case, and it is necessary to identify dominant processes and verify them by suitable criteria, for example, pattern matchings of relevant microstructural images retrospectively retrieved in each semantic capacity. Final self-consistency of explanations is to be achieved by human experts.

As long as the expression of technical terms is useful in representing such dynamic phenomena, conventional techniques of translation can be applied effectively. However, the final stage of validation requires the creation of various "virtual worlds," totally equivalent to the "real world," in a computer. Such worlds are described in various ways, for example, phenomenologically, through reasoning by terms, or as qualitative equations, continuous elastic approximations, rate equations, or a set of images obtained by electron microscopy. These dynamic manipulations of worlds are required until a comprehensive explanation is obtained. This process serves as a conversion of the communication between humans and materials to software. It is a translation of a natural process to a system. In Fig. 4, this sort of natural interface is illustrated.

Concluding Remarks

The lack of a serious discussion of user needs causes major problems in the application of AI technologies to fill those needs. Many tools useful for developing comprehensive standards for materials data bases have been developed, but they need refining and restructuring to make an elegant user interface for materials application incorporating user experiences. A clearer defini-

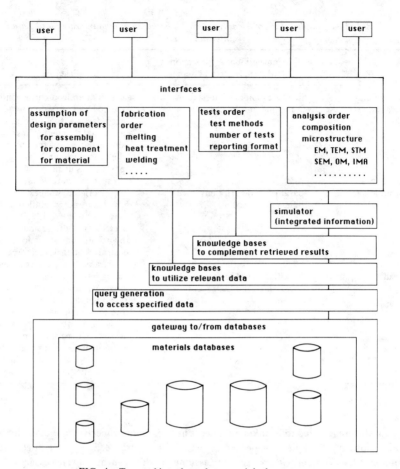

FIG. 4—*Types of interfaces for materials data systems.*

tion of information flow from data acquisition to design, is needed, as well as a clearer description of the need for intelligent help from computer data systems. Software tools designed to grasp realistic and quantitative images of microstructural changes and consequent property changes are needed to give theoretical background to the information flow.

Current AI technologies supply us with a new ability to handle logic. Many available applications take advantage of production systems, frames, and predicate calculus. However, such sophisticated problems as corrosion, environmental effects, and irradiation effects, are very difficult to handle with a universal procedure that explains everything. The production rule approach offers a good tool for representing phenomenological causality. It is possible to handle logic, numerical values, and procedures simultaneously. Default values and attached procedures in the frame approach widen the representation of expert knowledge. However, such a data set of knowledge should be prepared before application. With sophisticated problems, this preparation is very difficult. Identifying the problem comes first and is followed by the integration of the relevant information. One strategy for supporting this procedure is to prepare a complex software "simulator" that produces the equivalent of experimental information.

Robert A. Pilgrim,[1] *Phil M. Johnson,*[1] *and Patrick M. Falco, Jr.*[2]

An Interactive Inquiry System for Materials Data Bases Using Natural Clustering

REFERENCE: Pilgrim, R. A., Johnson, P. M., and Falco, P. M., Jr., **"An Interactive Inquiry System for Materials Data Bases Using Natural Clustering,"** *Computerization and Networking of Materials Data Bases, ASTM STP 1017,* J. S. Glazman and J. R. Rumble, Jr., Eds., American Society for Testing and Materials, Philadelphia, 1989, pp. 185–193.

ABSTRACT: User inquiry in a data-base management system (DBMS) takes on patterns that can be recognized by a monitoring system using hierarchical conceptual clustering. The application of natural clustering techniques to interactive query sessions permits the automatic development of predictive tools to assist the unsophisticated data base user. This paper defines the desired attributes of such a monitoring inquiry support system, discusses system design considerations, and provides an algorithm for the natural clustering of both numeric and nonnumeric property information. The implementation of such a system is described based on its application to the Spacecraft Materials Data Base SRS Technologies is currently developing.

KEY WORDS: natural clustering, hierarchical agglomerative clustering, interactive inquiry, materials data bases

As the use of data-base management systems (DBMS) increases, the need for more efficient and effective inquiry and decision support capabilities is recognized. Several solutions, such as those presented by Lagomasino and Sage [1], address the general problems of decision support in an interactive inquiry system. While these general methods are extremely useful in providing solutions for a broad spectrum of problems, general methods typically result in suboptimal solutions to a particular problem. In this paper, we note some of the special characteristics of materials data bases that may be exploited to reduce the time and effort required of a user in the search for candidate materials with particular material properties. We describe an approach to the development of a inquiry support monitor for a data-base management system and discuss a specific algorithm for effective acquisition of relevant information that builds on an approach to conceptual clustering described by Fisher and Langley [2]. We offer a methodology for clustering numeric and nonnumeric property data simultaneously. In addition, we derive a scaling function that can provide completely independent material property clustering criteria in a hierarchical system of comparison.

Baseline DBMS

We have chosen a relational data-base structure as a baseline for the implementation of our hierarchical conceptual clustering model. The standard row/column data structure is used in a

---

[1]Senior staff scientists, SRS Technologies, Huntsville, AL.
[2]Materials development engineer, Air Force Wright Aeronautical Laboratories, Wright-Patterson AFB, OH.

two-level format to represent both material sources and property data as shown in Table 1. The data used in this paper are from a preselected subset of a materials data base and cover data for specific types of adhesives. While we intend to eventually interface our monitoring system to a commercial DBMS, we have developed a simple proof-of-principle testbed DBMS in Pascal accessing a materials data base installed on a VAX-11/780. In our prototype relational database management system, we have provided basic relational operators such as SELECTION, PROJECTION, and JOINING to support the testing and evaluation of conceptual clustering techniques.

Advantages of Adaptive Inquiry Monitoring

In a query session with the DBMS, a user exhibits patterns of interaction that establish relative levels of importance for the various properties under consideration. While such patterns are difficult to define in the general case, the relatively predictable structures characteristic of user inquiry into materials data bases suggest straightforward approaches to efficient and effective search operations. These patterns can be entered directly by the user as relational operators or automatically derived through interpretation of session entries by the clustering model to provide hierarchical parameters for comparison. Systems such as HEARSAY III [3] make infer-

TABLE 1—*Excerpt from a two-level relational data base for material properties.*

Preselected Table (Adhesives) From Spacecraft Materials Data Base

DBMS

No.	Name	Category	Location	Source
1	Tra-bond2151	Adhesive	Medford, MA	TRA-CON, Inc.
2	Tra-duct2901	Adhesive	Medford, MA	TRA-CON, Inc.
3	Tra-bondJ1156/E8	Adhesive	Medford, MA	TRA-CON, Inc.
4	Tru-Cast111M/901	Adhesive	Ashland, MA	Fenwal, Inc.
5	TY-PLY-BN	Adhesive	Erie, PA	LordCorp.
6	U (CuringAgent)	Adhesive	Houston, TX	ShellChem. Co.
7	Ultra-Temp516	Adhesive	Ossinging, NY	AremcoProd. Inc.
8	Uniset906-25	Adhesive	Lexington, MA	Amicon
9	Uniset909-60	Adhesive	Lexington, MA	Amicon
10	UnisetA-316-7	Adhesive	Lexington, MA	Amicon

TY-PLY-BN Adhesive Example Properties Table

TY-PLY-BN ADHESIVE ERIE, PA

Category	Name	Value	Units
Mechanical	Viscosity	0.02–0.035	
Physical	StorageLife	12	mth.
Mechanical	Viscosity	0.05	N.s/m2
Optical	Color	Black	
Physical	Color	Black	
Chemical	Composition	Polymers	
Chemical	Composition	35–39	% by weight
Physical	Density	7.7–8.1	lb/gal.
Chemical	FlashPoint	42	deg. F

ences based on triggering patterns provided by the user, which can then be used to reduce the size and complexity of the problem domain. We use similar information provided by the user as well as these special characteristics of materials data bases to reduce the computational load and search time for data-base queries.

Common Problems in Materials Data Base Inquiry

The naive user encounters two common classes of problems in interactions with automated data base management systems. The first is a lack of familiarity with the nomenclature, which typically results in null responses to SELECT commands. The second is ineffective PROJECTION and JOINING operations, which result in acquisition of either the wrong set (the surviving candidate data set does not contain relevant information) or one which is too large to be efficiently scanned. Both of these problems can be mitigated through the use of the on-line conceptual clustering methods described below. Briefly, clustering provides a measure of similarity between all items of a group based on a selected set of characteristics. Hierarchical clustering establishes a relative level of importance for two or more classes of characteristics. We describe the theory of both numeric and nonnumeric clustering in detail in the section that follows.

Model Description

Natural Clustering Definition

A concise definition of natural clustering is provided by Geveden [4]. It is paraphrased here. Given a set X of n samples as described by the vectors $x_1, \ldots x_n$, which is to be partitioned into c disjoint subsets $X_1, \ldots , X_c$, we define each subset as the set of samples that are more similar (based on a selected set of criteria) to each other than they are to the members of any other subsets. The partition, as defined above, has the following properties:

$$X_1 \cup \cdots \cup X_c = X$$

$$X_i X_k = \emptyset, (i \neq k)$$

$$n_i \geq 1 \, (i = 1, \ldots , c)$$

$$1 \leq n_i \leq n \tag{1}$$

That is, each object is contained in only one cluster and each cluster contains at least one member. The measure of similarity used for partitioning is arbitrary and can be based on either numerical or nonnumerical criteria or a combination of both. The only requirement for pure clustering is that the same partitioning rule be used throughout the clustering process.

Various forms of natural clustering algorithms and their respective partitioning criteria are available in the pattern recognition literature. A fairly comprehensive review of the subject is provided in a monograph by Everitt [5] in which he divides clustering techniques into four basic types (hierarchical, optimization-partitioning, mode-seeking, and clumping) with a collection of other miscellaneous methods limited to special applications. Within the hierarchical type of clustering, Everitt makes a further division between *agglomerative* methods, in which individual elements are successively grouped into larger subsets, and *divisive* methods, in which a single set is divided into separate partitions. Our focus is on the agglomerative methods, since we usually start with a preselected set of individual candidate materials within which we wish to know the mutual similarities and differences with respect to certain properties of interest.

A Simple Numerical Example

We provide here a simple example of the agglomerative clustering method for a set of five elements. Assume for now that we have established a function $S(X_i, X_j)$ to generate a numerical value representing the similarity between X_i and X_j, which are elements of our set. Using the function S, we can calculate the similarities between every pair of elements in the set, expressed as the adjacency matrix, **A**

$$\mathbf{A} = \begin{array}{c|ccccc} & (1) & (2) & (3) & (4) & (5) \\ \hline (1) & 0 & 14 & 9 & 12 & 27 \\ (2) & 14 & 0 & 56 & 8 & 17 \\ (3) & 9 & 56 & 0 & 13 & 33 \\ (4) & 12 & 8 & 13 & 0 & 49 \\ (5) & 27 & 17 & 33 & 49 & 0 \end{array} \qquad (2)$$

where $\mathbf{A}(i, j) = S(X_i, X_j)$. Excluding the main diagonal, the minimum value $\mathbf{A}(2, 4) = 8$ (corresponding to the most similar pair of elements in this case) establishes the first cluster pair. The elements (X_2, X_4) are now combined and **A** is collapsed to a 4 by 4 matrix. At this point, a decision is necessary to determine how to treat the combined values for the (2, 4) element. In this example, we choose to use the minimum **A** values for each entry of the cluster pair. That is,

$$\mathbf{A}((2, 4), j) = \min(\mathbf{A}(2, j), \mathbf{A}(4, j)) \quad \text{and,}$$

$$\mathbf{A}(i, (2, 4)) = \min(\mathbf{A}(i, 2), \mathbf{A}(i, 4)). \qquad (3)$$

Note that the value leading to the clustering in each stage is replaced by a zero of the main diagonal in the operation of Eq 3, and that the clustered pairs are treated as single row/column entries in successive stages. This process is repeated until a single cluster remains, as shown in Eq 4 below

$$\mathbf{A} = \begin{array}{c|cccc} & (1) & (2,4) & (3) & (5) \\ \hline (1) & 0 & 12 & 9 & 27 \\ (2,4) & 12 & 0 & 13 & 17 \\ (3) & 9 & 13 & 0 & 33 \\ (4) & 27 & 17 & 33 & 0 \end{array} \qquad \mathbf{A} = \begin{array}{c|ccc} & (1,3) & (2,4) & (5) \\ \hline (1,3) & 0 & 12 & 27 \\ (2,4) & 12 & 0 & 17 \\ (5) & 27 & 17 & 0 \end{array}$$

$$\mathbf{A} = \begin{array}{c|cc} & (1, 3, 2, 4) & (5) \\ \hline (1, 3, 2, 4) & 0 & 17 \\ (5) & 17 & 0 \end{array} \qquad (4)$$

The results of this agglomerative hierarchical clustering can be represented graphically by a special type of tree diagram called a *dendrogram* as shown in Fig. 1. The length of the respective branches in the dendrogram indicates the level of difference between the connected elements. These difference measures are simply the A values that originally initiated each cluster. As

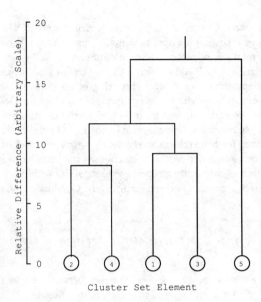

FIG. 1—*Dendrogram for sample clustering of a five-element set, based on single-element pair similarity values of adjacency matrix.*

illustrated for our simple example, X_2 and X_4 are closely related to one another, as are X_1 and X_3. These cluster pairs are in turn clustered since they are more similar to each other than to X_5, which is the last to be clustered.

The clustering algorithm described is not globally optimal. That is, the sum of difference values for all clusters is not minimized. First, this is typically not necessary or desirable, since we are more concerned with the close matches determined early in the clustering process and are less concerned with the optimality of the lower similarity scores. Second, the number of possible partitionings required to ensure global optimality for n elements divided into groups of length c is the Sterling Number

$$S_n(n, c) = (c!)^{-1} \sum_{j=1}^{n} (-1)^{c-j} \binom{c}{j} j^n \qquad (5)$$

For a 60-object example, global optimality requires the investigation of roughly 10^{18} clustering combinations. The compromise between optimality and processing efficiency is clear and the detrimental effects of a suboptimal clustering algorithm in this application are minimal.

Application

Occasionally the establishment of multiple levels of importance in the clustering similarity measures is desired. For example, in choosing an adhesive, we may decide that its color is the most important property and that although secondary properties must also be considered, their total measure must be subordinate to the single property of color. For natural clustering to be an effective tool in aiding the materials database user, some means of scoring one or more properties higher than others is needed. A simple static weighting scheme is not always effective, since multiple matches in a lower-priority criteria can accumulate to exceed the weighting for a higher-priority property. Conversely, separately preselecting (PROJECTION) each higher priority criterion requires revisiting the data base many times and can become computationally expensive for complex relations or large candidate data sets.

An alternative approach is provided in which a single accessing of the preselected group produces a completely independent hierarchical clustering of all elements and for all clustering criteria. Let n be the number of entries in the current tableau (domain), let p be the number of hierarchical levels being considered, and let q be the maximum number of properties to be matched at each level. We may now define a scale factor F_k to be multiplied by the cumulative similarity value calculated for hierarchical level k as $F_k = (qn + 1)^{p-k}$ for $k = 1, p$.

We can now define our similarity criterion function as the number of string (or substring) matches plus the number of numerical property values falling within a predefined range for nonnumerical and numerical property values, respectively. Combined with our scale factor definition we have

$$\mathbf{A}(i,j) = \sum_{k=1}^{p} [C_k(i,j)/(M(i) \times M(j))](qn + 1)^{p-k} \qquad (6)$$

where C_k represents the total number of k-level matches between cluster subset i and j of the adjacency matrix, and $M(r)$ is the number of materials represented in a subset r. In this clustering method, a larger $\mathbf{A}$ value represents greater similarity (the opposite of the previous example). Note that all subsets are single elements before clustering begins; hence $M(r) = 1$ for all $r = 1, n$ in the original adjacency matrix. Setting the base of the scale factor to $qn + 1$ ensures that even a single match at a given level, $(p - k)$ will result in a scaled value which will exceed the value obtained by maximal matching at all lower levels $(p - r)$ where $r > k$.

Absolute independence is not necessarily the best prioritization approach. If, for example, some cumulative score from matching lower hierarchical levels must eventually surpass a single match at a higher level, the base of the scale factor can be set to a value less than $qn + 1$. The amount of overlap in prioritization levels can be controlled by the percent the scale factor base is reduced by. For example, setting the base to 80% of $qn + 1$ results in a 20% overlap in hierarchical levels.

Results

In our application, the criteria for partitioning are taken from the set of properties associated with each material data base entry. The typical property categories and property names are listed in Table 2. As we take up our materials data base clustering example, we assume that some preselection has already occurred to obtain the group of candidate adhesives (Table 1). We also assume that during the query session, a hierarchy of property categories, names, and ranges of values has been established, either as the result of direct user entry or as part of an automatic inference monitoring system. As described in Table 3, four hierarchical levels have been established, each with a number of relevant property categories. A match occurs between two nonnumeric property values when their respective strings are the same (or nearly the same based on a predetermined substring matching criterion), while numeric property matches occur when the difference between two values is less than the acceptable variance specified for that property. The highest priority properties for this clustering include dielectric strength, dielectric constant, electrical dissipation factor, and volume resistivity. The next level priority properties are composition, specific gravity, and color, with some thermal properties and finally water absorption and flashpoint being the least important properties considered. No other properties were evaluated in this clustering.

Figure 2 illustrates the resultant dendrogram generated from the hierarchical clustering of the selected adhesives based on these properties. The interpretation of this graphic representa-

TABLE 2—*Property categories and names for materials data base.*

Category	Property Name	
Chemical	adhesive strength	color
Electrical	composition	compressive strength
Mechanical	density	dielectric constant
Optical	dielectric strength	dissipation factor
Physical	flashpoint	gel time
Thermal	hardness	head deflection temper
	heat distortion	impact
	temper	strength
	linear shrinkage	operating temperature
	peel strength	performance-high temperature
	pot life	specific gravity
	stability-vacuum	storage life
	temperature limit	tensile shear
	thermal conductivity	viscosity
	volume resistivity	water absorption

TABLE 3—*Clustering criteria and prioritization factors for comparison of candidate adhesives.*

Priority	Property Category	Property Name	Acceptable Variance (Numeric)
1	2	6	1
	2	7	25
	2	8	0.005
	2	27	1000
2	1	3	...
	4	2	...
	5	20	0.5
3	6	12	20
	6	18	...
	6	25	250
4	1	9	28
	1	28	0.1

tion is that candidate adhesives 1, 2, and 9 are very similar (with respect to the chosen hierarchy), and that if one of these items is chosen for an application, the other two should be considered as possible alternatives. Item 4 is the next most similar material to this group. Depending on the preselected threshold of similarity, it may also be a candidate. Materials 3, 6, and 8 are also related, but only moderately so, since their similarity factors are somewhat smaller. While having some common characteristics, these adhesives probably would not make good substitutes for one another.

To illustrate that clustering provides much more information than is obtainable from a single linear ranking of candidates, we contrast the above results with the clustering obtained by reversing the priority on the top two hierarchical property sets of Table 3 while keeping every other factor constant. By setting the properties of composition, specific gravity, and color to the top priority, we obtain a completely different clustering, as shown by the dendrogram of Fig. 3. In this case, candidates 1, 3, and 4 are very similar, while essentially no relationship exists between

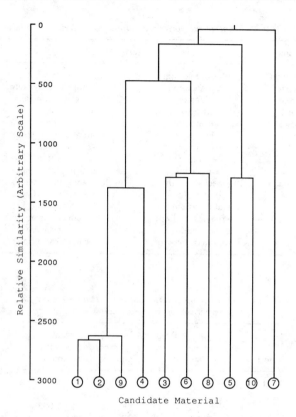

FIG. 2—*Dendrogram for hierarchical clustering of candidate materials with highest priority on selected electrical properties.*

the remaining candidates. Even though all the candidates were eventually clustered, their low similarity values (with respect to the selected set of criteria and priorities) show that they do not have much in common.

Conclusions and Future Work

Based on a preliminary analysis, we conclude that natural clustering can be useful in the reduction of the amount of user interaction with DBMSs needed to conduct successful query sessions (efficiency). Also, special characteristics of materials data bases promote the establishment of simple criteria functions for supporting hierarchical clustering algorithms (effectiveness). We encourage the use of clustering methods and the incorporation of graphical representations for data relationships such as dendrograms to enhance user comprehension.

We are currently evaluating commercial DBMS packages to determine their applicability to an interactive inquiry system for a spacecraft materials data base. If this research is successful, a natural clustering query support monitoring system based on the principles described in this paper will be interfaced with a commercially available relational DBMS in a multiuser environment.

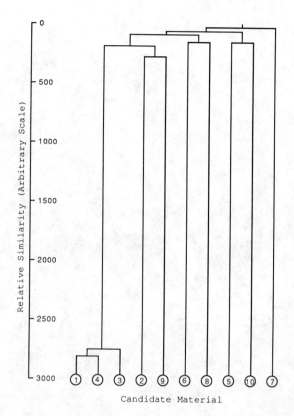

FIG. 3—*Dendrogram for hierarchical clustering of candidate materials with highest priority on selected physical properties.*

References

[1] Lagomasino, A. and Sage, A. P., "An Interactive Inquiry System," *Large Scale Systems,* Elsevier Science Publishers B. V. (North-Holland), Sept. 1985, p. 231-244.
[2] Fisher, D. and Langley, P., "Conceptual Clustering and Its Relation to Numerical Taxonomy," reprinted in *Artificial Intelligence & Statistics,* W. A. Gale, Ed., Addison Wesley, MA, 1986.
[3] Erman, L. D., et al., "Languages and Tools for Knowledge Engineering," reprinted in *Building Expert Systems,* F. Hayes-Roth, D. A. Waterman, and D. B. Lenat, Eds., Addison-Wesley, London, 1983.
[4] Geveden, R. D., "Minimum Variance Partitioning Applied to Classification," unpublished paper, Aug. 1987.
[5] Everitt, B., *Cluster Analysis,* Social Science Research Council, Heinemann Educational Books, 1974.

Impact of Materials Data Bases

Harris M. Burte[1] and Clayton L. Harmsworth[1]

Data Base R&D for Unified Life Cycle Engineering

REFERENCE: Burte, H. M. and Harmsworth, C. L., **"Data Base R&D for Unified Life Cycle Engineering,"** *Computerization and Networking of Materials Data Bases, ASTM STP 1017,* J. S. Glazman and J. R. Rumble, Jr., Eds., American Society for Testing and Materials, Philadelphia, 1989, pp. 197-199.

ABSTRACT: Unified life cycle engineering should encompass designing for performance, producibility, and supportability in an interactive mode. It will combine computer aided design (CAD) and computer aided manufacturing (CAM) with supportability concerns such as inspection, corrosion protection, repair, and replacement. This new engineering approach will deal with these parameters in the design process from a total life cycle standpoint.

The development of such an approach will require the integration of existing data bases and the development of new information sources. Today's data bases are not sufficient in scope nor do they meet the statistical requirements for such a concept. No system automatically prompts the user with the warning notes and "do's and don't's" that must be considered to prevent an unintentioned misuse of materials property data. Rapidly locating and accessing the needed range of diverse information and making it available for intelligent, interactive use by the designer presents a significant challenge.

KEY WORDS: reliability, data processing, manufacturing, design

Unified life cycle engineering (ULCE) is an integrated process that encompasses designing a component for performance, manufacturing, and maintenance throughout its service life. An illustrated introduction to the ULCE concept can be found in the references [1-2]. This paper briefly describes progress toward the ULCE concept, and indicates where additional R&D will be necessary, and particularly that involving data bases.

Status of ULCE and Some Directions for Data Base R&D

Computer aided design as limited to drafting has been widely applied. Further analysis, primarily of a few performance aspects of a design as permitted by finite element analysis, for example, is coming into use in some industries. The further growth of these performance modeling systems seems assured. However, most users must input materials property data for each problem; there is little automatic input from materials property data bases. Furthermore, they cannot easily ask "what if" questions such as those relating to material substitution, optimum heat-treatment, and so forth.

Simulating the manufacturing of components or even assemblies with computerized models is less advanced than analysis of performance, but is making rapid progress. For example, research supported by the Air Force leads to a simulation of metal flow during the forging process that can tell the engineer whether a particular preform-die combination would be suitable or if

---

[1]Chief scientist and technical manager for engineering and design data, respectively, AFWAL Materials Laboratory, Wright Patterson Air Force Base, OH 45433.

the part should be redesigned to avoid forging defects. Unfortunately, computer-usable data bases of materials behavior under the processing conditions required for such models do not exist.

In addition, ULCE should be able to assist in the selection of competing manufacturing processes—for example, a computerized trade-off between a forged, machined, and built-up component versus a one-piece, cast component. Such performance-manufacturing trade-offs should be considered as a part of the design process. A one-piece, cast configuration might differ from a machined and assembled configuration, and its design should consider the usually lower strengths that can be assured in a casting and the application of a casting factor. Although the technology for such a computerized process can be developed, the interactive manufacturing process data bases to support it are not available.

Those aspects of a product such as reliability, inspectability, maintainability, and other related phenomena that impact in-service use have received relatively little attention from the viewpoint of computerized simulation models or data bases. R&D towards this deficiency is now building up. For example, stress corrosion resistance is often a major factor in component life. Tables of stress corrosion resistances have appeared in MIL-HDBK-5 for a number of years. Obviously there are some combinations of conditions that should be avoided unless one is willing to increase maintenance costs to provide performance. From a combination of laboratory data and heuristics ("lessons learned" from service problems), the Air Force has formulated a set of "do's and don't's" that have been published as MIL-STD-1568 and MIL-STD-1587. These standards prohibit or control the use of certain combinations of materials, heat-treatments, thicknesses, and coatings. It should be possible to enter such information into a computerized data base, perhaps in the form of an expert system, to prompt the designer, and the Air Force has work under way to this end.

In another example, some capability to model nondestructive inspection for a given component and to predict the probability of detecting a flaw as a function of component geometry and flaw characteristics is emerging for ultrasonic and eddy current inspection. Given this, the inspectability of a design can be traded off against performance and manufacturing limitations. A set of rules about building inspectability into design can also be developed and could be applied in somewhat the same way as those previously given for corrosion, for example, to eliminate or redesign critical areas that would be uninspectable.

Handbook Computerization

A key element in the ULCE concept will be computerizing materials property data bases such as those contained in MIL-HDBK-5 for metals and MIL-HDBK-17 for composites. As early as 1976, the Air Force considered a computerized version of MIL-HDBK-5. At that time, the motivation was primarily to reduce the time and cost for the yearly insertion of several hundred pages of changes in the more than 10,000 copies of MIL-HDBK-5 in circulation throughout the country. A MIL-HDBK-5 users' survey that was conducted showed a preference for a centralized data base rather than disks or tapes that could be mailed to individual handbook holders, but at the time of this initial survey, there were three issues that produced concern.

First was the cost of transferring the tens of thousands of data points plus the charts and graphs to a computerized format and updating these values once a year. The closure of the Mechanical Properties Data Center owing to data development and maintenance costs did not increase confidence relative to establishing a similar system for MIL-HDBK-5.

The second concern was that extensive guidance in the handbook in the form of footnotes and warnings in the text might be lost. For example, a typical table on 2024-T6 aluminum clad sheet and plate shows five conditions where the allowables may not be valid. Possible solutions being considered include electronic asterisks or even flashing values in a data display. How such guidance might be handled in an interactive CAD system where the system picks values from the handbook for part design has been a matter of serious concern.

The third issue raised by several of the larger aerospace firms was loss of control during the design process. These companies have materials and processes or allowables groups that are charged with seeing that the design values are properly selected and used in accordance with their own company design practices. They may reduce the handbook numbers depending on the environment, third-dimensional loading, high degrees of forming, difficulties in inspectability, and so forth. These companies do not want to lose hands-on control of such data.

While these issues are matters of serious concern, they can be overcome. Consequently, a green light has been given to the development of a computerized version of MIL-HDBK-5 that will be part of an "Expert System" for materials selection, which will in turn be a key element of our ULCE approach. While a hard-cover version of MIL-HDBK-5 will always be available, researchers propose the creation of an electronically assessable data base in approximately two years. This data base will be structured so that the materials expert will stay in the loop; the system is designed to aid and not replace him. Only after the materials expert makes his decisions and the "approval" command is given could the data be fed into any actual design.

Summary

Only a few of the data bases are in place. Those that are available are not necessarily computer compatible. By their very nature, they are operated and applied in different ways. Issues relating to how the overall system should operate have not been resolved. The role of man in a particularly automated system needs further definition. While the Air Force is addressing these issues, it will be some time before they are resolved and a ULCE capability emerges. In the meantime, the spinoffs are great and many of the individual systems, although not fully unified, are of considerable value in their separate states.

References

[1] Burte, H. M. and Harmsworth, C. L., "New Challenges for Engineers: Unified Life Cycle Engineering," in *Computerized Aerospace Materials Data*, J. H. Westbrook and L. R. McCreight, Eds., American Institute of Aeronautics and Astronautics, Inc., New York, June 1987, pp. 19-29.

[2] Burte, H. M. and Coppola, "Unified Life Cycle Engineering," *Proceedings of the Annual Reliability and Maintainability Symposium*, Institute of Electrical and Electronics Engineers, Inc., Jan. 1987.

C. Dale Little[1] and Thomas E. Coyle[1]

Computerized Materials Data in Aerospace Applications

REFERENCE: Little, C. D. and Coyle, T. E., **"Computerized Materials Data in Aerospace Applications,"** *Computerization and Networking of Materials Data Bases, ASTM STP 1017,* J. S. Glazman and J. R. Rumble, Jr., Eds., American Society for Testing and Materials, Philadelphia, 1989, pp. 200–210.

ABSTRACT: The impact of computerized and networked materials data in the aerospace industry is conveyed by a description of its introduction, operation, and expansion plans in the Fort Worth Division of General Dynamics Corporation (GDFW). Like most other companies involved in the design and manufacture of aerospace systems and vehicles, GDFW was committed to an orderly buildup of Computer Aided Design (CAD) and Computer Aided Manufacturing (CAM) capabilities in the early 1970s. This milestone was followed in the 1980s by a commitment to integrate all computerized analysis and other support functions into the design process. This included the development and implementation of a computerized Engineering Materials Property Data Base (EMPDB), accessible by designers and analysts from their electronic workstations. The scope, development, and operation of the EMPDB is discussed here. Current and future expansion of computerized materials data bases through networking is addressed. Materials engineering files and other resources are now being shared through Local Area Networks (LANs). A computerized Laboratory Information Management System (LIMS) is now in the procurement cycle and will be operational in 1988. This information will be shared with other General Dynamics divisions and offices throughout the nation via Electronic Mail Office System (EMOS) through corporate-wide networks.

KEY WORDS: computerized networking, computer aided design, computer aided manufacturing, engineering materials property data base, local area networks, laboratory information management systems, electronic mail office systems

Computer aided design (CAD) and computer aided manufacturing (CAM) systems were established in the aerospace industry between the mid 1970s and the present. All support areas for CAD engineers at their workstations are being computerized. This paper presents the impact of on-line, computerized Engineering Materials Properties Data Bases (EMPDB) in the aerospace industry. The impact of computerized and networked data in the aerospace industry can be conveyed by a description of its introduction, operation, and expansion plans in one division of a large aerospace corporation. The Fort Worth Division of the General Dynamics Corporation (GDFW) is used as a model. Its EMPDB development may be ahead of some aerospace companies and may be behind others, but all are headed in the same direction.

GDFW, like most other companies involved in the design and manufacture of aerospace systems and vehicles, committed to an orderly buildup of CAD/CAM capabilities in the 1970s. CAD use at GDFW began in 1976. Now, most drafting tables and machines are gone from engineering design departments. Instead, most designs are developed using CAD systems and are represented by electronic data sets.

---

[1]Manager of materials and processes technology and assistant project engineer for computer aided engineering, respectively, General Dynamics, Fort Worth Division, Fort Worth, TX 76101

Indications of EMPDB Development in the Aerospace Industry

Before 1982, most materials property data were in handbooks and reports. Any data that were in a computer were certainly not available to support CAD. In 1982, the National Materials Advisory Board committee investigating materials information used in computerized design and manufacturing processes found that although much CAD development had taken place, very little had been done toward materials data base development for CAD use. Only one aerospace company was doing significant data base development for CAD use. This was the General Electric Company's aircraft engine division, which was establishing a data base of materials property curves for designers to use [1]. Therefore, in 1982, most of the aerospace industry was consequently without a materials properties data base for CAD use.

That 1982 survey showed a critical need in the United States for a computerized materials properties data base or a network of data bases with validated data and design analysis software for CAD use. J. G. Kaufman has recounted the chronological history of later events that resulted in the incorporation of The National Materials Property Data Network, Inc., a network for high-quality, well-documented data [2]. These events included the 1982 Fairfield Glade, TN, meeting; a feasibility study by Dr. J. H. Westbrook; and support-seeking presentations or workshops for a number of industries and professional societies.

Between 1982 and 1984, a number of aerospace companies began to develop materials properties data bases for CAD use. In April of 1984 at two symposiums, nine presentations reported milestones in EMPDB work and computerized analysis of parts in the United States, Great Britain, and the Federal Republic of Germany [3]. These papers indicated that more computerized data bases were being developed and placed in service. Besides the known computerized materials properties data bases of NASA, the reported papers indicated data bases at Battelle Columbus [4], the Federal Republic of Germany [5], and AiResearch [6]. Continued data base loading and development was underway at General Electric's aircraft engine division [7]. The ASM/NBS data-base program on alloy phase diagrams was underway [8], as well as a Materials and Processes (M&P) data base at Lockheed/Georgia [9] and a pilot data base at Martin Marietta/Denver Aerospace [10]. Other data-base papers dealt with the analysis of engineering components by Dr. A. J. Barrett [11], the use of an expert system to select materials by Dr. C. Edeleanu [12], and a new logic programming language (LOGLISP) by Dr. Volker Weiss for use with material properties data bases in a number of ways [13].

Another opportunity for assessing the status of other aerospace companies in their data-base progress occurred at the workshop on computerized materials property and design data for the aerospace industry held in El Segundo, CA, 23–25 June 1986 [14]. In discussions with representatives from other aerospace companies, nearly every company represented was involved in some sort of feasibility study or had different levels of plans of EMPDBs.

GDFW EMPDB Development History

Data base development at GDFW followed patterns taking place throughout the aerospace industry. The GDFW development history is shown in Fig. 1. A decision was made in 1982 to support CAD with materials properties made available at the CAD workstation. This led to a survey of designers for materials properties needs. Then, a small team was formed to do a feasibility study. It had written an Engineering Requirements document by mid-1984. A data-base management system was selected and a small, prototype data base was evaluated in the first half of 1985. The system was developed in phases so that loading could begin on some parts while design and development was still underway on other parts. Then, the EMPDB system was interfaced with computerized Application Lists/Parts Lists (AL/PL). Plans included interfacing it with the GDFW Manufacturing Resources Planning (MRP) system in 1987 and the computerized Class II Change system in 1988.

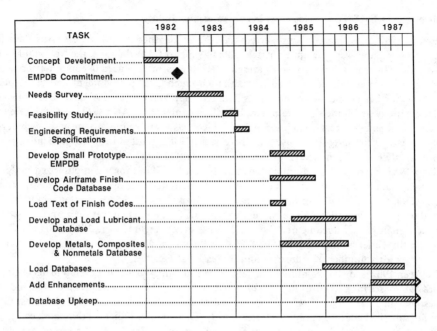

FIG. 1—*History of development of General Dynamics, Fort Worth Division engineering materials properties data base.*

Contents of the GDFW EMPDB

As indicated by Fig. 2, the EMPDB system is composed of five, on-line IBM Information Management System (IMS) data bases connected by a top menu plus eight supporting auxiliary data bases. Figure 3 shows the top menu that ties the various data bases together. EMPDB can be entered directly on IMS or it can be entered through the GDFW IMS Product Definition Systems Interface (PDSI). The latter system ties the EMPDB, the Standard Parts Data Base (SPDB), the two computerized AL/PL systems, and the MRP system together so CAD engineers can visit these data bases on-line and transfer information to their workstations. The EMPDB system will also interface with the computerized Class II change program that is being developed. Engineering Change Notices (ECNs) will be signed off on the computer beginning early in 1988.

In the EMPDB system, data bases exist for composites, lubricants, metals, nonmetals materials, and airframe finish codes approved for the F-16 program. Special search capability exists for bend radii, fracture toughness, weldable metals listings, and other items. Property curves, ground support finish codes, and access to MRP cost and lead time were added as 1987 update tasks. On-line viewing is available for material properties and selection aids.

Two automated AL/PL worksheet systems (one on IMS for non-CADAM®, and the CADAM® AL/PL) are aided by the on-line selection of materials specifications, F-16 airframe finish codes, drawing and PL material-related notes and documents. The two computerized worksheet systems provide automated verification before release of material specifications, F-16 airframe finish code, data on the compatibility of material and finish code, and the standard thickness of metals. Support systems are the three 1987 additions mentioned earlier, the M&P Notes Data Base, the Documents Data Base, the K-Standards Data Base (aluminum extrusions), the Special Searches Data Base, and an electronic mail system called Material Process

Principal Databases
  Composites
  Lubricants
  Metals
  Nonmetals
  F-16 Airframe Finish Codes

Support Databases
  Documents
  Ground Support Equipment Finish Codes
  Interface to MRP
  K-Standards
  M & P Notes
  M & P Request
  Property Curves
  Special Searches

FIG. 2—*Data-base system of the GDFW EMPDB.*

ENGINEERING MATERIALS PROPERTIES

SELECTION	PFKEY
COMPOSITES	1
FINISH CODES	2
LUBRICANTS	3
METALS	4
NONMETALS	5
SPECIAL SEARCHES	6
MATERIAL AND PROCESS REQUEST	7
EXIT	8
FOR HELP	9
PDSI MENU	11

FIG. 3—*Top menu for the GDFW EMPDB.*

Request (MPR). In the latter, engineers can request new materials or finish codes to be added to EMPDB or a new process specification to be written.

Movement through the EMPDB can be by menu choice, PF-key, or overtype of a key field (for example, change specification number in a field on the screen). Screens of the metals data base resemble MIL-HDBK-5 pages. Besides mechanical and physical properties for material specifications, design usage information is provided. Data base information also includes general use, processing, corrosion protection, necessary notes, hazardous or toxic considerations, and allowables source. Consideration for environmental protection was given to each material included in EMPDB.

Report generators exist for the F-16 Airframe Finish Codes Report, and the metals, nonmetals, and notes data bases. Other report generators are planned for listing specifications requiring qualification and listing the latest revision/amendment of M&P specifications.

One of the big cost savers is predicted to be the automated verification of the AL/PL before release. Many ECNs should be saved with that system by catching mistakes better than human drawing checkers can. Of course, this system has some drawbacks. Designers will be forced to either select specifications from EMPDB (and thereby get their selections transferred to the automated AL/PL), or they must input specification call-outs into the 18-character materials specification field correctly to keep an exception report from occurring.

All dashes and spaces must be entered just as the computer has them stored. For instance, a specification for aluminum sheet might be the characters "QQ-A-250/4-T81". A misplaced or omitted dash would get an exception report. Another problem exists in which class, type, grade, and so forth, are needed in combination. Their abbreviations are two characters each, but which comes first? To cope with this situation, the data base has the abbreviations in alphabetical order (for example, FMS-1020-CLBTYII).

Future Enhancements to GDFW EMPDB

Several additions are being considered at GDFW for the EMPDB. Process specification information needs to be added. It would cost very little to add a flag on specifications of materials that are included in the Air Force Technical Order (TO) #25-00-113 for "Critical Materials." A computer could then easily determine drawing part numbers and effectivities causing a report to print in support of that TO. Materials stocking numbers in MRP that cover materials on the "Controlled Materials List" will be flagged for ease in obtaining a computer printout when needed.

Probably the largest time saver for the CAD designer and stress engineer would be an interface between the EMPDB and computerized analyses programs performed by a material selection Expert System (E-S). Several preliminary investigations in this area are underway at GDFW.

Expert systems are a branch of applied Artificial Intelligence (AI). They are knowledge-intensive computer programs that contain much knowledge about their speciality. Rules of thumb, or *heuristics*, are used to focus on key aspects of specific problems and manipulate symbolic descriptions to reason about the knowledge contained. The rules of thumb used to solve a problem come from human experts in the narrow domain covered by the particular E-S [15]. In the past, a large E-S was built by "knowledge engineers," who would query the domain expert and translate the expert's reasoning into rules. Then code was written in a symbolic programming language that searched through the rules when a user queried the E-S. Such systems can be very costly owing to the time required by the knowledge engineer to build and test the system.

Software vendors are attempting to resolve the time bottleneck and eliminate the knowledge engineer. Expert system "shells" or "tool kits" are an answer to this problem. These tool kits are available for a broad range of computers, and allow domain experts and average programmers to develop knowledge-based systems without the assistance of AI specialists. The E-S shells are software packages that contain the code for the "inference engine" (for use in backward or forward chaining or both through the rules for a solution) and an empty knowledge base framework that the user fills in with rules and data [16].

The majority of these E-S shells are available for small systems. At the last ASTM E-49 Committee meeting, three expert systems that had been built using a commercially available E-S shell were demonstrated on a Personal Computer (PC) by the American Welding Institute.

The more rules the system contains, the more it can react like one or more experts at solving a problem. Whether there is a large E-S shell available for the large GDFW EMPDB is not known. If analyses programs are written in Fortran, can the symbolic languages of shells (mainly a dialect of LISP) communicate? Some E-S shells use PROLOG and others written in C language are becoming available. Perhaps C could be an interface between Fortran and the symbolic language. Finding the answers will be part of a planned investigation.

One must not overlook the possible uses of one of the PC E-S shells as a low cost tool to develop the series of rules needed for a larger E-S. There would be groups of selection rules for every one of the materials used in designing aerospace components. The adhesive selection rules could be developed independently of aluminum selection rules and so forth. With the groups of rules already developed on low cost PC systems by the domain experts, a large E-S would be less costly to develop.

As more professional PC workstations are purchased and placed in engineering departments, small PC expert systems will become more useful. E.I. Du Pont de Nemours had an AI team investigate more than 40 E-S shells that ran on VAX® and personal computers before selecting two to receive concentration [16]. They now have over 300 employees proficient in the use of these shells and are predicting a 10% increase in profits because of their expert systems in all areas.

Impact of Computerized Data Bases on the Materials Engineer

The materials and processes information contained in the EMPDB is one of the more formalized outputs of the materials engineer's work. This output, used in the design and manufacture of the company's products, is the result of collecting, screening, generating, analyzing, displaying, communicating, and storing materials and processes data. Computers, with the exception of word processors used in writing specifications, had played a relatively minor role in this process until recently.

This scenario is rapidly changing as the use of computers is being integrated into the materials engineers' functions. The era of voluminous file cabinets of materials and process data and specifications is coming to a close as old files are being converted to computerized data bases and new files are being generated within these data bases.

One important function of the materials engineer is to maintain an awareness of the characteristics of, and new developments in, the engineering materials classes for which he or she has been assigned responsibility or has professional interest. Recent estimates in the literature [17] place the number of existing engineering materials at greater than 100,000.

Economic considerations demand less than 1000 materials for which adequate design data and specifications are generated and maintained for the design and manufacture of the division's product lines. Figure 4 illustrates the partially implemented plan for maximizing the use of computers and networking to achieve this ongoing goal within GDFW more efficiently than has been possible by using the noncomputerized methods of the past.

Local Area Network (LAN) Demonstration Program

A LAN (Fig. 5) currently connects the Materials and Processes Technology (M&PT) office-based groups, staff, and the manager's office. In addition to building individual group material data bases, each group, as well as section staff and management, can share resources such as a hard-disk with a file-server and common data files, a dial-up modem, and a laser printer. A demonstration program consisting of several experiments has recently been conducted using the LAN as a prototype. In addition to other experiments concerned with section administration,

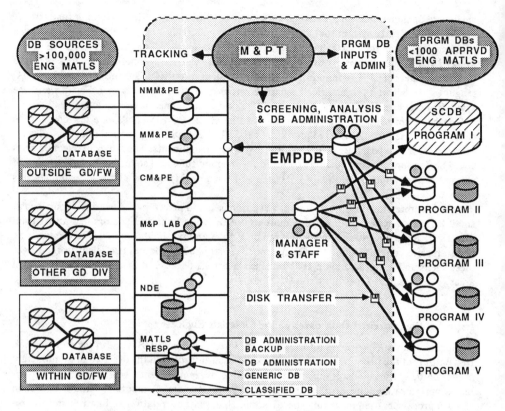

FIG. 4—*M&PT engineering materials data in networked computerized data bases.*

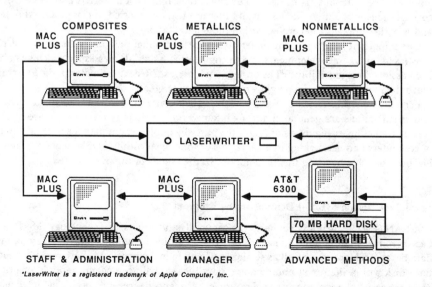

FIG. 5—*Materials and processes technology local area network (LAN).*

project management, and personnel management, those involving materials and process data had the following specific objectives (in italics), with excellent results:

(1) To demonstrate the compatibility of different hardware systems for the transfer and storage of data files.

Result—Apple® Computer, Inc. hardware (Macintosh® Plus computers and peripherals) was selected for the majority of the LAN based on its claimed ability to interface with AT&T, IBM, and DEC equipment for data file transfer, its outstanding graphics capabilities, the availability of integrated software programs for it, and its user-friendliness—all characteristics required for successful accomplishment of other objectives of the demonstration program. The required connectivity with the other systems was successfully demonstrated by the sharing of common data files, a hard-disk, printers, and the Electronic Mail/Office System (EM/OS).

(2) To promote and measure improved communication skills using graphic rather than tabular presentation of materials property data.

Result—Since the introduction of the LAN in mid-1986, materials engineers have been encouraged to present materials property data in graphical form in lieu of, or to compliment, tabular data. Although measurements of improved comprehension and retention of data have not been attempted, the improvements were readily apparent. The overwhelming favorable response to this presentation style insured its adoption at the Fort Worth Division.

(3) To evaluate an integrated spreadsheet, data base management, and graphics program as a depository of engineering materials property data.

Result—Microsoft's Excel program has been an excellent choice for this application. In addition to its ability to import data files from the popular Lotus 1-2-3® program, it is reputed to be the most powerful spreadsheet program available for personal computers. With its large number of cells, built-in and user-defined functions, macros, and capabilities for data base management, linkage to other spreadsheets, and integrated graphics, it provides the materials engineer with a powerful tool for storing, retrieving, analyzing, displaying, and communicating materials data.

(4) To communicate with and evaluate the use of outside data bases and services.

Result—With a dial-up modem, computers in the LAN have accessed outside services and data bases to read and download their files. Obviously, the amount of information made available by this is enormous. Since there are always costs associated with this method of data acquisition, the problem of selectivity confronts the user. The selectivity issue also includes the validity of the data from these sources. Several divisions of General Dynamics are observing with interest the progress of The National Materials Property Data Network, Inc., which is currently addressing these and related issues.

(5) To increase computer usage by materials engineers, especially among the more mature employees.

Result—Heretofore, material engineers at the Fort Worth Division have not been faced with the compelling incentives for computer literacy that their colleagues in some other disciplines have confronted; therefore, as a class, they were recently lagging somewhat in computer skills. This situation is rapidly changing in today's aerospace workplace. Although this change is of little consequence to the young engineers, the more mature materials engineers see their discipline being integrated into an overall computer-aided design and manufacturing process, and they are responding to the challenge of adapting to this relatively new computerized environment.

Fortunately, this process is greatly aided by the advent of today's user-friendly personal computers and software. The more experienced engineers are no longer intimidated by computers as they learn by hands-on experience, usually in the company of one or two other supportive colleagues who may have learned the same thing only the week before. The major problem remaining now is providing enough computers and software of this type for ready access by the many engineers and support persons wanting to use them.

Other Changes Attributable to Computerized Data Bases and Networks

The laboratory-based M&PT groups currently have and maintain individual group computerized data bases. A Laboratory Information Management System (LIMS) is now being procured. When the LIMS is installed and becomes operational in 1988, it will connect the M&PT laboratory-based groups, the office-based groups, section staff, and management. Additionally, as shown in Fig. 6, other departmental and divisional networking activities in progress will link the laboratory-based M&PT groups and the other departments within the division for which laboratory services are provided.

Since the laboratory-based groups are not in close proximity to many other groups and departments with which they interface, the LIMS is expected to greatly improve communications between these various groups. The laboratory-based engineers and the office-based engineers can interactively plan, schedule, and denote the status of laboratory projects using information stored in and readily available from the LIMS. The office-based engineer can monitor the progress of experiments and can receive experimental results in near-real-time by accessing the LIMS. Improved personnel relations and teamwork are anticipated as the various groups work more interactively through the LIMS. The LIMS is an integral part of an ongoing plan to computerize laboratory operations to the fullest extent possible with available technology and financial resources. This movement has created new career opportunities for computer-oriented engineers and scientists in the laboratory environment.

Computerized materials and processes data is changing the role of the materials engineer in

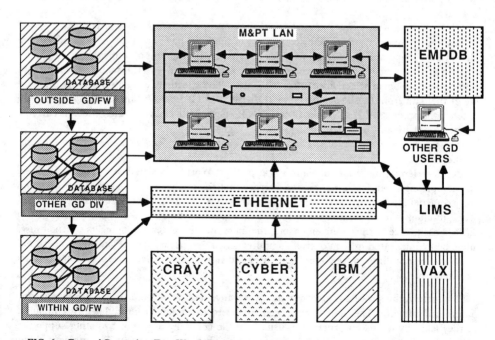

FIG. 6—*General Dynamics, Fort Worth Division, materials and processes computerized data bases.*

other ways. The typical materials engineer is a specialist in one or a few classes of materials, for example, steel, aluminum, adhesives, polymeric composites, and so forth. Until recently, unpublished materials data were usually in the possession of the specialist with only limited availability to others. To provide some redundancy, this materials engineer is usually backed-up by another specialist in a few other classes of materials. These material engineers are assigned to a functional home group, and either work in their home group or on a project team. Simultaneously, many projects, ranging from small to large, are ongoing, each under the direction of a project manager and staffed by an interdisciplinary team drawn from various functional groups. The budgets of smaller projects usually permit only one materials engineer working full or part-time. This requires the materials engineer to exercise responsibility for all materials instead of just some of the materials. This also requires the materials engineer to either make judgments in some situations for which he or she may have only limited qualifications, or else seek help from colleagues knowledgeable in the other material classes. Budget or schedule pressures imposed by the project can sometimes lead to nonoptimum decisions on the part of the materials engineer.

When material data resides in computerized data bases, and these data bases are connected by networks, at least two positive results can occur: (1) the data is readily available to any knowledgeable person with access to the network, and (2) the data was entered into a format and with help instructions previously documented and approved by knowledgeable and responsible persons. This means that a materials engineer trained to access computerized materials data through networks can have quick access to all data and on-line helps. Although this will not assure optimum decisions, it will certainly increase the speed and the quality of the materials engineer's decision process.

Each home group can assign material engineers as data-base administrators and back-ups. These persons would become knowledgeable about the formatted data and on-line help instructions residing in their data bases. With training, they could also become familiar with the contents and operations of other group materials data bases. These data-base administrators would then become candidates for the materials engineering representation on project teams. Through the network, the project materials engineers would then have access to all computerized materials and processes data and on-line consultation with the data-base administrators.

Conclusions

The EMPDB has made great strides in making materials information available to the designer. A networking concept linking the materials engineer's EMPDB, the Laboratory-Based Engineer's LIMS, and the Materials & Processes Technology's LAN is about to become a reality at the Fort Worth Division. Also, the EMPDB could be transportable to other divisions. In addition, by early 1988 this computerized technology will be used with other divisions within the corporation through the use of EM/OS. This mail system can be used to transfer both verbal data and graphics. The technology is available to extend the network concept to link all interacting groups, departments, divisions, corporate offices and headquarters, suppliers, and customer organizations. This movement is underway at General Dynamics' Fort Worth Division, and other corporate entities, not only for materials and processes data, but many other types of computerized information.

Acknowledgments

Appreciation is expressed to Mr. Marvin S. Howeth, engineering specialist, at General Dynamics, Fort Worth Division, for his support and assistance, which was essential in the timely completion and preparation of this paper and its figures.

References

[1] Brown, William F., Ed., *Materials Properties Data Management—Approach to a Critical National Need*, Report NMAB-405, National Materials Advisory Board, National Academy of Sciences, Washington, DC, National Academy Press, 1983.

[2] Kaufman, J. G., "The National Materials Property Data Network, Inc.," *Computerized Aerospace Materials Data*, American Institute of Aeronautics and Astronautics, Inc., New York, 1987, pp. 49-61.

[3] Coyle, Tom E., "Investigation of Symposium Papers on the Subject of Materials Properties Data Bases for Computer-Aided Design," unpublished paper for computer science class CSCI-590, North Texas State University, Summer term, 1984.

[4] Mindlin, Harold, "Data Dissemination and On-line Numeric Data Base Systems," paper presented in Las Vegas at the 29th National SAMPE Symposium on April 3-5, 1984.

[5] Dathe, Gert, "Two Data-Base Systems for the Access to Materials Data and for Their Statistical Analysis," paper presented in Las Vegas at the 29th National SAMPE Symposium April 3-5, 1984.

[6] Schwab, David E., "AMDADS: AiResearch Materials Data Analysis and Dissemination System," paper presented in Las Vegas at the 29th National SAMPE Symposium on April 3-5, 1984.

[7] Baur, R. G., Donthnier, J. L., Moran, M. C., Mortman, I., and Pinter, R. S., "Materials Properties Data Base Computerization," paper presented in Las Vegas at the 29th National SAMPE Symposium on April 3-5, 1984, and at the Denver IPAD meeting on April 17-19, 1984.

[8] Bhansoli, K. J., Redmiles, D. F., Murray, J. L., and Sims, J., "Data Base Development Under the ASM/NBS Program on Alloy Phase Diagrams," paper presented in Las Vegas at the 29th National SAMPE Symposium April 3-5, 1984.

[9] Howard, J. W. and Maddock, B. C., "Specification Information Handling for the M&P Engineer," paper presented in Las Vegas at the 29th National SAMPE Symposium on April 3-5, 1984.

[10] Karr, Patricia H. and Wilson, David J., "RIM As the Data Base Management System for a Materials Properties Data Base," paper presented in Denver at the IRAD meeting on April 17-19, 1984.

[11] Barrett, Dr. A. J., "Data and Systems Requirements in Computer Analysis of Engineering Components," paper presented in Las Vegas at the 29th National SAMPE Symposium on April 3-5, 1984.

[12] Edeleanu, C., "The Incorporation of Information on Deteriorating Processes in Data Bases," paper presented in Las Vegas at the 29th National SAMPE Symposium on April 3-5, 1984.

[13] Weiss, Volker, "Materials Selection With Logic Programming," paper presented in Las Vegas at the 29th National SAMPE Symposium on April 3-5, 1984.

[14] Westbrook, J. H., and McCreight, L. R., Eds., *Computerized Aerospace Materials Data*, American Institute of Aeronautics and Astronautics, Inc., New York, 1987.

[15] Harmon, Paul and King, David, *Expert Systems: Artificial Intelligence in Business*, John Wiley & Sons, Inc., New York, 1985.

[16] Shannon, Terry C., "AI Gets Commercial," *Digital Review*, 23 Feb. 1987, pp. 63-65.

[17] Bittence, John C., Ed., "When Computers Select Materials," *Materials Engineering*, January 1983, pp. 38-42.

Jane E. Martini-Vvedensky[1]

The Business of Materials Data Banks

REFERENCE: Martini-Vvedensky, J. E., **"The Business of Materials Data Banks,"** *Computerization and Networking of Materials Data Bases, ASTM STP 1017,* J. S. Glazman and J. R. Rumble, Jr., Eds., American Society for Testing and Materials, Philadelphia, 1989, pp. 211–215.

ABSTRACT: This is an analysis of the business experiences of the Engineering Information Company Ltd. in the provision of MATUS, their on-line materials selection data bank. Market demand and data-base design and distribution are discussed.

KEY WORDS: data banks, engineering materials, materials selection data banks

The value of accumulated information was apparent to the ancient Byzantine emperors who amassed the libraries of their conquered lands in the city of Constantinople. This made Constantinople the hub of civilization and one of the greatest capital cities of all times.

Accumulated information is just as important in modern times as it was in the days of the Byzantine Empire. However, its accumulation is no longer financed by wealthy emperors. In the Western world of today, information sources are either compiled under the auspices of elected government or by private industry in response to specific needs.

This analysis reflects the experience of one private company whose business is accumulating and providing data on engineering materials. In response to the specific industrial need to source engineering materials, the London-based Engineering Information Company Ltd.[1] collects data sheets and trade literature from materials suppliers and provides that information through the on-line data base, MATUS.

The survival of such an undertaking now depends on market demand rather than the wisdom of an emperor. The specific needs of the market must be well understood. Great effort is required to collect the information and force it into a data-base format. Once complete, the data base must be successfully sold and distributed.

The Nature of the Market

The need for materials information is a curious subject. On one hand, manufacturing companies happen upon problems in materials technology that they must either solve or design around. These problems may be anything from the need for a material for a new engine design that will maintain high strengths at high temperatures, to a wear-resistant coating for manufacturing equipment. Faced with an urgent need for a material with particular properties, companies will spend heavily to find a solution.

However, such urgent materials requirements are rare. Part of the art of a design engineer is designing around the properties of the available materials, so only occasionally is a new material essential. Even so, there are times when new materials give important competitive advantages. Lower-cost materials will improve the profitability of high-production parts. Occasionally, a technology cannot advance until new materials are found. Materials which maintain strength at

---

[1]Engineering Information Company, Ltd., 15/17 Ingate Place, London SW8 3NS, United Kingdom.

higher temperatures will improve engine efficiency. There are also times when unexpected environments or use will cause a material employed in a design to fail. If this happens to a part already in production, finding a substitute becomes an important priority. No matter how infrequently a manufacturing company faces an urgent materials selection problem, its successful resolution is of the utmost importance. Forming and assembling the correctly chosen materials into the desired product is what manufacturing is all about.

However, the need for information about materials is greater than simply what these urgent requirements demand. It is important that design engineers are continually fed information about new materials so that they can make an informed choice about material-design tradeoffs. Design engineers should be as informed about new materials as medical doctors are about new drugs. This is becoming an increasingly difficult task as the variety of materials grows; a challenge which can be addressed by information technology. The use of data bases containing information about materials of engineering interest would change the act of designing. Materials could be optimally selected rather than "designed around."

Even so, it seems that to date, few engineers are using computerized data bases to expand their materials knowledge. We believe the reason for this is two-fold: the data bases are not yet sufficiently complete, and the barriers to access are still too high.

Designing the Databank for the Market

Required Features for Data Base Use

There are three factors required for the acceptance and use of a data base system.

(1) The engineers who need the information must be aware of the existence of the data base and they must remember it when they have a materials enquiry. This requires a significant awareness campaign, not aimed at just those who will purchase the system, but at design, materials, and manufacturing engineers as well.

The barrier for gaining access to the system must be perceived as less than that for other equivalent sources of information. The engineer may find it easier to ask the materials scientist in the next room for an appropriate material than to set up a link to a computerized data base. To use a computerized data system rather than a readily available handbook or an expert, the required equipment must be at hand, communications links should be easily established, and the software must be friendly.

Once the engineer makes an attempt to use the system, it must reward him with accurate answers. Receiving no information, irrelevant information, or inaccurate information will result in disappointment. The engineer will be hesitant to try a second time.

Providing the Solution

Unfortunately for the data-base creator, engineers turn to data bases for the hardest questions first. Because of the barriers inherent to using a system for the first time, engineers will approach a materials selection data base only if their accustomed methods had been unsuccessful. This means that the data base must contain information about a wide variety of materials. Unusual as well as common properties must be described, since the tougher materials selection problems often entail a combination of common and uncommon properties. The data for each material must be as complete as possible and they must be accurate.

Since engineers encounter materials selection problems on an occasional basis, software must be extremely friendly. The engineer should not have to remember any details and should be guided through the searching procedure (although optional shortcuts through a menu-based system are often appreciated). Often the most unfriendly part of a data-base session is connecting to the data base (or, in the case of floppy-based systems, initiating the system to run on a particular machine). Unfortunately, many of the problems encountered are hardware-oriented: which cable to use, how to set up a modem, and so forth.

Ideally, a materials selection data base would be part of a regularly used information system available through a personal computer residing on the engineer's desk. A number of relevant data bases should be available, forming a "library" of data bases. This is an approach already adopted by the large hosts, such as Dialog. Where relevant, the information system should be coupled to other computer activities, such as CAD-CAM or electronic mail systems.

Finance

Financiers have several reasons to be reluctant to finance materials data-base systems. The development time is long. Finance houses prefer to have projects that become profitable within three years. Few (if any) data bases have been developed with such speed; many remain unprofitable even after 6 or 7 years. The development time is of indeterminate length. Unfortunately, the amount of detail required to provide a successful service depends on the data base; the question is answered only with hindsight.

Building the data base poses difficult personnel problems. Data entry requires technically trained individuals who are enthusiastic about the subject and who pay good attention to detail. The person managing and coordinating data entry not only needs technical training and the ability to perform detailed work, but also must have a global perspective. He must be able to understand the needs of the market and the best presentation methods.

The data base requires substantial resources to develop, but in accounting terms, it is an intangible asset. The data base is of no value until people pay to use it. Should an established data base go out of date or be superceded by a competing system, the value will drop as people stop using it.

Finally, the data base is difficult to protect. In the opinion of some legal advisors,[2,3] material property data represent facts rather than art and therefore are not covered by copyright law. If the data were obtained by a competitor, he would be able to avoid substantial development costs in producing a similar system.

The Experience of MATUS

Probably for the reasons outlined above, MATUS did not initially attract a great deal of development financing from venture capitalists. The system was under development for 5 years (albeit with limited staff) until funding was received attracted from the European Economic Community and an investment was made by a private party. MATUS is now able to answer approximately 60 percent of the enquiries made of it and is increasingly attracting subscribers.

One of the important lessons learned was the necessity of attracting the best technical personnel available. The Engineering Information Company is a small company in a popular city. It has a youthful outlook, giving its staff the freedom to express themselves and to take on increasingly challenging tasks. Another advantage is its strong ties with the Imperial College of Science and Technology, which provides individuals with a sound (broad) scientific grounding.

Distribution

One task that remains a problem is providing easy access to MATUS. The bibliographic data bases are available through the major hosts (such as Dialog). However, the hosts are not yet able to provide friendly software capable of carrying out the structured searches needed to use a data

---

[2]McKenna & Co., Inveresk House, Aldwych, London WC2.
[3]Fairfields, Grindall House, Newgate Street, London EC1.

base such as MATUS. Our software is dynamic, changing in response to users' needs and to the data available. Often the changes are subtle, but immensely important. Fine-tuning of the user interface improves the engineer's patience with the system, allowing him or her to make more detailed searches. This fine-tuning is difficult to achieve with software that is not under our direct control. Also, we are unwilling to allow our data to reside on external machinery where it is more vulnerable to theft and where we cannot gather user comments. We hope that the EEC Materials Data Banks Demonstrator Programme[4] will be successful in providing an umbrella under which a number of data bases can be accessed.

Although there are few profitable engineering data bases, the data bases used by the financial sector have become profitable. In the City of London, the swing to profitability occurred during the "Big Bang." The Big Bang eliminated the restrictions of the financial community regarding foreign participants and computer transactions. Suddenly, terminals were available on every desk; information banks could be accessed directly. It is true that the participation of foreign firms brought more money and competition to the community, but individuals working in the City indicate that improved accessibility played a key role in promoting data-base use.

Work is ongoing to attempt to standardize materials data banks so that several information sources can be used in tandem to enrich each other. Hopefully these efforts will ultimately bring easier access to a more complete set of information. Committee E49 of ASTM,[5] CODATA,[6] and VAMAS[7] have all made contributions in this field. Descriptions of their efforts are contained in Refs 1 to 3. The Code of Practice for use in the Materials Data Base Demonstrator Programme [4] is gaining widespread acceptance. Hopefully, these organizations will also pressure for improved communications networks and will find ways to encourage corporations to equip their engineers with the terminals and modems needed to access large, on-line materials data banks.

Conclusions

Building a computerized materials selection data base is a long task that requires imagination and perseverence. The business of materials data banks involves using highly capable technical personnel to develop an immense store of information. It will not be used until the data collection is vast, relatively complete, accurate, and easily retrieved, yet it has no value until it is used.

Although many financiers are wary of such undertakings, a fully developed materials databank is an important asset. An easily accessible system with a full data set will not only generate good cash flow to its investors, but it will provide an important engineering tool also. The availability of such data bases will change the way design engineers work. Instead of designing around the properties of the materials they know, they will now be able to select the optimal materials for their designs.

Providing the resources needed to accumulate bodies of information may no longer be in the hands of enlightened emperors, but fortunately there are now corporations, individuals and governmental bodies who do recognize the value of such developments.

References

[1] Kaufman, J. G., "Towards Standards for Computerized Material Property Data and Intelligent Knowledge Systems," *ASTM Standardization News*, March 1987, pp. 38–43.

[4] Materials Data Banks Demonstrator Program, Commission of the European Communities, DGXIII, Bat J. Monnet, Plateau Du Kirchberg, L-2920, Luxembourg.
[5] American Society for Testing and Materials (ASTM), 1916 Race Street, Philadelphia, PA 19103, U.S.A.
[6] CODATA, 51, Boulevard de Montmorency, 75016 Paris, France.
[7] VAMAS, Versailles Project on Advanced Materials and Standards, Secretary: Dr. B. Steiner, National Bureau of Standards, Gaithersburg, MD 20399, U.S.A.

[2] Westbrook, J. H., Behrens, H., Dathe, G., and Iwata, S., "Materials Data Systems for Engineering," Proceedings of the CODATA Workshop at Schluchsee, Fachinformationszentrum Energie.Physik.Mathematik GmbH, Karlsruhe, West Germany, 1986.
[3] Kröckel, H., Reynard, K., and Rumble, J., "Factual Materials Data Banks: the Need for Standards," VAMAS Technical Working Area 10, available from J. Rumble, National Bureau of Standards, A323 Physics Building, Gaithersburg, MD 20899, 1987.
[4] "The Operation of Materials Property Data Systems in the EC [European Community]—A Code of Practice for Use in the Materials Data Base Demonstrator Programme," MPD (MAT-02)-OS-03, Commission of the European Communities DGXIII, Bat J. Monnet, Plateau du Kirchberg, L-2920, Nov. 1986.

John R. Rumble, Jr.[1]

Socioeconomic Barriers in Computerizing Materials Data

REFERENCE: Rumble, J. R., Jr., **"Socioeconomic Barriers in Computerizing Materials Data,"** *Computerization and Networking of Materials Data Bases, ASTM STP 1017,* J. S. Glazman and J. R. Rumble, Jr., Eds., American Society for Testing and Materials, Philadelphia, 1989, pp. 216-226.

ABSTRACT: While the computer technology, both hardware and software, exists to allow easy and comprehensive access to computerized data on engineering materials, only a few data bases are publicly available today. The reason for this is a set of barriers based on socioeconomic factors that are only partly recognized. In this paper, the use of materials data is examined to identify some of these socioeconomic barriers. They primarily arise from three factors: complex usage patterns, a relatively new computer audience combined with an exploding computer technology, and a lack of understanding of the positive impact of computerized materials data. Ways to overcome these barriers are suggested.

KEY WORDS: computerized data, data base, data base management, information systems, materials data, on-line

The awesome power of computers to hold, manipulate, and deliver technical information is obvious, and data on the properties of engineering materials are no exception. But in spite of the capability now available and the progress of the last 5 years, very few materials data are really available via computer today. Only a handful of publicly available data bases can be accessed. One major reason is a set of barriers, best described as socioeconomic, that have not yet been overcome.

Since activity revolving around manufactured products depends vitally on data about construction materials, it is surprising that there is not a well-accepted model of materials data usage. This is probably a reflection of the fact that materials data are used in many different ways, from "blue-skying" about possible materials to tracking down the properties of a material used in a part manufactured 20 or 30 years ago. The variety of usage has a profound impact on the computerization process for materials data and is one source of the socioeconomic barriers.

Unlike other areas of science, most potential users of materials data bases are relatively new computer users. Furthermore, since their interactions with materials data are usually just a small part of their routine work, the amount of time and effort that they wish to spend on mastering and maintaining materials data bases will be limited. In addition, the range of computer products is becoming even more complex, if that is possible. The combination of these factors contributes to problems related to the use of computers by materials data users. Again, only cursory attempts have been made to deal with these problems.

Finally, as with many underlying technologies, the impact of better or more easily accessed materials data cannot be easily quantified. This is further complicated by the drastically different impact from industry to industry. One reason for this is that the impact has been deter-

[1]National Bureau of Standards, Gaithersburg, MD 20899.

mined by quantifying improvements to final products rather than by measuring the increased efficiency and productivity of the data users. The difficulty in measuring the impact of computer access is a third major factor contributing to the socioeconomic barriers.

In discussing socioeconomic effects related to the computerization of materials data, the terms sociology, economics, and socio-economics need to be defined [1]

- sociology—the science of fundamental laws of social relations and institutions,
- economics—the science treating the production, distribution, and consumption of goods and services, and
- socioeconomic—pertaining to the combination or interaction of social and economic factors.

This paper concerns usage and access to materials data, both on an individual and on a corporate basis, as well as how data production and distribution are affected. We examine the interaction between the socioeconomics of materials data bases and progress in making computerized materials data available.

The primary thesis of this paper is that progress in materials data bases is being impeded *not* by technical problems, but rather by institutional barriers of socioeconomic origin that often are not even recognized.

The following section of this paper discusses the use of materials data by individuals and companies. The next section outlines several generic aspects of using materials data in diverse applications. The section after that identifies several socioeconomic barriers to the use of computer data bases by materials data users. The final section summarizes the conclusions of the paper.

The Use of Materials Data

Generally speaking, materials data are used for problem solving, that is, in the context of a larger engineering or scientific activity. This means that materials data are not the end of an engineering process, but rather a part. Consequently, the use of materials data often takes on a subordinate role when the engineering activity of a particular individual or organization is analyzed. Even when materials data are the major effort for an individual, such as for a materials information specialist or a materials expert, the role of that individual is presumed rather than highlighted.

We will approach our discussion of materials data usage in two ways. First, the types of users and applications are classified and how the data are used in each type is discussed. Second, a look will be taken at four aspects of usage that occur for all types of applications. In a previous paper [2], the author has discussed generic steps in using scientific data.

A recent study on standards for materials data banks [3] has identified several different types of applications (Table 1). This list is by no means exhaustive, and the detailed usage within a given category can be quite complex. In practical terms, this means that materials data are used by a wide variety of people for a wide variety of reasons. It is rare that in a given organization a consistent set of data are used in every application.

We are all familiar with the complaint that design engineers come up with materials that turns out to be difficult to maintain. Naturally, for situations where product liability and safety are paramount, some degree of control is exerted, but these are the exception rather than the norm.

Materials Data Generation

In this category, we are concerned with the data associated with the development and testing of new or improved materials. The data involved are the numerous test measurements needed to

TABLE 1—*Applications involving materials data* [3].

MATERIALS DATA GENERATION
Collection and storage of data measurements and calculations
Materials development
Data evaluation and validation
Materials characterization

MATERIALS USAGE
Design engineering
Materials selection
Materials performance
Production engineering
Product maintenance

MATERIALS ACQUISITION
Product information systems

REGULATIONS FOR MATERIALS
Legislation information systems

establish the property data to be used later. Test results are often reported in detail so that statistical and other analyses can be done. The data users are generally materials experts who are interested in the variation of properties as a function of many parameters and want to know the reasons behind the variability.

The data are generally reduced to more tractable amounts by producing design or nominal values that are the basis for using a material in an application. Within a given organization, the number of people involved in this process is usually quite small and they are often in a laboratory settting.

Materials experts often have a very high range of interest that can be characterized as a general curiosity about materials. Sometimes this is application-driven, but many times it is not. Because many materials data are remembered, these people often need to verify data or browse through them, rather than checking for what the specific properties of given materials are.

The data generators themselves often are faced with large volumes of raw test data that will be refined later. Rarely will someone such as a designer want to see all these data. The data, however, are sometimes needed long after the fact, in an unpredictable way, to determine some type of historical record for use in failure analysis or service life extension.

Materials data are often the central issue for materials experts, unlike other users of the data. Though data generation is fundamental in the use of materials data, it is sometimes overlooked because of its indirect impact on tangible products.

Materials data bases can have a profound impact on the data generation process. Data bases allow for easy collection, storage, and manipulation of large amounts of data. Analysis can be done quicker and better. Numerous data sources can be browsed. Archiving data for historical purposes can be done. But because of the indirect nature of this impact, the benefits of materials data bases can be difficult to justify on this basis.

Engineering materials often change slightly as time passes. Improvements in manufacturing, for quality or economic reasons, occur. Batch to batch variations are common, especially when the material is not being used in a critical application. The characterization of engineering materials is therefore important. Statistical analysis of data spanning these changes or variations is needed so that the "average" properties of these materials are well known and documented.

Materials Usage

People involved in using materials usually do so in the context of making a product. Designers, materials selectors, production engineers, and maintenance engineers all use materials data to help achieve some tangible outcome. Their materials data interests are very specific and often quite secondary to other parts of their technical activity. Their efforts are evaluated not on the basis of the quality of information behind a result but rather on the tangible result itself.

The data needed are specific values for a given set of conditions that the users specify. The variations, subtleties, and nuances are often secondary concerns. The approach is "I need to know these data now" and "If measurements are not available, what are the best estimates."

While traceability is often required, for instance, in the aerospace industry, often it is not. Sometimes that has bad consequences when unpredicted failures occur. When such failures are intolerable, the degree of interaction with the materials data generators and experts is appreciable. When tolerable, the interaction is less. Often, there is a remarkable lack of interaction between different users, such as between designer and production engineer or designer and maintenance engineer. Again this degree of interaction depends quite strongly on economic and other consequences. However, the degree of communication more often results from tradition than the real needs of the work.

In the use of materials, materials data bases can provide immediate benefits as well as a degree of consistency and control not easily achieved in a paper environment. But realizing these benefits is yet to come.

Materials Acquisition

This third category of data usage refers to the actual acquisition of materials as specified. The process involves translating a materials specification with a given set of properties into a purchasing requirement that can be satisfied by a materials supplier. For bulk purchases or situations when the supplier and materials expert or designer work together, this process may be trivial. In many cases, however, off-the-shelf purchasing must be done, and matching becomes more problematical.

The data here are often used by persons not expert in materials and the opportunity for problems is real. Good companies have procedures for minimizing these problems. The data required involve not only the specifications but also the supplier data and a determination of the equivalency. Consequently, though supplier data are often less documented than desired, they must be accepted.

For newer materials, especially new polymers, composites, clad materials, and so forth, the amount of independent generic test data is much less than for traditional metals, ceramics, and polymers. In many cases, the manufacturer's data are all that are really available. The positive consequence of this is that products are specified by their trade name rather than generically. The negative side is that the requirements become manufacturer-specific.

Legal Use of Materials Data

The last category to be mentioned has two aspects. The first is the data related to government control of the use of materials. Here the requirements are two-fold. Test results must be reliable enough to support regulatory action and be accepted by the community at large. In addition, the data must be specific enough to cover the situation accurately. This is necessary to make sure that loopholes do not allow avoidance of compliance.

The second aspect is the use of materials data for legal reasons, such as product liability suits. It is in this process that the deficiencies of materials data use in the previous categories become obvious.

For both regulatory and legal actions, there are well-accepted procedures for dealing with materials data issues in the present-day paper information systems. Computerization of these data will change those procedures in a way that is not yet well understood.

Aspects of Using Materials Data Bases

We can distinguish four levels of activities associated with the use of materials data, namely:

- awareness of existence,
- understanding the range of possibilities,
- decision-making, and
- assessing the correctness of decisions.

Awareness of data existence is essential. A materials data user must be aware that various materials exist and what the sources of information about them are. For an expert immersed in a narrow field, little help may be needed beyond one's own memory. But when confronted with a problem such as "What materials could we consider in replacing our chrome trim in this product?" the level of awareness must be deeper.

Keeping this awareness current is a labor-intensive job because of the diffuse nature of the range of materials and because the number of data sources is so large that the task is impractical in every sense. But it must be done to a degree satisfactory for the level of problem-solving encountered. The difficulty described above is of course the reason that traditional materials are so often used, and innovative, better, or less expensive materials are underused. The traditional means for maintaining awareness certainly cannot cope with the flood of information available except in a piecemeal fashion, and data bases should greatly increase awareness. Often data exist in publications that have very limited circulation. Data base networks can greatly improve circulation and increase awareness.

In understanding the range of possibilities with respect to materials data, the user in the most ideal situation must know everything about the materials requirements for what is being designed. This is more than simply the performance characteristics for the material. However, users live in less-than-ideal worlds and must almost always be willing to accept a smaller range of choices with which they have some familiarity. The practical implication of this is that even though data exist to cover most possibilities, the user is not willing to invest the time or energy to ferret them out.

Indeed, a full range of materials choices frightens users, especially when their expertise often is strongly focused on one class of materials, for example, metals, with little or cursory knowledge of other classes. A user will recognize that the failure mechanisms for a part made out of a polymer may be significantly different from one made out of a metal, but may not have the time, resources, or inclination to explore the data necessary to account for the differences. Thus, the materials choice is severely restricted.

Consequently, the decision-making based on materials data covers far more than answering the question "Which material is best?" The economics of decision-making involves not only the quality and availability of materials data, but also the economics of the selection process itself. It is not "How much do I have to invest to make sure I am choosing a suitable polymer when I don't really know if it is readily available in the quantities I want?" It is instead, "If I choose a metal, I know that suppliers A, B, and C all make it."

Recognizing this problem, we must also state that the quality of the materials data does make a significant difference, especially in regard to assessing the correctness of the decision. Often, however, that assessment involves not the question "Is it the best?" but rather the question "Can we live with it?" This might include verification of manufacturer-supplied data, checking to see if the corrosion resistance of a material selected on the basis of mechanical properties is adequate, or determining if a reliable supply can be found.

Sometimes the correctness is assessed a long time after the fact, such as in failure analysis. Good engineering practice assumes the adequacy of documentation related to materials data decisions. In some industries, such as aircraft manufacturing, rigid controls are in place. For third-party construction projects, say of oil wells, less satisfactory procedures exist, and failure analysis can be complicated because of a lack of knowledge of the materials data used.

We have taken the time to present the above discussion of materials data use to provide a background for its computerization process. As shown, the use of these data usually occurs in the context of far greater activity of great diversity. Each of the uses places an additional burden on the computerization process, and much of the burden is not computer-related in nature. Rather, it reflects socioeconomic problems in materials data use, as we shall discuss now.

Barriers to Materials Data Bases

In spite of all the talk about the computer information revolution, materials data users would be hard pressed to find evidence that the revolution has changed their life. But aside from the fact that reality takes time to catch up to the hype, several problems have prevented more rapid progress.

The significant barriers to the computerization of materials data are not technical in nature, but rather reflect socioeconomic concerns. There are some technical barriers that must be overcome before the final form of computer access to materials is in place, but these do not hinder the building and maintaining of full-fledged systems today. Table 2 lists the status of some of these technical concerns.

As previously stated, the real barriers are more socioeconomic in nature and are given in Table 3. The barriers are grouped as shown.

Data Base Building

The first three problems impede the data base building process itself.

Lack of Standards and Conventions—The conventions and standards for the representation of technical information are almost nonexistent for computers. For every type of application, there are multiple computer possibilities for hardware, operating systems, data base management packages, graphics packages, word processors, and what have you. No one can keep up with all the developments, much less take advantage of them, because each one requires separate mastery.

One practical consequence of this for materials data bases is that no two data bases are compatible in any feature—input, output, search techniques, loading, editing, report generation, or graphic display. When you add the oncoming surge of expert system capability on top of all this, the situation appears hopeless. The other consequence is that each data base building effort must start from scratch, and there is no sharing of experience.

Two types of standards are needed—one set for computer issues, the other for handling mate-

TABLE 2—*Status of technical concerns for materials data bases.*

Not Barriers	Barriers
Data base management systems technology	interfaces to general engineering software
Identification of user needs	support of multiple hardware types
User interfaces	distributed data base systems
Mass storage	model of materials data

TABLE 3—*Socioeconomic barriers to materials data bases.*

DATA BASE BUILDING

Lack of standards and conventions
Bewildering computer choices
Redundant data entry

COMMITMENT TO DATA BASES

Lack of resources
Vested interests of technical data suppliers
Lack of understanding of impact
External versus internal data

DATA BASE USAGE

Lack of training in information systems
Need for multiple data bases
Personal information systems

---

rials data itself. The problems have been well defined [3] and activities have started in both areas (see other papers in this volume). Once the standards are in place, both building and using data bases will be much easier. However, until the groups involved realize the extra costs associated with the lack of standards, progress will be slow.

Bewildering Computer Choices—Rather than a lack of computer technology, it often appears that we have too much. In reality, we have not yet learned to harness the computer power already available. Any noncomputer expert who starts out on a data base building project is faced with such a bewildering set of choices related to hardware and software selection that it is difficult to know where to begin.

Depending on the intended distribution of the data base, certain choices may be easy to make. Even in these cases, simply knowing the true alternatives is hard because of conflicting advice from the vendors. Often even the data processing staff of a company, though skilled in many computer techniques, simply do not have the background in data base usage to be helpful.

There are, of course, techniques that have been developed to assess what the true requirements are, resulting in a requirements analysis that provides the basis for intelligent choices. But those techniques are often unknown to the neophyte materials data base builder.

A further complication arises when a data base is built for local use and later a decision is made to expand distribution. What were suitable choices for the data base on a local level may be entirely inappropriate for a larger scale system. Given the investment already made, a change will be made only reluctantly.

Since scientists and engineers are problem solvers by nature, there is a tendency to approach the data base building process as an intellectual exercise never confronted by anyone else. One result is that a significant number of groups have chosen to build their own data base management system, an expensive process. The cost of maintaining in-house software is frighteningly high, and the do-it-yourself approach is usually a very expensive one.

As progress in building data bases grows, there will be more emphasis on timely and efficient completion of these projects, and data base builders will make better choices, though still with difficulty.

Redundant Data Entry—Since most organizations use internally generated data with external data sources, several questions related to data entry arise. While obviously internal data must be entered by the organization itself, whose responsibility are the external data sources? If all the external handbooks, reports, and data collections were available in computer-readable form, this would not be a problem except for the cost of purchase or lease.

However, they are not, so organizations are confronted with the necessity of entering someone else's data. Given the high cost of data entry, this course of action is usually met with some resistance. As a consequence, only part of the external data now used are available, and the total data needs cannot be met by turning to the information system. As more published data and commercial data base systems become available in computerized form, this problem will disappear, but it is very real now.

Commitment to Data Bases

The next four areas involve reasons for lack of commitment to computerized materials data systems.

Lack of Resources—It is expensive to build, distribute, and maintain a materials data base. We can easily project that in 25 years, all significant materials data will be available on a computer, but we cannot easily identify where the needed investment is going to come from. So far, government, industry, and information vendors have each put up modest amounts, but an order of magnitude of additional funding is needed. A positive note, however, is that once other socioeconomic barriers are overcome, groups will be more willing to provide adequate support.

Vested Interests of Materials Information Suppliers—A substantial amount of economic activity related to materials data presently exists. The suppliers of paper-based data sources as well as information specialists in industry have a vested interest in computerization. Computer-based information systems clearly will greatly perturb that activity. Several reactions have already become evident. Publishers, whether private or technical, are staking out their turf. But because the rules of the games are changing and the technology is new, this also involves barriers to progress until a more obvious course of action appears. These barriers take the form of excessive study by staff and committees, quick and dirty products, and objections to other activity that might prove threatening.

As long as these delaying actions are accompanied by a long-range planning process and eventually adequate investment, they are acceptable. Fortunately, that appears to be the case, at least at the present.

Lack of Understanding of Impact—The wide range of impact for materials data presents a problem because of the corresponding lack of focused successes. Rarely do we encounter a situation where the impact of bad data or the lack of good data can be documented. Field trials in which engineer Y designed with data sources A, B, C, and D while engineer Z used only sources A, B, and C are just not done. Often, the impact would not be easily measurable, as in reduced maintenance costs over the lifetime of a product.

Yes, we can say that improved engineering has reduced the recall rate of American-made automobiles, but can that be directly traced to better materials performance, better quality assurance and control, better manufacturing, or something else not easily measured.

Yet that is the level of justification that is often placed on the technical end-user in a company when he or she desires to participate in the creation of computerized materials data sources. The common good is rarely justification enough. The lack of a measure of impact of materials data on an industry is a barrier because a significant investment must be made by industry before the U. S. will have a large enough data base available.

This is not to say that this barrier is insurmountable. Industry has contributed to many materials data programs in the last decade, many of which are described in this volume. These include the following:

- alloy phase diagrams,
- corrosion data,

- ceramics phase diagrams,
- tribology data,
- materials property data network, and
- Military Specification Handbooks 5 and 17.

In the same way, government agencies have also invested many general and programmatic funds to achieve the same results. But as any of the participants in these programs can attest, the lack of minimum resources has been a real problem and the positive impact of the programs on an individual company virtually impossible to demonstrate.

External Versus Internal Data

Many companies have a poor awareness of how many nonproprietary (external) data are used in their engineering work. Even in industries such as airframe manufacturing, nonproprietary data are important. They may be used by only a limited number of people who are responsible for incorporating the data into proprietary (internal) sources, but the data are used.

For most other industries, these external data sources are more widely used, and copies of handbooks sit on every engineer's desk. When these handbooks are available in computer form, they too will be widely used.

Data Base Usage

The final set of barriers hampers the use of data bases, thus reducing the momentum toward a more complete computerization of materials data.

Lack of Training in Information Systems—In the present, paper-based information system, most technical people receive at least a 4-year formal education in using printed technical information sources. From the very first day of college, we must read, interpret, and create manuscripts, tables, and graphs of technical information. Inherent in this education is training in how to recognize and, when necessary, extend the conventions, explicit and implicit, involved.

Though younger technical people have a full dose of computers in their education, this almost never extends to computerized information sources. The result is that we are untrained to take advantage of computerized technical information sources and companies rarely allow such training by end-users. Most materials data bases are not user-friendly enough to be used without training, and the full power of computerized information cannot be realized.

Need for Multiple Data Bases—Because of the very nature of their work, users solving technical problems need to access multiple data sources. When the data sources are computerized, the investment in time and money needed to master more than one or two is prohibitive because the data bases are now imcompatible in every respect (as discussed before).

The incompatibilities are on two levels. First, the expression of the data content itself is unstandardized. What is your "modulus of elasticity" is someone else's "Young's modulus." While this has been well documented [3-6], the problems have not been resolved.

The other level of incompatibility concerns computer formats themselves. Ideally, the user will one day have interfaces with systems that do not depend explicitly on the underlying software and hardware. At present, interfaces are so disparate as to present a major hurdle. My "locate" is your "search" and your "exit" is my "bye." Trivial perhaps, but as the number of systems grows, this becomes much more than annoying.

To correct this problem, the same standards and guidelines needed for building materials data bases will facilitate the use of multiple data bases concurrently. The results include greatly reducing training costs and producing better user acceptance.

Personal Information Systems—The development of the personal computer/workstation is a mixed blessing. While potentially it can free the user of reliance on huge, nonresponsive, monolithic entities such as data processing (DP) shops, MIS departments, or on-line systems, there are downsides also. Merely being aware of the existence of a data source is a problem, as well as determining the relevance of that source to your problem-solving needs. Once identified as of value, just the acquisition of multiple personal computer (PC) data bases can be a problem. Can each engineer have 50 data bases, each costing $200 to $2000? With books, that is not allowed now. There is no reason to expect that the corporate policy will be any different with computerized data sources.

The broader problem concerns maintaining a single-person information center. These costs can be impossibly large, as seen by the problems that people have in keeping the standard four software packages—word processing, spread sheets, data base management, and telecommunications—all up and running on a personal computer. The potential cost of maintaining PC hardware also diminishes the attractiveness of a personal information system.

However, the traditional library and DP service functions are equally inappropriate. The computer affords the end-user the opportunity to sift through the information him/herself and to maintain the type of computing environment appropriate to her/his situation. The tyranny of information specialists and DP departments is real, and PCs afford real relief.

The solution lies in the amalgamation of the information sources and support function in a comprehensive but unobtrusive way. Collections of data sources, such as those provided by on-line systems with individual, identifiable data bases, are one answer. Standards for software and hardware, on a discipline basis or corporate basis, are possible. The landscape in this area is changing rapidly, and the final lay of the land is very indeterminate. Until it is clear, personal information systems will be a barrier.

Conclusion

In a paper such as this, the heavy emphasis on discussing barriers and problems can leave too negative a picture. The progress made in the 7 years since the National Academy of Sciences/National Research Council study [7] on Materials Data Bases has been impressive.

First, standards activity on a national and international scale has become vigorous (see other papers in this volume). Second, as other parts of the engineering process are computerized, the need for computerized materials data grows more evident. This in turn will result in better resources and a greater commitment. Third, the commercial value of computerized information has not only been recognized, but also has begun to be realized. Last, as patterns in data base usage, though on a small scale now, are recognized, better solutions to these problems are being developed.

The socioeconomic barriers discussed in the paper have arisen because of the newness of computerized technology in information handling. Once we learn to appreciate and use this new technology, most of these types of barriers will fade. At that stage, computer access to materials data will be not a goal but a reality.

References

[1] Random House College Dictionary, Random House Inc., New York, 1980.
[2] Rumble, J. R., Jr., Bullock, E. et al., "Data Systems for Engineering Materials, The Materials Engineer Point of View," *Materials Data Systems for Engineering—Proceedings of a CODATA Workshop*, J. H. Westbrook, Ed., Fachinformationszentrum Karlsruhe, Karlsruhe, FRG, 1986.
[3] Kröckel, H., Regward, K., and Rumble, J. Jr., Eds., *Factual Materials Data Banks—The Need for Standards*, VAMAS, 1987. Limited distribution available from Standard Reference Data, National Bureau of Standards, A323 Physics Bldg, Gaithersburg, MD 20899.

[4] Westbrook, J. H., *Standards and Metadata Requirements for Computerization of Mechanical Properties of Metallic Materials*, NBS Special Publication 702, U.S. Department of Commerce, Washington, DC, 1985.
[5] Kaufman, J. G., *ASTM Standardization News*, March 1987.
[6] Ambler, E., *ASTM Standardization News*, Aug. 1987.
[7] *Mechanical Properties Data for Metals and Alloys: Status of Data Reporting, Collecting, Appraising, and Disseminating*, National Academy Press, Washington, DC, 1980.

Materials Data Base Projects

Ligaya S. Petrisko[1]

Engineering Plastics via the Dow MEC Data Base

REFERENCE: Petrisko, L. S., **"Engineering Plastics via the Dow MEC Data Base,"** *Computerization and Networking of Materials Data Bases, ASTM STP 1017,* J. S. Glazman and J. R. Rumble, Jr., Eds., American Society for Testing and Materials, Philadelphia, 1989, pp. 229-238.

ABSTRACT: The Materials Engineering Center (MEC) of the Dow Chemical Company has developed a computerized engineering data base for plastics. The properties and parameters needed to specify materials for engineering design were defined and a system was developed to correctly link test data results to specific specimens. Validation of the engineering data base entry is achieved by combining knowledge of the laboratory source reliability and of statistical records generated within the system.
Information can be retrieved by selecting a combination of properties based on a specific set of conditions or by specifying a property of various environmental or processing conditions or both.
In this fashion, the MEC data base supplies the information needed by design, materials, and process engineers during the development of new products, applications, and in the modification of existing systems. The data base response speed and simplicity of information retrieval, combined with its well-defined data, make this data base a significant technical development in the materials engineering world.

KEY WORDS: MEC data base, plastics, engineering design data, material specification, property data reporting, material selection

Engineers and designers confronted by the myriad plastics available today need quick access to accurate materials property data when selecting polymers for durable goods applications. Therefore, the Materials Engineering Center (MEC) of the Dow Chemical Company has developed a computerized engineering data base for plastics that supplies the end-user with the engineering data needed to design parts, develop products and processes, and compare competitive materials. The data are available in an easily accessible and concise format, and are finding widespread use among our R&D, TS&D, marketing and sales, and production and quality control staffs and, of course, by our customers. Thus the data base serves as a central focus for most of the MEC activities. A typical approach we use for new application development is displayed in Fig. 1.

The scope of the data base is diagrammed in Fig. 2, which covers: (1) thermoplastics and thermosets produced by the Dow Chemical Co. and its competitors; (2) physical, thermal, rheological, mechanical, ignition, optical, permanence, and electrical end-use properties; (3) the characterization used to evaluate and control end-use properties; (4) processing parameters, including specifics about the equipment and conditions of operation; (5) the bulk list price in $/lb and $/cu in; and (6) external files containing the raw data for further analysis, for example, stress-strain curves.

We have found that the data base is extremely useful in preventing inappropriate selections of materials, and has been invaluable in assessing the strengths and weaknesses of competitive

[1] Development leader, Materials Engineering Center, Dow Chemical Company, Midland, MI 48667.

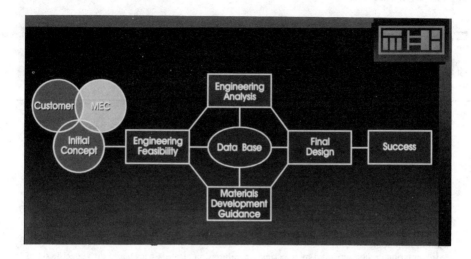

FIG. 1—*MEC application development approach.*

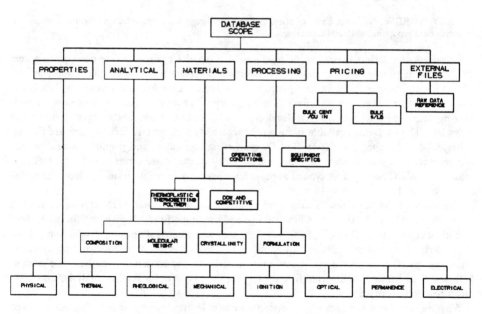

FIG. 2—*MEC engineering plastics data base.*

design approaches. We believe that the data base provides economic and strategic benefits to its users.

Elements of the MEC Data Base

Material Specification

Most of the existing data bases for plastics consist of single data point properties that are useful only for quality control and specification purposes. There is a need to access other engineering design properties like those listed in Table 1 (entries in capital letters are not commonly available for most plastics). Lot numbers, process/product modifications, and data on how the test specimens were prepared are necessary to unambiguously identify a material. Once identified, the material must then be linked to generic groupings and classifications to assess its competitive advantages. Our efforts to meet these needs have been guided by the findings of ASTM task group E49.01.04, which is providing guidelines for the identification of polymers in computerized materials property data bases. These guidelines identify those elements that are essential for describing polymers, yet can be efficiently searched and sorted with a computer. Table 2 shows the 13 classifications needed to properly specify a polymer.

The generic family classification is a nonstandard one and has been employed to help group different materials according to their monomeric compositions. The preferred naming system is based on poly(monomers) for homopolymers and monomers A-B-C, and so forth, for heteropolymers. Unfortunately, common names such as nylon (polyamides) and saran (polyvinylidene chloride) are also used in the industry. This ambiguity presents a problem in all data bases for plastics, as does the extensive use of abbreviations.

Fortunately, ASTM Standard Abbreviations of Terms Relating to Plastics (D 1600) contains standard abbreviations for both homopolymers and heteropolymers. These abbreviations have come into widespread use, having evolved from the industry's need for convenient, readily comprehended shorthand notations. For example, consider that rubber modified polystyrene has been referred to as super high-impact polystyrene (SHIPS), high-impact polystyrene (HIPS), or

TABLE 1—*Mechanical design protocol—MEC data base.*

Properties	Independent Variables
Tensile Properties	RATE, TEMP.
Flexural Properties	TEMP.
Compressive Properties	TEMP.
Izod Impact	Temp., GEOMETRY
Dart Impact	Temp., RATE, GEOMETRY
Shear Strength	TEMP.
SHEAR MODULUS	TEMP., FREQUENCY
POISSON RATIO	RATE, TEMP.
CREEP	STRESS, TEMP.
STRESS RELAXATION	TEMP.
FATIGUE	STRESS, TEMP., FREQUENCY
BUCKLING	STRESS, TEMP.
FRACTURE ENERGY	RATE, TEMP., GEOMETRY
EFFECTS ON STRENGTH, STIFFNESS & IMPACT	
AGING	TEMP.
ENVIRONMENTAL RESISTANCE	MOISTURE, CHEMICALS, UV
ORIENTATION	GEOMETRY, FABRICATION

TABLE 2—*Guideline for polymer identification.*

Classifications

1. Material class (designated as polymer to differentiate it from metals, ceramics, and composites.
2. Polymer class, that is, thermoplastics, or thermosets.
3. Generic family.
4. Manufacturer's designation, including trade name and the material grade.
5. Material supplier.
6. Production process.
7. Date and place of manufacture.
8. Condition of the product as tested.
9. Description of the modifiers.
10. Specifications, classifications, or regulations covering the material or some combination thereof.
11. Form of the material used for testing.
12. Processing used to prepare the test specimens.
13. Conditions of fabrication.

medium-impact polystyrene (MIPS). It is indeed a challenge to keep up with the number of new materials developed and the constant modifications of existing materials. The MEC data base accomplishes this by continually updating the abbreviations list shown in Appendix A. The data base also has a dictionary of other terms used by the industry that will help the user during searches.

Property Data Reporting

Once a material is uniquely identified, reporting its property data requires definition of all test and specimen variables affecting the results. Standard test methods established by ASTM have been a great help to the industry for generating property data. Unfortunately, reference to the ASTM number alone does not make the data meaningful, as plastic suppliers often do not follow the test method fully or fail to report important variables such as temperature, rate of testing, conditioning procedure, strain measuring device, gage length, specimen thickness, or direction of test for anisotropic materials. This frustrates end-users, as accurate materials comparisons are nearly impossible.

Consider the data in Table 3. Six major manufacturers report deflection temperature under load (DTUL) data for a polymer obtained by using the ASTM Standard Test Method for Deflection Temperature of Plastics Under Flexural Load D 648. Manufacturers A and B were the only ones that reported the fabrication method, annealing condition, and specimen thickness. The rest ignored these key variables, which are known to significantly affect the DTUL value. No definite conclusion on the superiority of one material over another can be made because of

TABLE 3—*Data reporting—deflection temperature.*

Manufacturer	Fabrication Method	Annealed (Yes/No)	Thickness (in.)
A	Injection Molding	N	1/8
B	Compression Molding	N	1/2
C	...	N	...
D	...	...	...
E	...	...	...
F	...	Y	1/4

the uncertainty regarding how thickness and stress in the specimen affected the reported DTUL. In contrast, when run conditions are standardized (see Table 4), the results indicate that for a minimum stressed part, resins 2 and 3 from manufacturer A have the highest heat resistance, while for a highly stressed part, resin 1 from manufacturer A and resin 1 from manufacturer B show the best retention of heat resistance.

Access Methods

The MEC data base currently offers two query user formats (Fig. 3). One method allows the selection of materials based on combinations of different properties. Each property is defined under a specific set of conditions. This provides the user with a quick way of screening different materials based on various property requirements.

An example is shown in Fig. 4. Suppose a materials engineer is interested in all materials having a heat resistance greater than or equal to 149°C (300°F), a flexural modulus greater than or equal to 3448 MPa (500 000 psi), an Izod impact greater than or equal to 107 J/m (2.0 ft-lb/in.), and a UL 94 flammability rating of V-0. The selection can be accomplished in one command, such as

SELECT DUT GE 149 AND FMA GE 3448 AND IZA GE 107 AND ULA = V − 0

All data elements are identified by three letter codes obtainable through a help system. This user output mode can be designed to access any set of properties and materials desired.

The second mode of access allows the user to address one property at a time. With this method, the user is able to examine the effects of the test and the fabrication conditions on the data and is supplied with the measurement statistics. The source and date of data generation are also provided.

A view (property) is selected, aided by a help system, which guides the user into the different groupings of properties (Fig. 5). If the mechanical grouping was selected, the next screen tells

TABLE 4—*Deflection temperature at 1820 kPa (264 psi), °C (°F).*

	Injection Molded, 1/8"	
	Unannealed	Annealed
MANUFACTURER A		
Resin 1	87 (189)	102 (215)
Resin 2	84 (184)	112 (234)
Resin 3	87 (188)	113 (236)
MANUFACTURER B		
Resin 1	86 (187)	102 (215)
Resin 2	76 (168)	104 (220)
HIGHEST HEAT RESISTANCE		
Based on Minimum Stressed Part	Manufacturer A Resins 2 & 3	
Based on Highly Stressed Part	Manufacturer A Resin 1 Manufacturer B Resin 1	

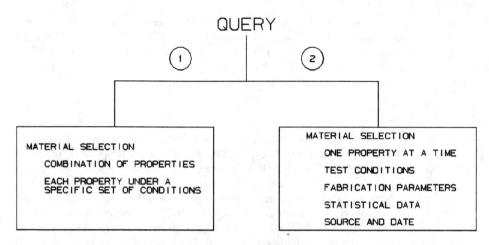

FIG. 3—*Methods of access.*

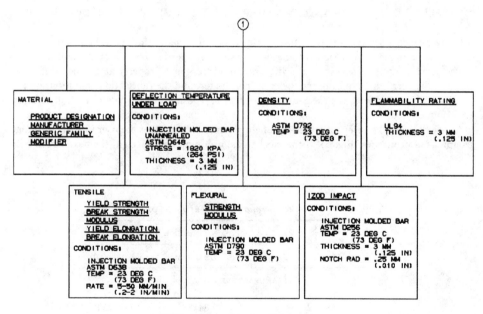

FIG. 4—*Example of type 1 query.*

the user the different types of properties accessible under the "mechanical" designation (Fig. 6). After choosing a property, in this case "tensile," the user can start with the query selections. Each view includes material identification, fabrication conditions, test data, and conditions. The result of such a search is shown in Fig. 7 and the desired property obtained by a simple command

VIEW IS [4 letter code for property name]

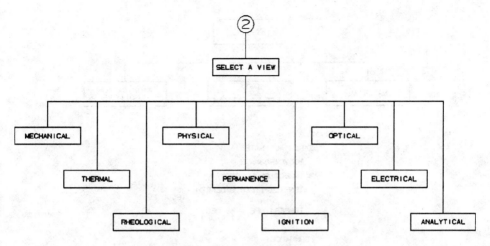

FIG. 5—*Help for type 2 query property classifications.*

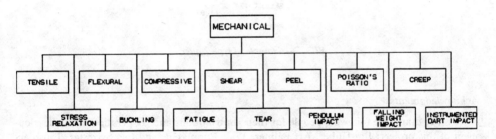

FIG. 6—*Help for type 2 query mechanical properties.*

Then the SELECT command is employed to search for the generic material and the property at the desired temperature and fabrication condition.

Conclusion

Dow's MEC data base offers a well-developed capability for selecting, evaluating, and standardizing the data used by researchers and engineers working with structural polymeric materials. The data base operates on a mainframe computer, and has accessibility through a modern or a network link.

The data base presently covers about 100 records and 1000 data and information fields. It has a dynamic system of operation and maintenance to satisfy the needs of new polymer, process, and application technologies. Future capabilities may include data analyses or integration with CAE programs or both.

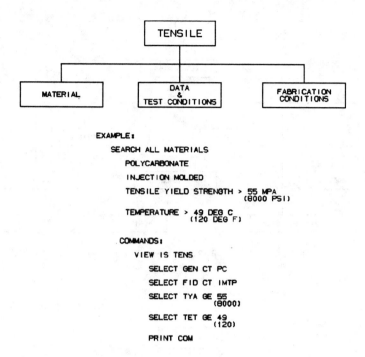

FIG. 7—*Example of type 2 query.*

Acknowledgments

I wish to thank the following people for their contributions and cooperation in the development of this project: Jim Huber, Art Palmer, Walt Rupprecht, and Leo Sylvester. I also acknowledge the endless support of Bob Cleereman, Senior Research Manager, Dow Materials Engineering Center.

APPENDIX A

STANDARD CLASSIFICATIONS FOR FAMILY OF POLYMERS

ABBREVIATION	FULL NAME
*ABS	ACRYLONITRILE-BUTADIENE-STYRENE
ABS/CPE	ACRYLONITRILE-BUTADIENE-STYRENE/CHLORINATED POLYETHYLENE
ABS/PA	ACRYLONITRILE-BUTADIENE-STYRENE/POLYAMIDE
ABS/PC	ACRYLONITRILE-BUTADIENE-STYRENE/POLYCARBONATE
ABS/PVC	ACRYLONITRILE-BUTADIENE-STYRENE/POLY(VINYL CHLORIDE)
ACS	ACRYLONITRILE-CPE-STYRENE
*ADC	ALLYL DIGLYCOL CARBONATE
AES	ACRYLONITRILE-EPDM-STYRENE
ASA	ACRYLONITRILE-STYRENE-ACRYLATE
*CA	CELLULOSE ACETATE
*CAB	CELLULOSE ACETATE-BUTYRATE
*CAP	CELLULOSE ACETATE-PROPIONATE
*CF	CRESOL FORMALDEHYDE
*CN	CELLULOSE NITRATE
*CP	CELLULOSE PROPIONATE
*CPE	CHLORINATED POLYETHYLENE
*CPVC	CHLORINATED POLY(VINYL CHLORIDE)

CSPE	CHLORSULFONATED POLYETHYLENE
*CTA	CELLULOSE TRIACETATE
*DAP	DIALLYL PHTHALATE
EAA	ETHYLENE-ACRYLIC ACID
*EC	ETHYL CELLULOSE
ECTFE	ETHYLENE-CHLOROTRIFLUOROETHYLENE
*EEA	ETHYLENE-ETHYL ACRYLATE
*EMA	ETHYLENE-METHYL ACRYLATE
EMAA	ETHYLENE-METHACRYLIC ACID
EP	EPOXY-EPOXIDE
*ETFE	ETHYLENE-TETRAFLUOROETHYLENE COPOLYMER
*EVA	ETHYLENE-VINYL ACETATE
*FEP	PERFLUORO(ETHYLENE-PROPYLENE) COPOLYMER
*FF	FURAN-FORMALDEHYDE
HDPE	HIGH DENSITY POLYETHYLENE
HIPS	HIGH IMPACT POLYSTYRENE
LDPE	LOW DENSITY POLYETHYLENE
LLDPE	LINEAR LOW DENSITY POLYETHYLENE
LMDPE	LINEAR MEDIUM DENSITY POLYETHYLENE
MDPE	MEDIUM DENSITY POLYETHYLENE
MF	MELAMINE-FORMALDEHYDE
PA-6	POLYAMIDE-6
PA-66	POLYAMIDE-66
PA-69	POLYAMIDE-69
PA-610	POLYAMIDE-610
PA-612	POLYAMIDE-612
PA-11	POLYAMIDE-11
PA-12	POLYAMIDE-12
*PAA	POLY(ACRYLIC ACID)
PAR	POLYARYLATE
PAE	POLYARYLETHER
PAI	POLYAMIDE-IMIDE
PAK	POLYESTER ALKYD
*PAN	POLYACRYLONITRILE
PASU	POLYARYLSULFONE
*PB	POLYBUTENE-1
PBT	POLYBUTYLENE TEREPHTHALATE
*PC	POLYCARBONATE
PC/PBT	POLYCARBONATE/POLYBUTYLENE TEREPHTHALATE
PC/SMA	POLYCARBONATE/STYRENE MALEIC ANHYDRIDE
*PCTFE	POLYMONOCHLOROTRIFLUOROETHYLENE
*PEEK	POLYETHERETHERKETONE
*PEI	POLYETHERIMIDE
PESU	POLYETHERSULFONE
*PET	POLYETHYLENE TEREPHTHALATE
*PETG	POLYETHYLENE TEREPHTHALATE-GLYCOL COMONOMER
*PF	PHENOL FORMALDEHYHE
*PFA	PERFLUORO(ALKOXY ALKANE)
*PFF	PHENOL FURFURAL
*PI	POLYIMIDE
*PMMA	POLY(METHYL METHACRYLATE)
*PMP	POLY(4-METHYLPENTENE-1)
*POM	POLYOXYMETHYLENE
POMCO	POLYOXYMETHYLENE COPOLYMER
*PP	POLYPROPYLENE
PPCO	POLYPROPYLENE COPOLYMER
PPC	POLYPHTHALATE CARBONATE
PPE/HIPS	POLYPHENYLENE ETHER/HIGH IMPACT POLYSTYRENE
PPE/PA	POLYPHENYLENE ETHER/POLYAMIDE
PPE/PBT	POLYPHENYLENE ETHER/POLYBUTYLENE TEREPHTHALATE
*PPS	POLY(PHENYLENE SULFIDE)
*PS	POLYSTYRENE
*PSU	POLYSULFONE
PSU/ABS	POLYSULFONE/ABS
*PTFE	POLYTETRAFLUOROETHYLENE
PU	POLYUREA
*PUR	POLYURETHANE
*PVAC	POLY(VINYL ACETATE)
*PVAL	POLY(VINYL ALCOHOL)
*PVB	POLY(VINYL BUTYRAL)
*PVC	POLY(VINYL CHLORIDE)
*PVCA	POLY(VINYL CHLORIDE-ACETATE)
*PVDC	POLY(VINYLIDENE CHLORIDE)
*PVDF	POLY(VINYLIDENE FLUORIDE)
*PVF	POLY(VINYL FLUORIDE)
*PVFM	POLY(VINYL FORMAL)
*SAN	STYRENE-ACRYLONITRILE

*SI	SILICONE-PLASTICS	
*SMA	STYRENE-MALEIC ANHYDRIDE	
SMA/ABS	STYRENE-MALEIC ANHYDRIDE/ACRYLONITRILE-BUTADIENE-STYRENE	
SMMA	STYRENE-METHYL METHACRYLATE	
*TEEE	THERMOPLASTIC ELASTOMER, ETHER-ESTER	
*TEO	THERMOPLASTIC ELASTOMER, OLEFINIC	
*TES	THERMOPLASTIC ELASTOMER, STYRENIC	
TPUR	THERMOPLASTIC POLYURETHANE	
UHMWPE	ULTRAHIGH MOLECULAR WEIGHT POLYETHYLENE	
*UF	UREA-FORMALDEHYDE	
*UP	UNSATURATED POLYESTER	
VE	VINYL ESTER	

NOTE: 1. ELASTOMERS WILL BE ADDED TO THIS LIST
2. SOURCES OF ABBREVIATIONS WILL BE REFERENCED

*SOURCE - ASTM D1600

B. J. Moniz[1] and T. C. Wool[1]

RUST: A Coupon Corrosion Test Data Base for Metals and Nonmetals

REFERENCE: Moniz, B. J. and Wool, T. C., **"RUST: A Coupon Corrosion Test Data Base for Metals and Nonmetals,"** *Computerization and Networking of Materials Data Bases, ASTM STP 1017,* J. S. Glazman and J. R. Rumble, Jr., Eds., American Society for Testing and Materials, Philadelphia, 1989, pp. 239–252.

ABSTRACT: Du Pont has devised an interactive computerized corrosion test data base called RUST, which calculates, stores, and allows retrieval of corrosion data generated at many Company sites and test laboratories. This paper describes Version 3 of RUST, which has been expanded to include data on nonmetallic and metallic coupons. It describes the formats, the computer aspects, and the methods employed to increase user acceptance.

KEY WORDS: coupon corrosion testing, computer, metals, UNS, nonmetals, data bases

Data Base Needs

Corrosion test data derived from the exposure of coupons to a corrosive environment is the most common basis for selecting materials for corrosive environments. Although the industry has recognized this need [1], there are few commercial or public formats that allow the building of corrosion test data bases. RUST is an internal coupon test data base developed by the Du Pont Company to satisfy these needs. RUST has undergone three main revisions, or versions, since its inception in 1983.

Uses for Coupon Corrosion Test Data

Coupon corrosion testing is the most common method of developing information used to select materials for corrosive service. Coupon corrosion testing is also used to monitor material quality for corrosive applications.

Coupon corrosion testing is carried out using small coupons (samples such as discs, rectangles, tubes, or cylinders) of the materials under consideration, which are exposed singly or in groups to the corrosive environment. In some cases the environment may be tailored to simulate specific conditions, such as intermittent exposure. In other cases, the coupons are tailored to simulate equipment history, such as with stressed samples. To facilitate comparison between tests, coupon fabrication procedures are often standardized. When coupon testing is used as a quality control tool, the test conditions are also standardized.

Many standard coupon corrosion tests have been developed to meet specific end-user requirements. By contrast, there is little or no guidance on how to standardize the information obtained from coupon test data for computerization. Recently, a joint venture between NACE and ASTM was initiated to develop a format for computerizing corrosion test data. This format is available from ASTM Subcommittee E49.02 or NACE Technical Committee T-3-2.

[1]Consultants, Engineering Department, E. I. du Pont de Nemours & Company, P.O. Box 6090, Newark, DE 19714-6090.

Some commercial corrosion data software exists [2,3]. The best known of these have been developed from two established source books; however, the pedigree of the data in these source books varies, so this information is used as a first cut in conjunction with other sources of information, such as textbooks and experience.

Several expert systems on corrosion are under development. These systems do not include corrosion data bases that could be used as an ongoing supplement to the rule base of the expert system [4,5].

RUST Development

The RUST coupon corrosion test data base represents an attempt to share corrosion test data across Du Pont Company lines by encouraging participation between diverse groups of field sites and laboratories within the Company. In addition to storing data as it is generated, RUST can also be used to collect existing data useful to Du Pont Company business from any source.

Du Pont carries out an enormous amount of corrosion testing. Test data is used for such diverse purposes as process materials selection, classification of shipped products, marketing services, and product development. The vast majority of this corrosion test data is not commonly accessible. Even if it were, it is often written in highly specific terms that suit the immediate needs of the developer of the information. For example, important pieces of data such as concentrations of corrosives or temperature range may have been omitted from the information. As such, the reuse of the information is difficult, especially in another application.

RUST has experienced three major revisions or versions.

Version 1 was first organized in 1983 as a field coupon data base for several Du Pont plants. This initial program attempted to coordinate information among a variety of petrochemical and specialty chemical plants. It was used to calculate corrosion rates for metal coupons and store data from tests. Version 1 allowed comparisons to be made, for example, between the corrosiveness of processes at different sites that made identical products. Version 1 also allowed historical information to be preserved. This information is highly useful for engineers who must recommend materials of construction.

Version 2 was developed in 1985. It was primarily an attempt to include laboratory data. The format was revised and this was reported in a previous paper [6].

Version 3 is a continuation of the development process. It also reflects the changing needs of the corporation. The data base was expanded to include nonmetallic materials. Polymers represent a significant portion of Du Pont Company business and there are new ventures into advanced composites and advanced ceramics. There is no reliable way of documenting the behavior of these and other nonmetallic materials, such as coatings and linings, in Version 2 of the data base.

RUST Aspects

RUST uses a software package called DATATRIEVE from Digital Equipment Corporation. This permits the program to be accessed in two modes—shared and protected. The protected mode is menu driven and simpler to use. It is the preferred method.

The program consists of three domains in which data is stored and manipulated. These are the Process Environment Domain, the Metal Coupon Domain, and the Nonmetal Coupon Domain. The three domains are linked by the test number.

Software

RUST Version 3 uses a software package called DATATRIEVE, Version 4.0 (Digital Equipment Corporation). This is a more powerful version of the earlier versions of DATATRIEVE in which RUST was written. And with DATATRIEVE, updates and modifications have been relatively easy to make.

DATATRIEVE allows two types of access to RUST, the protected mode and the shared mode. The protected mode is menu driven. The menu driven mode suits the needs of the great majority of end users and requires no knowledge of the programming language to input or access data. The shared mode allows more flexibility by permitting the user to employ DATATRIEVE commands to access information. However, use of the shared mode is currently restricted to those who have sufficient knowledge of the programming language to avoid corrupting the data.

The most commonly performed tasks are tabulated on a menu (Fig. 1). From the menu, the user is directed to various submenus that allow specific tasks to be performed.

Hardware

To use the RUST program, the user must have the necessary equipment and resources to access the VAX computer in which the program is stored:

- a DEC compatible video display terminal,
- a modem equivalent to the Bell 212, 1200 baud modem (if the user is not hard wired to the computer), and
- a printer to print reports and other data.

Process Environment Domain

The Process Environment Domain uses the Process Data Form, which contains all fields relevant to the corrosion test environment (Fig. 2). It is linked to the other domains via the test number. In RUST, specific fields are designated as validated. This means the input must be exactly as required by the program. Lists of the contents of these fields indicating the exact spelling are accessed via Menu Option 9 (Display Lists). The number of validated fields has grown with the data base to facilitate searching for information.

Test Time—In this field the actual test time in days is entered to two decimal places. Since the test time is used in the corrosion rate calculation, it is important to obtain an accurate estimate, especially in field tests where downtime must be accounted for.

RUST Corrosion Data Base

```
( 0) Exit from RUST
( 1) Input Process Data
( 2) Input Metal Coupon Data
( 3) Input Nonmetal Coupon Data
( 4) Delete Process Data
( 5) Delete Metal Coupon Data
( 6) Delete Nonmetal Coupon Data
( 7) Generate Corrosion Reports
( 8) Display Site Test Numbers
( 9) Display Lists
(10) System Maintenance
```

Enter the number of the function desired: ■

FIG. 1—*The Basic Options Menu tabulates the most commonly performed tasks.*

```
                RUST Process Data Entry
 Test No.                   Site
 Test Time                  Product
 Test Type                  Equipment
 Start Date                 Prod'n Area
 Job No.                    Materials
 Department                 Engineer
 Process


 Test Loc.
    First Corrosive                                     X Wt
    Second Corrosive                                    X Wt
    Third Corrosive                                     X Wt
    Fourth Corrosive                                    X Wt
 Minimum Temperature      (Deg C)       Turbulence
 Maximum Temperature      (Deg C)       Aeration
 Average Temperature      (Deg C)       Pressure
    Remarks
_____
 Enter the next test number or leave blank to end data entry ▇▇▇▇
```

FIG. 2—*Information on the corrosive environment is entered in the Process Data Form.*

Test Type—This field identifies the type of test conducted, whether field (F) or laboratory (L). The test type quickly alerts the end user to the broad level of the test. A laboratory test is usually quick and carried out in a small volume of test solution. A field test is usually the opposite.

Start Date—The month, day, and year the test started are entered here. This information is probably of greatest value for field tests. The test start date indicates the approximate period when the exposure was conducted, which may be important. For example, a check of the production records may show that the process was operated at a higher than normal temperature during the test period, possibly explaining anomalous corrosion behavior.

Job Number—This field is used to relate a series of laboratory tests to a common project number. It enables data from these tests to be manipulated and displayed in specific ways.

Department—The name of the operating department that sponsored the test is entered in this field.

Site—This is a validated field in which the name of the test site is entered.

Product—This is a validated field. The name of the product made is entered. Several products are usually made at one site and some products are made at more than one site.

Production Area—This is a validated field. A process is usually manufactured in a series of discrete steps, such as feed preparation, reaction, or separation. The production area indicates which process step the test is being conducted at or is simulating.

Equipment—This field is used to identify the equipment item or test vessel in which the coupons were installed. It is not a validated field because often identical pieces of equipment, such as the air-cooled heat exchanger and the fin-fan cooler, are given different names according to industry preference.

Materials—This field is used to describe the materials of construction of the equipment item in which the test is being conducted or simulated. It could also be used to describe any problems with the equipment, such as abnormally high maintenance, that justify a corrosion testing program.

Engineer—This field is used to indicate the names of the people acquainted with the test and the background of the problem. It is useful when follow-up information is required at a later date.

Process—This field is used to describe the characteristics of the environment or process in the test location (for a field test) or simulation (for a laboratory test). Such information provides useful historical background for comparing tests when conducting data searches.

Test Location—This field is used to describe the location of the coupons in the test environment, such as fully immersed, or at the splash zone, vapor space, or inlet nozzle. This provides important information on how to interpret the corrosion data. For example, splash zones are usually more corrosive than full immersion.

First, Second, Third and Fourth Corrosives—These are validated fields. They are reserved for what are considered the major corrosives in the test environment. Up to four are allowed and it is not necessary to put them in any order of corrosiveness. The weight percentage of each corrosive may be entered to four decimal places, that is, parts per million.

Maximum, Minimum, and Average Temperature—These fields are used to document the temperature range during the test. Temperature is an extremely important parameter in a corrosion test. Hence, the use of three fields to bracket it.

Turbulence—Turbulence (agitation) or the absence of it often plays a key role in corrosion reactions. This field allows documentation of turbulence or the lack thereof, for example, 30 rpm, 3 fps, laminar flow, and so forth. The descriptor "turbulence" was chosen for this field over the commonly used term "agitation" because it was felt that this word is more recognizable.

Aeration—This field is used to indicate the degree of oxygen or air entrainment in the process solution. Aeration, or the absence of it, plays a significant part in corrosion reactions.

Pressure—This field is used to document the operating pressure during the test.

Remarks—This field is used to input any relevant information, such as the reason for the test, that cannot be input elsewhere on the form.

Metal Coupon Domain

The Metal Coupon Domain uses the Metal Coupon Data Entry Form, which contains all the fields required for documenting the measurement and appearance data of metal coupons, to develop a concise statement of the corrosion behavior of the coupon in a test.

Coupon Number—This field records the identification stamped or inscribed on the coupon. A coupon number cannot be duplicated within a particular test; otherwise, corrosion rates cannot be calculated. For example, if a test contains two coupons stamped 4, say Monel and 316 stainless, the user cannot enter the number 4 to identify both coupons. If this is attempted, only the first will be identified with that number, and it will not be possible to enter data on the second coupon. In this case, the coupons could be identified as MON-4 and 316-4, respectively.

UNS Code—This is a validated field. When the UNS code is entered, the computer matches the code to the corresponding alloy name and density. Most commonly used metals have official UNS designations [7]. An unofficial group having the prefix X has been added to the UNS codes, which is for alloys not having UNS codes. When a non-UNS alloy acquires an official code it is shifted to the appropriate UNS section. All alloys having official or unofficial UNS codes are stored in the computer and may be displayed via Menu Option 9 (Display Lists) (Fig. 3). The use of UNS codes ensures there is only one way of entering the description of an alloy. The common name of the alloy is also indicated in the list and is printed out alongside the UNS code in corrosion reports to improve understanding.

Original Condition Code—This validated field uses a series of codes to indicate the original surface condition and thermomechanical history of the coupon. The original condition codes consist of a letter followed by a number and may be displayed via Menu Option 9 (Display Lists) (Fig. 4). Up to four original condition codes are permitted.

Code	Name	Density
A03560	356 ALUMINUM	2.68 g/cc
A04430	443 ALUMINUM	2.69 g/cc
A91060	1060 ALUMINUM	2.70 g/cc
A91100	1100 ALUMINUM	2.71 g/cc
A92024	2024 ALUMINUM	2.77 g/cc
A93003	3003 ALUMINUM	2.73 g/cc
A95005	5005 ALUMINUM	2.70 g/cc
A95052	5052 ALUMINUM	2.68 g/cc
A95083	5083 ALUMINUM	2.66 g/cc
A95086	5086 ALUMINUM	2.65 g/cc
A95154	5154 ALUMINUM	2.66 g/cc
A95454	5454 ALUMINUM	2.68 g/cc
A95456	5456 ALUMINUM	2.66 g/cc
A96061	6061 ALUMINUM	2.70 g/cc
A97075	7075 ALUMINUM	2.80 g/cc
C10100	OFE COPPER	8.89 g/cc
C10200	OF COPPER	8.89 g/cc
C11000	ETP COPPER	8.89 g/cc
C11400	STP COPPER	8.91 g/cc
C12200	DHP COPPER	8.94 g/cc
C12210	PHOS DEOX HI P	8.94 g/cc
C17200	BE COPPER 25	8.23 g/cc
C19400	HSM COPPER	8.78 g/cc
C22000	COMMERCIAL BRONZE	8.80 g/cc
C23000	RED BRASS	8.75 g/cc
C26000	CARTRIDGE BRASS	8.53 g/cc
C26800	YELLOW BRASS	8.47 g/cc
C28000	MUNTZ METAL	8.39 g/cc
C31600	LEADED BRONZE	8.83 g/cc
C36000	FREE CUTTING BRASS	8.49 g/cc
C44300	ADMIRALTY BRASS	8.52 g/cc
C46400	NAVAL BRASS	8.41 g/cc
C48500	LEADED NAVAL BRASS	8.44 g/cc
C51000	PHOSPHOR BRONZE	8.86 g/cc
C61400	ALUMINUM BRONZE D	8.33 g/cc
C63000	NI ALUMINUM BRONZE	7.58 g/cc
C64200	ALUMINUM BRONZE	7.69 g/cc
C65500	HIGH SILICON BRONZE	8.52 g/cc
C67500	MANGANESE BRONZE A	8.63 g/cc
C68700	AS ALUMINUM BRASS	8.33 g/cc
C70610	90-10 CUPRONICKEL	8.94 g/cc
C71000	80-20 CUPRONICKEL	8.94 g/cc
C71500	70-30 CUPRONICKEL	8.94 g/cc
C83600	85-5-5-5 BRONZE	8.83 g/cc
C86300	MANGANESE BRONZE	7.70 g/cc
C90300	TIN BRONZE	8.80 g/cc
C95400	ALUMINUM BRONZE 9L	7.45 g/cc
F12801	CLASS 40 GRAY IRON	7.28 g/cc
F32800	60-40-18 DUCT. IRON	7.10 g/cc
F33100	65-45-12 DUCT. IRON	7.10 g/cc
F41002	TYPE2 AUST.GRAY IRON	7.31 g/cc
G10080	1008 CS	7.85 g/cc
G10100	1010 CS	7.85 g/cc
G10180	1018 CS	7.85 g/cc

FIG. 3—*A partial listing of the UNS/Alloy Name/Density Table.*

Filler Metal—This validated field is used to record the weld filler metal, braze, or solder (if any) on the coupon. Filler metals, brazes, and solders are usually described by their AWS designations, for example, ER309, ERNiCr-3, and so forth. An AWS publication is available that cross references commercial filler metals against their AWS designations [8].

Shape—This validated field consists of a number that directs the computer to the appropriate

Code	Condition
C1	FLAT WASHER
C2	MULTIPLE CREVICE WASHER
C9	SEE REMARKS
F1	80 GRIT
F9	SEE REMARKS
H1	SOLUTION ANNEALED
H2	SENSITIZED
H3	PRECIPITATION HARDENED, SEE REMARKS
H9	SEE REMARKS
M1	METALLIZED
M2	PLATED
M3	WELD OVERLAY
M9	SEE REMARKS
S1	U BEND
S2	C RING
S3	TEARDROP
S9	SEE REMARKS
W1	WELDED
W2	AUTOGENOUS WELDED
W9	SEE REMARKS

FIG. 4—*The Original Condition Codes describe the surface finish and thermomechanical treatment of metal coupons.*

surface area and volume formula to calculate the density and corrosion rate. There are five shape codes, which are: 1-rectangular, 2-disc shaped, 3-tubular, and 4-cylindrical.

Initial Weight—The initial weight of the coupon (to 0.0001 g) is entered in this field.

Initial Measurements—These fields are used to record the initial dimensions of the coupon (to 0.025 mm or 0.001 in.) according to their shape.

Final Weight—This field is used to record the final weight of the coupon (to 0.0001 g) after exposure and cleaning.

Area—This field is used to record the surface area (in square inches) of coupons that cannot be calculated using the basic dimensions recorded in the various measurement fields. Examples include extra large coupons, coupons with more than two holes, or irregularly shaped coupons.

Corrosion Rate—This field may be used in two ways. First, when RUST is used to input existing corrosion data (for which the corrosion rate calculation is bypassed), the corrosion rate is entered in this field in mpy (mils per year). One mil equals 0.025 mm or 0.001 in. Alternatively, when RUST is employed to calculate corrosion rates, data is not entered in this field. It will be automatically stored once the corrosion rate is calculated by RUST.

Crevice Depth—This field is used to enter the maximum depth of crevice corrosion measured on the coupon in mils.

Maximum Pitting—This field is used to record the maximum depth of pitting measured on the coupon in mils.

Attack Types—This is a validated field that uses code numbers for attack types. Attack type codes consist of two numbers displayed by Menu Option 9 (Display Lists) (Fig. 5). Up to five attack type codes are allowed to record the appearance or type of corrosion on each coupon. Attempts to standardize observations of attack types are described in the section entitled Data Quality.

Nonmetal Coupon Domain

The Nonmetal Coupon Domain uses the Nonmetal Coupon Data Entry Form, which is comprised of all the fields required for documenting the behavior of nonmetallic coupons.

Coupon Number—This field records the identification of the coupon.

Code	Attack Type
1	General corrosion
2	Unattacked
3	Superficial corrosion
4	Uniform general corrosion
5	Non-uniform general corrosion
6	Accelerated corrosion at punch marks
7	Partly corroded away
8	Missing -rate probably higher
10	Crevice corrosion
11	Superficial (unmeasurable) corrosion under spacer
12	Superficial (unmeasurable) corrosion under deposits
13	Crevice corrosion under spacer
14	Crevice corrosion under deposits
20	Pitting
21	Superficial (unmeasurable) pitting
22	Single or very few pits (ASTM G46 density A1)
23	A few scattered pits (ASTM G46 density A2)
24	Profuse pitting (ASTM G46 density A3-A4)
25	Heavily pitted (ASTM G46 density A5)
26	Broad pits (ASTM G46 size >B2 or 1/d >1/3)
27	Undercut pitting
28	Pitting of weld
29	Pitting of heat-affected zone of weld
30	Intergranular or interdendritic attack
31	Weld unattacked
32	Attack described in parent metal
33	Attack described in weld
34	Attack described in weld heat-affected zone
35	Knife line attack
36	Slight intergranular attack
37	Moderate intergranular attack
38	Severe intergranular attack
39	Slight end grain attack
40	Moderate end grain attack
41	Severe end grain attack
50	Cracking
51	Stress-corrosion cracking at bend
52	Stress-corrosion cracking at weld
53	Stress-corrosion cracking at punch mark
54	Cracked when coupon bent
60	Other categories
61	Selective phase attack
62	Layer type dealloying
63	Plug type dealloying
64	Exfoliation
65	Erosion/corrosion
67	Unattacked under spacer
68	Unattacked under deposits
70	Corrosion in specific location
71	Attack described at edge
72	Attack described at hole
73	Attack described on one side only
80	Scaling
81	Tight hard scale
82	Loose hard scale
83	Soft scale
84	Coupon plated
90	General Remarks
91	Missing from causes other than corrosion
92	Mechanically damaged
93	Weight gain
94	Incomplete scale removal
95	See remarks

FIG. 5—*The Attack Types List allows for two levels of sophistication in interpreting the appearance of coupons after the test, by the use of the headings for specific groups of attack types.*

Trade Name—This field is used to record the commercial name of the coupon material.

Group—This validated field is used to indicate in which of three main application groups the nonmetal belongs. These groups are C&L (Coatings and Linings), S&G (Seals and Gaskets), and SNM (Structural Nonmetals).

Subgroup—This validated field is used to indicate the materials type within the group. Certain materials types may be present in all three groups, for example, elastomeric materials. In this case, ELAS would be entered in its application group division (Fig. 6).

Change Degree/Type—This validated field is used to enter information on the change in appearance of the coupon as a result of the test. Up to five sets of information are allowed, each consisting of a set of numbers, for example, X/XX. The degree of change is given a qualitative ranking from none (1) to severe (4). The type of change is divided into two groups—visual changes and mechanical or physical property changes. All these are given different numbers (Fig. 7).

Comments—This field documents information that enhances the quality of the test data for the coupon.

Cooperation and Quality

A key aspect of an interactive data base whose growth and quality depends on the cooperation of a large group of contributors is the level of motivation and training of the contributors. Ac-

Group	Subgroup	Description
C&L		
	CEMENT	CEMENTITIOUS
	CERAM	CERAMIC
	ELAS	ELASTOMERIC
	GLASS	GLASS
	METAL	METALLIZED
	MORTAR	MORTAR
	OTHER	OTHER
	THIN	THIN (<20 MILS)
	TPLAS	THERMOPLASTIC
	TSET	THERMOSET
S&G		
	ELAS	ELASTOMERIC
	INORG	INORGANIC (INC. CARBON)
	OTHER	OTHER
	REINF	REINFORCED (ALL)
	TPLAS	THERMOPLASTIC
SNM		
	CARBON	CARBON PRODUCT
	CERAM	CERAMIC
	ELAS	ELASTOMERIC
	GLASS	GLASS
	MORTAR	MORTAR
	OTHER	OTHER
	REFRAC	REFRACTORY (>1000F)
	RELAS	REINFORCED ELASTOMERIC
	RINORG	REINFORCED INORGANIC
	RTPLAS	REINFORCED THERMOPLASTIC
	RTSET	REINFORCED THERMOSET
	TPLAS	THERMOPLASTIC
	TSET	THERMOSET
	WOOD	WOOD

FIG. 6—*Nonmetals are grouped into three main divisions, according to their end use.*

Code	Degree of Change
1	NONE
2	SLIGHT
3	MODERATE
4	SEVERE

Nonmetal Coupon
Types of Change

Code	Type of Change
1	VISUAL CHANGE
2	ABRASION/EROSION
3	BLISTERING
4	CHALKING
5	CHARRING
6	COLLAPSE
7	COLOR CHANGE
8	CRACKING
9	CRAZING
10	CREEP
11	DELAMINATION
12	DISSOLUTION
13	FLAKING
14	ORANGE PEEL
15	PITTING
16	POPCORNING
17	SINTERING/DENSIFICATION
18	SLAG ATTACK/FLUXING
19	SPALLING
20	TACKINESS
21	WARPING
50	PROPERTY CHANGE
51	SHRINKAGE
52	SWELLING
53	WEIGHT LOSS
54	WEIGHT GAIN
55	TENSILE STRENGTH LOSS
56	TENSILE STRENGTH GAIN
57	COMPRESSIVE STRENGTH LOSS
58	COMPRESSIVE STRENGTH GAIN
59	ULTIMATE ELONGATION LOSS
60	ULTIMATE ELONGATION GAIN
61	HARDNESS LOSS
62	HARDNESS GAIN

FIG. 7—*Nonmetals may be described by the degree of change and type of change in their appearance and properties as a result of exposure.*

companying the technical development of RUST are motivational aspects, such as the feeling of ownership in the user, the reliability or quality of the information, and the ease of searching for information.

Data Ownership

Encouraging ownership of the data is extremely important. RUST must satisfy the needs of a diverse group of users who must be encouraged to treat the data base as their own. The data base is personalized through the test number and training programs to enable the user to earn a password to use the system.

Test Number—When setting up a data base dependent on voluntary contributions, it is important that each contributor be made to feel part owner of the system. One important method of achieving this in RUST is via the test number. The test number consists of up to four characters and six numbers. The test number links the three domains.

The first four characters of the test number are used to identify the site responsible for the test, for example, VCT (Victoria) or WASH (Washington Works) or XSTN (Experimental Station). The last six numbers identify the test uniquely, so that each site could enter 999 999 tests before having to look for a new identifier.

"Personalizing" the test number with an identifying prefix makes the contributor more willing to share data, because it does not lose its identity within the data base.

Menu Option 8 (Display Site Test Numbers) allows the user to locate all the test numbers for a particular identifying prefix or obtain the highest number currently used. The user is thus able to select the next highest number for a new test. The source of all test number identifiers (for example, VCT, WASH, or XSTN) is stored in the computer and may be accessed through Menu Option 9 (Display Lists).

Password—A password is required to use RUST. The password is only given to qualified users, whether they are using RUST for searching or for data entry. Before being given the password, users must demonstrate their satisfactory credentials via training (in the case of data entry) or an interview (in the case of data searching).

Data Quality

As data is stored for continual reuse, quality becomes an important issue. The human aspects of gathering data from a large number of sources and levels of aptitude require the setting of a quality level. This is also necessary because a computer demands a degree of discipline that was not required in our previous methods of storing data (books and reports).

We recognized that to maintain the reliability of data entry between a diverse group of users, a comprehensive guidebook would be required to support the data base. This guidebook will include the program manual as well as technical sections on eliminating errors in weighing, measuring, and cleaning coupons, and photo albums of attack types for metallic and nonmetallic coupons.

The data base is supported by a core user group representing various interests, whose members can be contacted if there are any problems. The user group is responsible for maintaining the quality of input and for auditing and training other users.

Errors in Weighing and Measuring Coupons—When metal coupons are initially weighed and measured, the computer calculates their density values. These are compared with their standard density values stored in the UNS/Alloy Name/Density Table. The result is flagged if the calculated density differs from the theoretical density by more than ±5%. This alerts the technician to possible misweighing or mismeasurement.

Other Sources of Error—The calculated corrosion rate is subject to three main sources of error [9]. The most important source of error is due to the exposure time estimate in field coupon tests. The second most important source is the variation in coupon weight arising from the coupon cleaning operation. The third is the variation between technicians making the measurements. For all these reasons, the corrosion rate in the printout is reported to no more than two significant figures.

Validated Fields—To facilitate searching, key fields in each of the three domains are validated and have standardized entries. The correct spellings or designations for validated fields are given in specific lists displayed on Menu Option 9 (Display Lists).

Delete Function—RUST contains a delete function to allow the removal of corrosion test information that is not considered worthy of permanent storage. An example might be where RUST was used to calculate corrosion rates and print a report for a routine chemical cleaning operation.

Training—Users of RUST are encouraged to attend a training session to improve their understanding of coupon corrosion testing and the interpretation of the appearance of coupons after testing. For example, the attack types list (Fig. 5) is arranged in groups. The title of each group, for example, pitting, crevice corrosion, and so forth, constitute attack types, and may be used to describe the appearance of the coupon (for example, pitting) rather than a more subtle description such as profuse pitting (for example, in ASTM Recommended Practice for Examination and Evaluation of Pitting Corrosion G 46, density A3-A4), which would require a more sophisticated level of interpretation.

On-Line Help—On-line help is available at all stages of entry.

Data Searching—In the shared mode, data searching is done using menu Item 7 (Generate Corrosion Reports). This leads the user to a submenu that shows four search modes: Print a Single Test (where the test number is known), Retrieve Data by Equipment, Retrieve Data by Corrosive, and Tabulate Data for a Job Number (Fig. 8).

When searching for data by equipment or by corrosive, the user is led to a form that allows various "windows" to be set on the data search fields. For example, a user may set concentration ranges for a particular combination of corrosives. If no ranges are set, all the data on those corrosives will be obtained. Similarly, if the same product is made at more than one site, all corrosion data for a specific product may be obtained by not specifying a site name (Fig. 9).

To sort data, the user may request a full or brief report. A brief report contains a summary of the information requested, such as test site, product, and equipment, for the information obtained. This allows the user to fine tune the data search parameters before printing out all relevant data.

The output may be (1) printed, (2) displayed on the terminal screen, or (3) stored on a file for later printing, inclusion in a report, or for transmittal to another party.

Conclusion

RUST has demonstrated that corrosion test data for metals and alloys can be computerized in a way that retains flexibility. Rules for computerizing data on the degradation of nonmetals are less formalized and the challenge of making RUST work for nonmetals remains.

RUST Corrosion Reports

```
( 0) Return to the Main Menu
( 1) Calculate Corrosion Rates
( 2) Print a Single Test
( 3) Retrieve Data by Equipment
( 4) Retrieve Data by Corrosive
```

Enter the number of the function desired: ■

FIG. 8—*Corrosion data may be retrieved via a submenu.*

```
                                                              Part I
     RUST Corrosion Report Selection Menu
                         Site
                       Product
                      Equipment
                     Prod'n Area
                       UNS Code
                     Filler Metal


                 Full or Brief Report ▮
                 Print on Attached Printer
                 Output Location for Report
```

> Complete both sections of this form. In the first section, enter a value
> for each field that you want to use to identify which tests are to be
> included. If you want all values for a field, then leave the field blank.
> Use the second section to identify the type of report you want and where
> you want the report printed. For more help on this section, hit the HELP
> key or PF2 key twice.

```
                                                             PART II
     RUST Corrosion Report Selection Menu
                  Corrosive              Low        High
       First
       Second
       Third
       Fourth
                    Average Temperature Range

                         Site
                       Product
                       UNS Code
                     Filler Metal


                 Full or Brief Report ▮
                 Print on Attached Printer
                 Output Location for Report
```

> To get help on filling out this form press the HELP key or the PF2 key twice

FIG. 9—*Corrosion data may be searched by equipment and corrosive.*

Acknowledgments

The authors acknowledge the following Du Pont and Conoco personnel who assisted during the development of RUST Version 3: J. L. Cooney, W. F. Gentile, T. Pugh, F. O. Rinebold, and R. E. Tatnall. Valuable discussions with the Du Pont Materials Engineering consultants are also acknowledged.

References

[1] Verink, E. D., Kolts, J., Rumble, J., and Ugiansky, G. M., *Materials Performance 26*, (4), pp. 55-60.
[2] "CorSur 1" (Metals), Corrosion Data Survey Software for Metals, NACE, Houston, TX.

[3] "CorSur 22" (Nonmetals), Corrosion Data Survey Software for Nonmetals, NACE.
[4] Wescott, W., Williams, D. E., Croall, I. F., Patel, S., and Bernie, J. A., "The Development and Application of Integrated Expert Systems and Data Bases for Corrosion Consultancy," *Computers in Corrosion Control*, NACE, Houston, TX, 1986.
[5] "Expert Systems," MTI Project No. 53, Materials Technology Institute of the Chemical Process Industry.
[6] "RUST, A Company-Wide Computerized Corrosion Test Data Base," *Computers in Corrosion Control*, NACE, Houston, TX, 1986.
[7] Metals and Alloys in the Unified Numbering System—Fourth Edition, ASTM Publication DS-56C, ASTM, Philadelphia, 1986.
[8] "Filler Metal Comparison Charts," Publication AWS-FMC-86, AWS, 1986.
[9] Hess, J. D., Internal Report, Du Pont Company, Wilmington, DE, 1986.

Kenneth Ranger[1]

Generation and Use of Composite Material Data Bases

REFERENCE: Ranger, K., **"Generation and Use of Composite Material Data Bases,"** *Computerization and Networking of Materials Data Bases, ASTM STP 1017,* J. S. Glazman and J. R. Rumble, Jr., Eds., American Society for Testing and Materials, Philadelphia, 1989, pp. 253-264.

ABSTRACT: The use of composite material data bases in the structural design process is important due to the great number of material systems that are currently available. In preliminary and final design, it is efficient to have computer programs that use data bases containing fiber, matrix, ply, and laminate information. Definitions and methods of generation (estimation techniques and test methods) for these data bases are included. The AS1/3501-5 graphite/epoxy material system is used in the examples. This paper is intended as a guide for implementation of a composite material data base system for a design group.

KEY WORDS: composite material data bases, computers, structural design

Because designing and manufacturing with composites can be more complicated than it is with "isotropic" materials, there have been attempts to make the task easier by using computer automation. This paper presents an automation technique that employs the use of computer material data bases.

Three types of data bases will be presented: constituent, laminae, and laminate. Content, method of generation, and usage will be described for each. HERCULES AS1/3501-5 graphite epoxy tape [1,2] will be used in the examples. This paper shows what type of data should be in the data bases and how they should be used; it is not a presentation of data. The data where included are typical.

Composite material data bases organize the materials, design, stress analysis, and quality assurance groups on composite allowable related issues. They also increase the efficiency of the design cycle, especially if many different composite material systems are used, if the material properties are likely to change, or if the material systems will change during the design cycles.

Constituent Data Base

Composites are made of fibers "glued" in place by a matrix. The fibers and matrix are the constituent materials of a composite material. A constituent data base contains the material properties for the fibers and matrices. Table 1 shows typical contents of a data base in which AS4 graphite fiber and 3501-6 epoxy matrix properties are the entries [1]. The AS4 fiber and the 3501-6 matrix properties are assumed similar to the AS1 fiber and 3501-5 matrix. A full-sized data base would be divided into fiber and matrix sections and contain many fibers (graphite, glass, KEVLAR®, quartz) and matrices (epoxy, bismaleimide, polyimide, phenolic) with various properties at different temperatures. The properties of a particular fiber or matrix can be accessed by using a computer program by specifying its material number.

[1]Engineer, The MacNeal-Schwendler Corp., 815 Colorado Blvd., Los Angeles, CA 90041.

TABLE 1—*An example of two entries in a constituent properties data base.*

	FIBERS		
TITLE	AS4 GRAPHITE		
TEMPERATURE (°C)	24.0		
E1 (nt/m^2)	2.20^{11}		
E2 (nt/m^2)	1.4^{10}		
G (nt/m^2)	9.14^{10}		*
U12	0.2		
STRN1	0.013		*
FT1 (nt/m^2)	2.90^{9}		*
ALPHA1 (1/C)	-0.4^{-6}		
ALPHA2 (1/C)	18.0^{-6}		
DIA(m)	7.1^{-6}		*
DENSITY (kg/m^3)	1.8E3		*

	MATRICES		
TITLE	3501-6 EPOXY		
TEMPERATURE (°C)	24.0		
MOISTURE (%wt.)	1.0		
E (nt/m^2)	4.3^{9}		
G (nt/m^2)	...		
U	0.34		
STRNT YLD	...	STRNT ULT	...
STRNC YLD	...	STRNC ULT	...
STRNS YLD	...	STRNS ULT	...
FT YLD (nt/m^2)	...	FT ULT (nt/m^2)	...
FC YLD (nt/m^2)	...	FC ULT (nt/m^2)	...
FS YLD (nt/m^2)	...	FS ULT (nt/m^2)	...
ALPHA (1/C)	40.0^{-6}		
BETA (1/M)	2.0^{-3}		
DENSITY (kg/m^3)	1.3^{3}		*

NOTE: Data is from Ref *1* except where marked. Values marked with an * are typical.
TEMPERATURE—Temperature at which property measurements were made (C);
        E1—Tangent Young's Modulus in the fiber direction (nt/m^2);
        E2—Tangent Young's Modulus in the transverse direction (nt/m^2);
        G—Tangent Shear modulus (nt/m^2);
     U12—Poissons's ratio;
   STRN1—Strain to failure;
     FT1—Failure stress (nt/m^2);
 ALPHA1—Thermal coefficient of expansion in the fiber direction (1/C);
 ALPHA2—Thermal coefficient of expansion in the transverse direction (1/C);
     DIA—Diameter of fiber (m);
DENSITY—Density (kg/m^3);
MOISTURE—Level of moisture when properties were measured (%wt);
STRNT YLD—Tensile yield strain. Defined as either the departure from linearity or offset exceedence;
STRNC YLD—Compressive yield strain. Same definition as above;
STRNS YLD—Shear yield strain. Same definition as above;
STRNT ULT—Tensile strain to failure;
STRNC ULT—Compressive strain to failure;
STRNS ULT—Shear strain to failure;
  FT YLD—Tensile yield stress. Defined as either the departure from linearity or offset exceedence (nt/m^2);
  FC YLD—Compressive yield stress. Same definition as above (nt/m^2);
  FS YLD—Shear yield stress. Same definition as above (nt/m^2);
  FT ULT—Tensile ultimate stress. Defined as either the departure from linearity or offset exceedence (nt/m^2);
  FC ULT—Compressive ultimate stress. Same definition as above (nt/m^2);
  FS ULT—Shear ultimate stress. Same definition as above (nt/m^2);
    BETA—Moisture coefficient of expansion (1/M);
       M—Percent moisture absorbed (%wt).

Fiber longitudinal Young's modulus (E1), strain-to-failure (STRN1), failure stress (FT1), longitudinal thermal coefficient of expansion (ALPHA1), diameter, and density can usually be obtained from the manufacturer. The transverse Young's modulus (E2) and transverse thermal coefficient of expansion (ALPHA2) may have to be backed out of laminae data using micromechanics theory [3], as they are difficult to obtain or measure experimentally.

Matrix properties (neat resin) are sometimes harder to obtain from manufacturers than fiber properties are, but material tests can be run. The matrix values in the sample data base are mostly blank because of the lack of public information, but these properties are needed because they enhance our understanding of the complex failure mechanisms that occur in composite materials.

Since the matrix Young's modulus and shear modulus (E and G) depend on the stress level, a stress strain curve is necessary to describe them. If a bilinear curve can approximate the nonlinear behavior and the analysis programs can use it, then the bilinear points can be included in the data base. Otherwise, they can be a linear value that represent (conservatively) the strain the design will experience. In the sample data base, the Young's modulus value is the initial straight-line portion of the stress/strain curve. However, the stresses and strains in tension (FT, STRNT), compression (FC, STRNC), and shear (FS, STRNS) are more accurately listed as yield and ultimate values. If these yield and ultimate stress/strain points are plotted, the result is a bilinear matrix stress/strain curve. It is called bilinear because it is formed by two, straight-line portions. The definition of the yield point can either be the point of departure from linearity or the point where an offset amount is exceeded.

Resin expands significantly when it absorbs moisture, so the moisture coefficient of expansion (BETA) is included [1].

Fiber and matrix properties can be used by a composite micromechanics program to predict laminae properties. This is done if little or no laminae or laminate material property test data are available. This can occur during preliminary material selection, or if material tests will occur later, or if test data are not available for certain properties. Table 2 shows an example of tape laminae properties calculated from the fiber and matrix properties shown in Table 1.

Micromechanics theory [3] can be used to calculate only tape laminae moduli, the thermal coefficient of expansion, and density. But if another similar system's properties are known, an estimate of the new systems longitudinal stress and strain allowables can be made by comparing the fiber properties of the two systems.

For instance, suppose a search is being made for a higher-strength system. Some new fibers on the market are found to have higher strains to failure and higher Young's moduli than the current system's fibers. We can infer that the new system's tension strength will be higher by some fraction of the fiber's increased strength. The laminae's matrix-dependent properties

TABLE 2—*Example of tape laminae properties generated by micromechanics theory (Ref 3).*

E1 (nt/m^2)	1.33^{11}
E2 (nt/m^2)	8.2^{9}
G (nt/m^2)	5.9^{9}
U12	0.26
ALPHA1 (1/C)	0.0
ALPHA2 (1/C)	23.4^{-6}
BETA1 (1/M)	$.02^{-3}$
BETA2 (1/M)	1.07^{-3}

NOTE: (From Table 1 fiber and matrix properties). See Table 1 for a definition of property abbreviations.

(G and STRNS) can be similarly inferred by comparing the matrix properties at various temperatures. One unknown in this process is the degree to which the fibers and the matrix match each other (fiber-matrix interface strength). This cannot be known until laminates are tested, especially under hot-wet conditions.

While estimates of tension laminae properties can be made relatively accurately, compression strength and strain-to-failure are more difficult to estimate because the initial failure mechanism is not fiber breakage but fiber microbuckling. Buckling is a function of the diameter of the fiber and the amount of lateral support the matrix gives it. Thus, if, in the above example, the increase in modulus and strain of a new laminae are brought about by the use of a smaller diameter fiber, there is a chance that while the tensile strength increases, the compressive strength may suffer. Also, while a matrix with a slightly lower Young's modulus may have increased tensile strength at room temperature under dry conditions, it may experience decreased compression strength at a high temperature under wet conditions. This is because the loss of lateral support to the fibers causes a lower fiber buckling strength. If an analytic model for microbuckling is available, then the effect of the fiber stiffness, fiber diameter, and matrix stiffness on microbuckling can be quantified.

Estimates of cloth laminae properties can be made by analytically cross plying the tape estimates and decreasing the value by a tape-cloth comparison from a similar system. Cloth laminae typically have lower moduli and strength than equivalent cross plied tapes due to the warp of the weaving process.

Two analysis programs that use the constituent properties are

1. Adhesive joint analysis programs that need matrix properties and
2. Laminate physical property programs that need accurate fiber and matrix densities for a prediction of ply thicknesses.

Figure 1 shows the uses for the constituent property data bases.

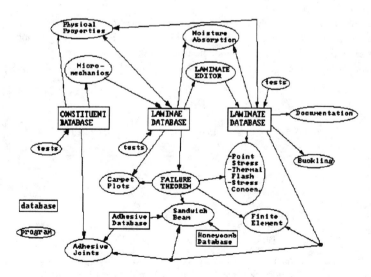

FIG. 1—*Laminate/program flowchart.*

Laminae Data Base

Composite laminates are made of layers of matrix-impregnated fibers called laminae or plies. In a laminae, all the fibers lie in the same direction in a tape or are crossplied in a cloth. Laminae properties are the building blocks for calculating laminate properties. If an engineering company is using several composite materials in several design environments (temperature and moisture levels), storing the laminae properties in a computer data base is one way to organize them, keep them current, and make them available to the users. Table 3 is an example of an entry in a laminae data base for HERCULES AS1/3501-5 graphite epoxy tape [2]. This older system is given as an example because the properties of newer, more high-performance systems are not public. But it does show the properties necessary for a complete data base. An actual data base would contain many different material systems, each of which could be accessed by its data-base material number by analysis programs.

TABLE 3—*An example of one entry in a laminae data base.*

MATERIAL SYSTEM	AS\3501-5		
TYPE	GRAPHITE/EPOXY TAPE		
ENVIRONMENT	ROOM TEMPERATURE DRY OR WET		
E1 TENSION (nt/m^2)	1.39^{11}		
E2 TENSION (nt/m^2)	9.28^9		
E1 COMPRESSION (nt/m^2)	1.19^{11}		
E2 COMPRESSION (nt/m^2)	9.62^9		
G12 (nt/m^2)	0.5^9		
U12	0.2		
DENSITY (kg/m^3)	1.61^3		
ALPHA1 (1/C)	0.0		
ALPHA2 (1/C)	28.8^{-6}		
FT1 YLD (nt/m^2)	1.54^9	FT1 ULT (nt/m^2)	1.54^9
FT2 YLD (nt/m^2)	4.0^7	FT2 ULT (nt/m^2)	4.0^7
FC1 YLD (nt/m^2)	-9.07^8	FC1 ULT (nt/m^2)	-9.07^8
FC2 YLD (nt/m^2)	-1.37^8	FC2 ULT (nt/m^2)	-1.95^8
FS12 YLD ±(nt/m^2)	4.1^7	FS12 ULT ±(nt/m^2)	8.24^7
STRNT1 YLD	0.01126	STRNT1 ULT	0.01126
STRNT2 YLD	0.0047	STRNT2 ULT	0.0047
STRNC1 YLD	-0.00809	STRCN1 ULT	-0.00809
STRNC2 YLD	-0.01309	STRCN2 ULT	-0.02858
STRNS12 YLD	0.00976	STRSN12 ULT	0.04760
FAW (kg/m^2)	0.15		*
RC (%)	31		*
VV (%)	1		*
t (in)	1.5^{-3}		
MOISTURE ABSORPTION PARAMETERS (Ref 5)			
a (M/RH)	1.5		*
b	1.0		*
D1 (m/s-C)	0.04		*
D2 (K)	-5560		*

See Table 1 for property abbreviation definitions except for the following: RH = relative humidity, K = degrees rankine, FAW = fiber areal weight, RC = percent resin content by weight, VV = percent void volume, and t = laminae thickness.

Values are from reference 2 except where marked.

* = Typical values.

The laminae coordinate system is defined as: direction 1—the fiber direction for tape, the fill direction for cloth, direction 2—transverse to direction 1 in the plane of the ply, direction 12—in the plane of the ply.

All properties are in the laminae 1-2 coordinate system. The elastic, physical, and thermal properties are all average values. These include the Young's modulus and shear modulus (E1, E2, G12), Poisson's ratio (U12), density (RHO), fiber areal weight (FAW), resin content by weight (RC), void volume (VV), thickness, moisture absorption constants (a, b, D1, D2), and thermal coefficients of expansion (ALPHA1, ALPHA2). Although tests often show that tension and compression Young's moduli values are different, they are sometimes averaged if a single value is needed, such as for finite element loads analysis. The strength and strain allowables (F1, F2, F12, STRN1, STRN2, STRN12) are statistically reduced to account for the desired amount of scatter. In this case, they are reduced to a Weibull b-allowable. In cases in which the stress/strain curves are linear to failure, the yield (YLD) and the ultimate (ULT) points are the same value (F1 tension, STRN1 tension, F2 tension, STRN2 tension, F1 compression, STRN1 compression). Different yield and ultimate values indicate nonlinear behavior. The definition of the yield point can either be the point of departure from linearity, the point where an offset amount is exceeded, or the mid-point of the bilinear curve.

Composite material allowables are the properties used by stress analysis groups in margin of safety calculations. Laminae material allowables are preferably obtained by allowable tests, but in the early stages of design, it may be necessary to estimate them from other data sources, such as qualification test data, vendor information, or public literature. Also, they can be estimated from fiber and matrix properties using micromechanical theory [3]. This can be accomplished by a micromechanics computer program accessing the constituent data base.

Laminae allowables testing uses laminates with 100% of the plies in the 0-degree direction for longitudinal and transverse tension and compression tests, and laminates with 100% of the plies in the ±45-degree direction for shear tests. The test data is adjusted to the design physical properties, fiber areal weight (FAW), void volume (VV), resin content (RC), and thickness (t) (Table 3). Then it is statistically reduced and used to update the laminae data base.

Generally, FAW, VV, and RC are supplied by the material system manufacturer and updated by the materials and processes group and manufacturing and quality assurance group based on their experience with the material. Thickness (t) is calculated from these values and from fiber and matrix densities. Figure 2 is an example of a thickness versus RC, FAW, and VV plot drawn by a physical properties computer program. These physical properties are important because they are used by the materials and processes group when procuring the material, by stress analysis when allowables are generated, by design when drawings are made, by manufacturing when tools are made, and by quality assurance when parts are checked. If everyone gets values from the same source (the laminae data base), much confusion is eliminated.

Laminae data bases are primarily used to create laminates. A laminate editor computer program (discussed later) accesses the laminae data base to get the properties necessary for the analysis of laminates.

Carpet-plot programs also access laminae data bases. They are plots of materials properties or allowables versus either lay-up or loading condition. Computer programs can automatically generate a range of unimaterial laminate models or loads to produce plots for designers and stress analysts. Figures 3 and 4 are examples of elastic property carpet plots that use E1, E2, G12, and U12 from the laminae data base to produce laminate EX, EY, GXY, and UXY by varying the percentages of 0-, 45-, −45-, and 90-degree plies [2]. Thermal coefficient of expansion carpet-plots that use ALPHA1 and ALPHA2 laminae properties to produce laminate ALPHAX and ALPHAY properties can be similarly drawn. Carpet-plots usually assume that the laminates are symmetric and balanced.

If computer programs with laminae failure theorems are used for analysis, then laminate stress or strain to failure carpet plots can be generated using the laminae stress or strain to failure allowables. Figures 5 to 7 show examples of allowable stress and strain-to-failure carpet-plots [2]. The failure criteria used in these examples is ply maximum strain. Figure 5 gives uniaxial strain allowables with a 0.635-cm (¼-in.) hole for a range of laminates, Figure 6,

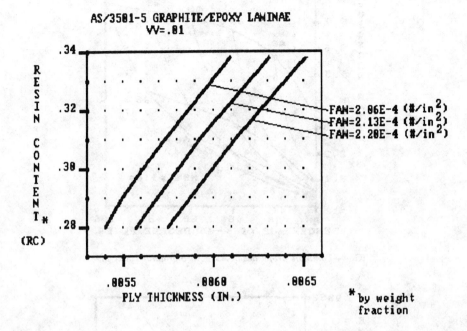

FIG. 2—*Physical properties of AS/3501-5 graphite/epoxy laminae, $W = 0.01$.*

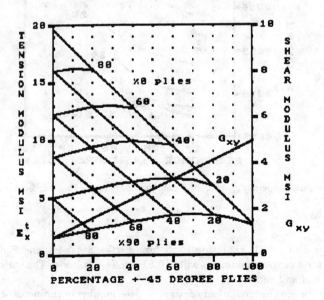

FIG. 3—*Laminate properties of AS/3501-5 graphite/epoxy tape.*

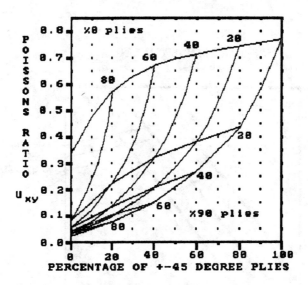

FIG. 4—*Laminate properties of AS/3501-5 graphite epoxy tape.*

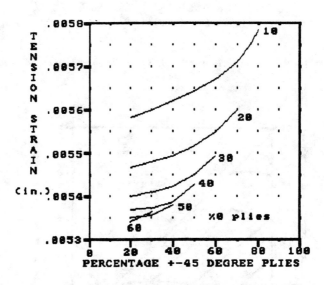

FIG. 5—*Laminate properties of AS/3501-5 graphite/epoxy tape with a 0.635-cm (1/4-inch) hole.*

biaxial loadings with a 0.635-cm (1/4-in.) hole for a particular lay-up, and Figure 7, the axial bearing loads for a range of laminates with a 0.635-cm (1/4-in.) bolt. The varieties of possible carpet-plots are almost endless.

Carpet-plots are good for preliminary design, or if the number of laminates or loading conditions in a design is few. They also might be adequate if a simple conservative failure criteria is used (such as maximum laminate strain), or if a single, well-tested material is used. But if a complex ply failure criteria is used or the number of unique laminates is large, then a computer-

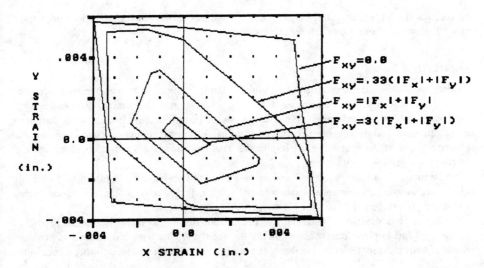

FIG. 6—*Laminate properties of AS/3501-5 graphite/epoxy tape RTW (25/50/25) under biaxial loading with a 0.635-cm (1/4-inch) hole.*

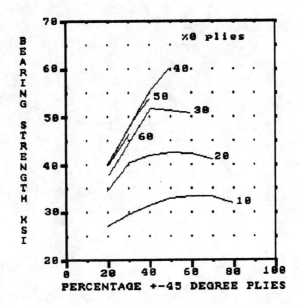

FIG. 7—*Laminate properties of AS/3501-5 graphite/epoxy tape with a 0.635-cm (1/4-inch) bolt.*

ized failure theorem program may be needed in addition to carpet-plots. In either case, the laminae data base would either be used to create the carpet-plots or would be accessed by the failure theorem program.

Any analysis program that uses ply failure theorems will access the laminae data base for ply properties. Two such programs are point stress analysis programs that use ply failure theorems and thermal flash programs that fail plies one by one as they exceed their temperature limits. Figure 1 shows a schematic of laminae data-base usage.

Laminate Data Base

A composite laminate is produced by "gluing" laminae together at high temperatures and pressure. It is described by the number, type, and orientation of the laminae that make it up. The descriptions of all the laminates in a design can be stored together in a laminate data base on a computer. So all the composite configurations can be viewed at once or a particular one can be analyzed quickly by specifying the laminate's data-base number. Figure 8 is an example of one entry in a laminate data base.

The laminate data base is created and updated by the designers during design cycles, and used by stress analysts during analysis and by manufacturing and quality assurance during production. It can be used later for product tracking and repair.

The laminate is described by listing the plies, from the top to the bottom, giving angles, thicknesses, and laminae data-base numbers for each ply. The data-base numbers are used instead of listing the material properties of the laminae. They are pointers defining where in the

```
TITLE          8 PLY (25/50/25) AS/3501-5 GRAPHITE/EPOXY TAPE RTW
TITLE          UPPER WING SKIN
# OF LAYERS             8
                                    DATABASE
                        LAYER #     MATERIAL #      ANGLE           THICKNESS
                                                    (degrees)       (m)

                            1           5            0.0            1.56E-4
                            2           5           45.0            1.56E-4
                            3           5          -45.0            1.56E-4
                            4           5           90.0            1.56E-4
                            5           5           90.0            1.56E-4
                            6           5          -45.0            1.56E-4
                            7           5           45.0            1.56E-4
                            8           5            0.0            1.56E-4

      EX (nt/m**2)  = 5.4E10              EY (nt/m**2)  = 5.4E10
      G  (nt/m**2)  = 2.1E10              U12(nt/m**2)  = .31
      ALPHAX (1/C)  = 2.10E-6             ALPHAY (1/C)  = 2.10E-6
      THICKNESS(m)  = 1.20E-3

A MATRIX (nt/m)           7.067E7            2.119E7            4.952E-2
                                             7.067E7            2.904E0
                                                                2.457E7

B MATRIX (nt/rad)         0.0                0.0                0.0
                                             0.0                0.0
                                                                0.0

D MATRIX (nt-m/rad)       1.418E1            2.113E0            8.809E-1
                                             3.604E0            8.809E-1
                                                                2.538E0

COEFFICIENTS OF
THERMAL FORCE (nt/C)      FX = 2.72E0       FY = 2.72E0         FZ = 0.0

COEFFICIENTS OF
THERMAL MOMENTS (nt-m/C)  MX = 0.0          MY = 0.0            MZ = 0.0
```

FIG. 8—*An example of one entry in a portion of a laminate data base.*

laminae data base the properties are located. In addition to fully describing the laminate, the data base contains lamination theory results so computer programs without lamination theory code can still use the data base.

Two types of plate information are included: a shorthand version and the full version. The shorthand version includes laminate moduli (EX, EY, G, UXY), coefficients of thermal expansion (ALPHAX and ALPHAY), and the nominal thickness of the laminate. These are balanced and symmetric values that do not account for membrane-bending coupling or axial-shear coupling, but they give an instant feel for what the laminate is. The full version is comprised of the point stiffness matrices that fully describe the laminate and are suitable for in-depth analysis. They include the A, B, and D matrices [4], which describe how stiff the laminate is for every plate direction, and coefficients of thermal force and moment, which tell how the laminate will move when heated or cooled. With very little change, this data can be incorporated into finite element codes to describe plate material properties.

A good method for generating or updating laminate models in the laminate data base is using an interactive laminate editor computer program. An editor allows laminates to be described ply by ply or by using editing features such copying, moving, and mirroring ply sections. It can display a list of current materials in the laminae data base so the user can pick the one wanted by number, removing the need for hand typing all the materials properties into the computer. After the laminate is described, the editor calculates the plate properties using lamination theory and saves everything in the data base.

Analysis programs can then access any available laminate by its data-base number for a quick, accurate analysis. Some programs that would use a laminate data base are listed below

- finite element codes,
- stress concentration programs,
- buckling programs,
- moisture absorption programs,
- sandwich beam programs (these also need honeycomb and adhesive data bases),
- bonded and bolted joint programs (these need an adhesive data base),
- quality assurance programs that access the effect of laminates, analyze non-specification, and
- documentation programs.

Figure 1 shows a schematic of laminate data-base usage.

Analysis/Data Base System

Data bases and analysis programs can be tied further together by a system level program in which everything from data bases through analysis computer programs is available through menus. Upon start-up, the following options are displayed to the user:

(1) laminate editor,
(2) analysis programs, and
(3) laminae editor.

The laminate editor would have suboptions for making a new laminates data base, accessing an existing laminates data base, or editing the laminate data base. The analysis programs option allows the user access to any program that uses any of the data bases. The laminae editor would have suboptions for laminae micromechanics, accessing existing laminae, or editing the laminae data base. Any new methods, options or analytic programs would be added as they become available.

Conclusion

Composite material data bases are valuable tools for insuring that all disciplines in a composite design organization have access to and use the same material properties. They also increase the efficiency of the structural analysts through the increased accessibility of laminate descriptions and material properties. They can help keep the design process in order in the volatile composite material industry, where new materials are appearing so fast that material tests struggle to keep up. Often material-allowable testing must be done concurrently with designing. Because of this the chances are good that before a design is completed, the current material will not live up to expectations or will be obsolete and the final allowables are invariably different from the preliminary values used. The way to handle this when data bases are used is to release updated data bases at every design cycle. The allowables group updates the laminae data base, the design group updates the laminate data base, and stress analysis uses them with their updated loads.

As an example, suppose an aircraft is being designed using the latest graphite/epoxy material system. It works well, but late into the design, tests show the material has little impact resistance. A quick industry search finds new resins that greatly improve impact resistance through a rubber additive. The new material has had little testing, but promises to be a direct substitute requiring no design or drawing changes. Preliminary quality assurance compression, short beam shear, and flex beam tests will be positive, so the material substitution is approved. The allowables are assumed to be the same as those in the old system until the allowables tests results are obtained in about 1 year. As the testing progresses, the initial results confirm that the material is generally good. But a pattern slowly emerges indicating that some of the odd matrix-dependent properties that are more difficult to test for—cold notched tension, hot/wet interlaminar tension, and hot/wet compression—are not as good as they were in the old system because the rubber additive causes the resin to be soft under hot/wet conditions, or creates a poor matrix/fiber interface bond. As this is discovered, the affected allowables are lowered, the laminae data base is updated, and the stress group is notified. The design is checked using the updated data base and design is notified of any negative margins. Design changes are made, the laminate data base is updated, and stress analysis performs final analysis.

References

[1] Adams, Donald F., and Monib, Mohamed M., "Moisture Expansion and Thermal Expansion Coefficients of a Polymer-Matrix Composite Material," from *Fibrous-Composites in Structural Design,* Edward Lenoe, Ed., 1978, p. 819.
[2] Grimes, G. C., Ranger, K. W., Brunner, M., "Element and Subcomponent Testing," Section 5, *ASM Engineered Materials Handbook,* Volume 1, Composites, T. R. Reinhart, Technical Chairman, Nov. 1987, p. 313.
[3] Whitney, James M., *Experimental Mechanics of Fiber-Reinforced Composite Materials,* Society for Experimental Stress Analysis, Brookfield Center, CT, SESA Monograph No. 4., 1982, Chapter 2, p. 10.
[4] Halpin, J. C., *Primer on Composite Materials: Analysis,* Technomic Publishing Company, Lancaster, PA, 1984, p. 46.
[5] Shen, C. H. and Springer, G. S., "Moisture Absorption and Desorption of Composite Materials," *Journal of Composite Materials,* Vol. 10, 1976.

Joseph K. Lees,[1] Beverly K. Roberts,[1] and Robert J. Michaud[1]

Designation and Characterization of Composite Materials

REFERENCE: Lees, J. K., Roberts, B. K., and Michaud, R. J., "**Designation and Characterization of Composite Materials**," *Computerization and Networking of Materials Data Bases, ASTM STP 1017*, J. S. Glazman and J. R. Rumble, Jr., Eds., American Society for Testing and Materials, Philadelphia, 1989, pp. 265–271.

ABSTRACT: Designing and using a computerized materials property data base for composites include the very difficult tasks of fully and uniquely identifying each material and characterizing the material properties in a way that will satisfy the needs of all potential users. Approaches to meeting this criteria are discussed, including abbreviated methods of identification, combining computer fields to fully and uniquely identify both research and commercial composites, structuring data so that different functional groups can access varied subsets, integrating test conditions into the data base, and presenting data in more effective graphic form, when applicable. A multi-tiered system for assessing data quality is also discussed.

KEY WORDS: composite materials, lamina, laminate, isotropic, lay-up, anisotropic, prepreg, preconditioning

Designing and using any computerized materials property data base involves solving two key questions:

1. how can the materials be designated so that they are fully and uniquely identified; and
2. how can the properties of the materials be characterized so that the needs of all potential users are satisfied?

Resolution of these issues is not necessarily straightforward for isotropic materials such as most metals and polymers. For anisotropic materials such as composites, the questions become much more complex. The lack of universal designation and characterization standards for composites further complicates these designation and characterization tasks. The development of comprehensive yet relatively simple data tables such as those that appear in *Metals Handbook* is a nearly impossible task for composites.

There are several crucial differences between composites, which are anisotropic, and isotropic homogeneous metals. Both metallic alloys and composites are combinations of two or more constituents. The difference is that composite constituents do not melt together and form a solution—they are two distinct phases. During curing or other processing, the active matrix phase will either chemically react or else melt, while the fiber will remain inactive.

A close metallic analog to a composite is the precipitation-hardened aluminum alloy or steel. But composites are different. Unlike a precipitation-hardened alloy in which the precipitates are dispersed isotropically, the reinforcements in composites are typically distributed anisotropically. A composite may have small particle reinforcements distributed relatively uniformly, but it could also have continuous fiber reinforcements parallel to each other (a unidirectional com-

---

[1]E. I. Du Pont de Nemours & Co., Inc., Composites Division, Wilmington, DE 19880-0702.

posite). Other reinforcement options are continuous fibers oriented at different angles with respect to one another, for example, a woven fabric. Not only are composites generically anisotropic, but there are varying anisotropies from system to system.

Another metallic system with a similarity to composites is a rolled steel. In this case the grain structure parallel to the rolling direction is elongated with respect to the grain structure perpendicular to the rolling direction. This difference can induce a slight amount of anisotropy in mechanical properties, but the difference may be neglected for nearly all purposes. A unidirectional composite is similar in that the continuous reinforcements are oriented parallel to each other. A cross-section parallel to the reinforcements has a very different microstructure from a cross-section perpendicular to the reinforcements. Composites are more complicated in that while a rolled steel sheet has only one rolling direction, a composite can be constructed of many layers of unidirectional plies with each ply oriented at a different angle with respect to an arbitrary specimen axis.

Geometric factors are also more important for composites than for metals. The geometric structure of a composite part strongly affects the performance of the part. In addition, there are no simple formulas to relate the properties of a unidirectional composite to the performance of a cylindrical part comprised of various layers oriented at different angles with respect to the cylindrical axis. The design of metal parts and prediction of their performance is more straightforward than for composites.

Manufacturing also plays a more critical role for composites than it does for metals. Here again, there are metallic analogs, but the effects are more complex in composites than in metals. Incorrect processing can adversely affect metallic microstructure, primarily grain size and inclusion content, which in turn degrades mechanical properties. When processing composites, many factors need attention, for example, prepreg quality, correct reinforcement orientation(s), uniform distribution of matrix and reinforcement, different processing requirements for amorphous versus crystalline matrices, and the extent of curing or polymerization. This complexity must be reflected in a comprehensive composites data base.

The Du Pont Composites Division has a Data Base Management Team chartered to develop and recommend how to manage a computerized composites data base for internal use. Eventually, selective access for external use will be possible after the security clearance issues have been resolved. The issues of designation and characterization of composites will be discussed based on the team's experience in executing its charter. Note that this paper does not give a detailed description of what we have done. Rather, it provides a clear, concise statement of the issues involved in developing a data base for composite materials. It is designed to provide a well-grounded starting point for those developing composite material data bases.

Composite Designation

As an example of the difficulty in designating and characterizing composites, consider the task of querying a computerized data base for the mechanical properties of the following laminar composite:

- Fiber: graphite,
- Matrix: epoxy,
- Prepreg form: unitape,
- Layup: (90/90/0/0/0/0/90/90)—designates the orientation angle of each layer, that is, lamina, in the composite (Fig. 1).

Some considerations in executing such a query would include: (1) a complete description of the material in the query; (2) the specific graphite fiber used and the associated filament count; (3) the specific epoxy resin used, including any additives used; (4) the prepreg batch used and

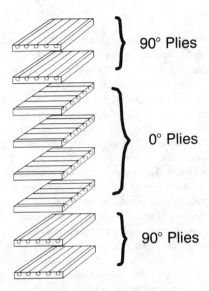

FIG. 1—*Construction of (90/90/0/0/0/0/90/90) laminate (adapted from Ref 1).*

how long was it stored before making the composite; (5) a lay-up configuration in the proper form for compatibility with the data-base lay-up designation; (6) the composite curing schedule; (7) the specimen orientation and configuration used; (8) a description of any specimen preconditioning used before testing; (9) the test procedure and apparatus used; (10) the number of specimens tested and who performed the tests; and (11) a description of any anomalies in fabricating the composite or testing the samples.

Although some of these considerations apply to queries about any material, many of them are unique to composites. For example, the layup referred to in Fig. 1 can also be expressed as $(90_2/0_2)_s$. Either of these lay-up configurations is correct, but the execution of a successful query will depend on using the description compatible with the data base. A related consideration is the necessity to identify each material in a form that is correct, but as abbreviated as possible so that the data base software can function as quickly as possible.

Depending on the composite being queried, these and other questions might be critical to obtaining the data the user really wants instead of just related data. The specific composites lay-ups described previously are examples of codes that attempt to fully describe composites in as few fields as possible. These codes may appear acceptable, but in fact don't represent a universal standard. They apply to laminar composites only, and do not allow for expansion into other types of composites. In addition, such codes are hard to computerize and make retrieval difficult if the user does not query exactly. An approach to this problem is to separate each part of the composite designation into a separate field. More complex composites can be added easily by simply adding more fields. Retrieval becomes dependent on content rather than form.

Note that we have chosen to designate composite materials by specifying the starting materials and processing history. Another approach would be to determine the chemical nature and microstructure of the finished material. This approach is similar to that used in designating many metallic materials. Unfortunately, this alternative is not viable for composite materials at this time. There is insufficient consistent data on how the starting materials and processing history correlate with the microstructure of the finished composite. As the composites industry matures, such data may become more readily available.

Composite Classification

Classification of composites is an equally difficult task. The complexity of composite materials is enormous. In addition to laminar or layered composites, there are four other general types—fiber, particle, flake, and filled (Fig. 2). This issue is further compounded by the many refinements of each composite type. For example, a laminar composite may be a flat plate fabricated from layers of prepreg, but it could also be a tube that was filament-wound from an impregnated tow. Still another aspect to be considered is reinforcement orientation (Fig. 3). The reinforcement orientation could range from one dimension (unitape) to two dimensions (fabric prepreg) to three dimensions (filled composite). A comprehensive data base needs a structure that can accommodate any of these possibilities.

Even if a data base only includes data on laminar composites, there are still more considerations. Laminar composite properties may be lamina properties, that is, generated from composites in which all layers are identical and have the same orientation, or laminate properties, that is, generated from composites in which each layer may either be different or have a different orientation or both. Depending on the design philosophy of an organization, the data base could include just lamina properties, just laminate properties, or lamina properties with only selected laminate properties. Some groups prefer to measure only lamina properties and then calculate laminate properties. Others prefer to measure laminate properties directly. The database structure must accommodate these varying needs.

Our approach to providing the needed structure was to develop a sound data-base shell, allowing for the addition of more and more complexity as the data base expanded to new types of composites. An important part of this approach was to obtain a clear vision of the final data base so that we were sure the initial simplified shell would meet the long-term needs. As each new level of complexity is added, we check it against the final vision to be sure that the structure is sufficiently flexible for subsequent expansion. The data base presently includes lamina properties. We are in the process of expanding the data base to include laminate prop-

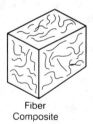
Fiber Composite

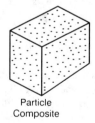
Particle Composite

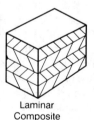
Laminar Composite

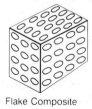
Flake Composite

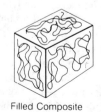
Filled Composite

FIG. 2—*Classes of composites (adapted from Ref 1).*

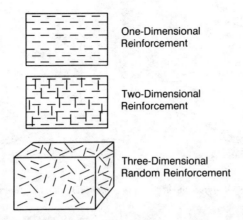

FIG. 3—*Types of reinforcement orientation (adapted from Ref 1).*

erties plus hybrid composites, and plan to add nonlaminar composite capability in the near future.

Different functional groups, for example, design, manufacturing, and research, have diverse needs at various stages in composites development and manufacture. A designer wants to know what statistically based, high-quality data, for example, "A" allowables, are available for a particular system. A manufacturing engineer might want to know how the properties of the latest batch of one type of prepreg compare with those of all similar batches received over the past 6 months. A researcher could be interested in what data are available for a new experimental composite system, even if the data are of questionable quality. We addressed these diverse needs by having each group represented on our team and by including stages and properties in a manner that allows each group to access the data base subset applicable to its concerns. Our goal was targeted data access allowing access to different portions of the data base by various groups, but not burdening any user with unnecessarily large amounts of data that do not relate to the task at hand. We also structured the data base to accommodate published data as well as data generated in-house.

Another issue concerning composites data bases is whether or not data on the composite constituents, that is, reinforcement and matrix, should be included. Researchers in particular might find it useful to know if the properties of one batch of experimental resin vary from those of another and how any difference translates into composite properties. Inclusion of constituent properties is also important from a manufacturing standpoint. It allows a composite to be tracked from start to finish, with properties recorded at all stages of production. Such a system provides an accurate log of product quality and traceability—desirable features for manufacturing operations.

A complication related to constituent properties is the wide variety of reinforcements and matrices possible (Fig. 4). A reinforcement could be a strong, high-modulus fiber such as graphite, but it could also be particles of tungsten carbide or glass microspheres. Matrices can vary similarly from a room temperature curing epoxy resin to stainless steel. The structure of a composites data base needs the flexibility to accommodate this variety.

Composites are relatively new materials as compared to metals, polymers, and ceramics. Less time has elapsed for development of comprehensive computerized composites data bases. In many cases, individuals or small groups maintain their own limited data base. We are developing the appropriate interfaces so that data in small, individual data bases can be downloaded easily into our new larger data base.

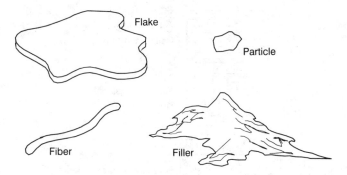

FIG. 4—*Reinforcement forms in composites (adapted from Ref 1).*

Composite Characterization

In most cases, the quality of data obtained from a material is a function of the quality of the specific samples used for testing. This is true for composites, but with a difference. Poor composite quality may result from a number of factors, such as poor adhesion between the fiber and matrix, curing problems, voids in the sample, and old prepreg. The composites data base structure must be able to accommodate this type of composite quality information. It also needs to accommodate information on measures taken to improve composite quality, for example, finishing or treating a fiber to improve adhesion to the matrix.

Since existing composites test standards do not describe all the tests being run on composites, for example, Iosipescu shear, test conditions must become an integral part of a composites data base. Specimen preconditioning and testing data are as important to the definition and characterization of a composite as are the properties of its constituents. The conditions prevalent during testing and preconditioning directly affect the properties of the composite. Each property result may be associated with many testing and preconditioning fields. Specimen size, configuration, orientation, test instrumentation, test fixturing, and loading parameters are examples of appropriate test conditions. The data base must include full descriptions of these conditions for every individual property and must be flexible enough to accept test codes as they become standard in the future. Comment text fields describing any new or unusual test circumstances are also useful. Implementing these features required us to develop concise methods to incorporate this data, including ways to effectively display it.

General Data Base Issues

Many properties cannot be represented adequately in tabular format, either because of the amount of data or their form. Fatigue data, for example, are much more readily displayed in a graphic format (Fig. 5). A comprehensive data-base system needs the capability to display data graphically. We plan to add this capability within the next few months.

Assuring data quality and integrity are critical aspects of any data base. Many questions must be answered before the data obtained can be assessed in terms of quality and used properly. Was the value obtained the result of one specimen or many replications? Were test conditions stable throughout? Was the data source fairly reliable? Was the test performed for design purposes, or was it just done for a quick, "ballpark" estimate? We developed data quality criteria to permit the categorizing of data before entry in the data base. It is important to know if data is of the highest quality, or is of suspect quality but the only available data on a given material.

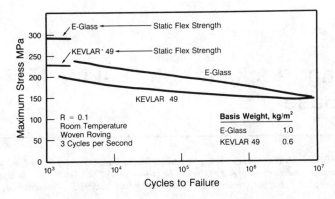

FIG. 5—*Flexural fatigue response of fabric reinforced composites—polyester matrix (Ref 2).*

Data integrity can be protected by addressing other important questions. Are the data consistent with data for similar materials already in the data base? If not, is there a mechanism to determine why the data are different and which data should be excluded from the data base? Are obsolete data removed from the data base? Are the data checked for accuracy after entry to be sure there are no typographical errors? It is essential that the data in the data base be verified or validated in this manner. Such verification is a key issue that cannot be emphasized too much. Our approach to these questions was to funnel data input through a gatekeeper function. The gatekeeper has both the responsibility to identify data that should not be included in the data base and the authority to exclude such data from the data base. In this manner, a central unit is responsible for maintaining data integrity.

Conclusions

We have raised many of the critical questions to be answered when setting up a materials property data base for composites. Thus far in our work to develop a composites data-base management system, we have addressed some of the questions and made plans to tackle the remainder. A key point is that these considerations are much more complex for composites than for isotropic materials such as most metals, polymers, and ceramics. Part of the complexity arises because composites represent an embryonic technology. As the technology becomes more mature, many of the complex questions may become moot points. For the present, these factors must be addressed so that the quality and integrity of a composites data base will sustain meaningful use.

References

[1] Metals Engineering Institute, *Composites: The Basics,* American Society for Metals, Metals Park, OH, 1984, Lessons 1 and 5.
[2] *Data Manual for Kevlar® 49 Aramid,* E. I. du Pont de Nemours & Co., Inc., 1986, Wilmington, DE, p. 58.

Hui H. Li[1] and Cho-yen Ho[1]

Building Blocks for an On-Line Materials Data Base

REFERENCE: Li, H. H., and Ho, C. Y., "**Building Blocks for an On-Line Materials Data Base**," *Computerization and Networking of Materials Data Bases, ASTM STP 1017*, J. S. Glazman and J. R. Rumble, Jr., Eds., American Society for Testing and Materials, Philadelphia, 1989, pp. 272–279.

ABSTRACT: Problems that surface in the development of materials property data bases are many and solutions are varied, with different data base applications and designs. This results in incompatibility among existing data bases. This paper discusses the essential requirements of on-line materials property data bases and the building blocks of various file systems. Areas covered include the types, basic structures, and various forms of the building blocks and the solutions to the problems encountered in the construction of the file systems. Under this scope, three file systems—namely, the materials files, variables files, and data files—are reviewed in terms of the basic building blocks. It is made clear that integration and transformation of data bases are possible if and only if the data bases under consideration are constructed with fully resolved building blocks. With the above considerations in mind, activities of standards establishment and data base development can be carried out in parallel.

KEY WORDS: data bases, basic structures, file systems

During this information explosion era, technical professionals are confronted with a broad range of options in specifying their needs and with a fast growing volume of available materials property data. Effective decision making requires gathering a great deal of data rapidly and completely. Ideally, these data should have a high level of reliability. Depending on the nature of the work, users may need data that meet specific criteria. In some cases, they may need data with a wide variety of supporting information for a host of selected parameters. To meet such demands, a well-designed and fully developed materials property data base is the answer.

Although modern-day computer technology is highly advanced and has been widely applied in many areas, the development of materials property data bases has unfortunately not kept pace with technology. The reasons for such slow progress are many. Aside from the societal and economical issues [1], the main reason is that materials property data are extremely varied and vast.

Many still believe that there must be a large number of data bases already computerized that can easily be integrated or accessed. In reality, however, this belief is rather over optimistic. There are many data bases in existence, but they were developed for different purposes, under different circumstances, and in an uncoordinated way. They are very likely of different quality, depth, and perhaps comprehensiveness.

It is certainly a good idea to integrate the existing data bases not only for cost effectiveness, but also to reveal the missing gaps. However, because these data bases were developed independently, incompatibility among them is expected. Eliminating the differences among them will be difficult and time consuming. Because many data bases are developed by individuals in the

---

[1]Deputy director and director, respectively, Center for Information and Numerical Data Analysis and Synthesis (CINDAS), Purdue University, West Lafayette, IN 47906.

same or closely related fields, considerable overlap of coverage is anticipated. Even if the differences can be resolved, the combination of two or more such data bases may be just slightly better than one. Furthermore, the unwillingness of the data base owners to participate in such a program is a major stumbling block.

It becomes increasingly evident that ground rules and standards for the computerization of materials property data should be planned so that developing new data bases and converting existing ones can be achieved in harmony. This paper addresses the fundamental requirements of an on-line materials property data base in general and discusses problems and solutions in dealing with the very basic unit—the building block—in particular. In the following discussion, the data base is assumed to be built according to the following hierarchy:

- data base,
- file systems in the data base,
- building blocks in the file system, and
- data elements in the building block.

The definitions of some terms used later are

- Data element—the smallest piece of information used to represent data, it can be a number, a character string, or paragraph(s) of text,
- Building block—the basic unit for file construction; it contains data element(s) and is unique, self-contained, and meaningful,
- Resolved block—the block that contains one or more simple, orderly arranged blocks,
- Degenerated block—the block that contains more than one simple, randomly stacked block, and
- Resolving power—a measure of the resolvability of the data base.

Essential File Systems of a Data Base

The backbone of a materials property data base is by definition a collection of building blocks. When a block is isolated, a host of questions are asked regarding how to establish the connections of this block with all the other related blocks within as well as across the file systems. Orderly answering of these questions is exactly the way to build various data base file systems. For example, when a set of data is given to the user, the user will naturally ask (but not be limited to) the following questions:

- For what material is this data set obtained?
- What property does this set represent?
- What is the primary variable of the data?
- What are the affecting parameters?
- What method is used in obtaining the data?
- How accurate is this data set?
- From what reference(s) is the data extracted or derived?

The list of questions can go on and on. To respond to the above questions, the data base must be equipped with four basic file systems, namely, materials, variables, references, and data. The data and references file systems mainly perform storage, while the materials and variables file systems act as data descriptors, do indexing, and search. In between there are interface files so that the data base can be searched and the qualified data and references can be retrieved.

For on-line operation of the data base, three additional, basic file systems are considered essential: help, menu, and executable interactive program. In designing these file systems, keep in mind that the system must be flexible and user-friendly. Equally important, the system must not require an extensive user's manual, tedious queries, and rigid spelling.

Building Blocks of a File System

The purpose of building a data base is to deliver useful data and information to users according to their specifications. The file system and building blocks have to be tailored to meet users' requirements. Since the requirements vary significantly among different user groups, the resolving power built into the data base varies, and so does the content of the building blocks. It is a common practice of a data base designer to resolve the most important data while lumping the secondary data together into a degenerated block. This is the area where the incompatibility among data bases creeps in.

To reduce the incompatibilities, the simplest way is to increase the resolving power by decomposing the degenerated block into resolved blocks. This is done if the data elements in the degenerated block are entered in an orderly manner with proper delimiters or traceable patterns. For randomly stacked ones, there is no other alternative but to separate them manually.

A careful study of the functions of various files finds that the building blocks of a data base can be classified into three types. These are the storage block, the interface block, and the function block.

The structure of a resolved storage block has the basic form

$$\text{identification} * \text{secondary identification} * \text{value}$$

where the asterisk above represents a delimiter separating the fields in the block.

In the above expression, the value in the block has a general meaning. It depends on the data type that the identification represents. The value can be a number, a character string, paragraph(s) of text, or several related data elements. The secondary identification is optional. Its presence depends on the requirements of the data base. By the same token, the identification can be single or a combination of several single identifications.

Deviations from the basic form can be considered resolved as long as they are done properly and uniformly in a given file. When a file can be divided into a number of groups and the members in each group stay in sequence and carry the same identification, the data elements can be organized into a block of the form

$$\text{identification} * \text{secondary identification} * \text{value} \ldots$$

where " . . . " indicates the repeating of the "* secondary identification * value" throughout the entire group. Or more clearly:

> identification
>
> secondary identification * value
>
> secondary identification * value
>
> secondary identification * value
>
> " . . . "

In the latter form, two types of delimiters are used and one of them is the new line character. In the extreme case, the file consists of a single group and the file name becomes the identification. The executable program, menu, and help information are of this type.

The structure of the interface block is perhaps the simplest. Whether it is created through file inversion, direct construction, or other means, its basic form is identification * equivalent identification

or

identification * address of interested data item

In case the position of the block in the file is implied by the identification, the interface block is simply of the form:

address of interested data item

with the understanding that the relationship between interfaced data items is implemented in the software.

The structure form of a function block is not much more complicated than those of other types, but the contents of the block vary widely depending on the application of the data base. The basic form of a function block

identification * func. 1 * func. 2 * . . . * func. n

The functions can be a

- data availability indicator,
- data item counter,
- data range indicator,
- data source identifier,
- data access pointer,
- data type indicator,
- and so forth.

Just how many functions should be included in a block and how the function blocks are divided into files is totally at the discretion of the data base designer.

Building Blocks in a Materials File System

It is a well-known fact that materials property data obtained under an identical experimental set up and on the same material from different suppliers can show significant differences [2]. In most cases the origin of such disagreements cannot be attributed to the experimental errors alone, but stem from the lack of an adequate material characterization as well [3].[2] Some properties are extremely sensitive to the variation of material characteristics. For example, thermal radiative properties vary vastly with surface conditions; the electrical resistivity of a semiconductor at low temperatures changes substantially with impurities. It becomes increasingly evident that the material characterization is as important as property values. The demand for adequate material characterization in connection with the experimentally determined data not only leads to better quality data, but also provides better clues for the study of materials behavior. An awareness of this importance is reflected in some of the technical reports published in recent years [4].

---

[2]The two volumes quoted here are typical examples of many materials property data compilation and evaluation series. A large number of data sets are extracted from a wide variety of data sources. Not all of the captured data are qualified for data evaluation because of insufficient information given for material characterization.

Although it is difficult to characterize a material completely, it is possible to incorporate those characteristics that have significant effects on the property as material descriptors such that a given set of data can be identified with a particular sample of a material. There are two kinds of material descriptors, intrinsic and extrinsic characteristics. The commonly used intrinsic material descriptors are:

- composition,
- composite construction or configuration,
- crystal structure,
- molecular structure,
- fabrication process,
- physical treatment,
- chemical treatment,
- characteristic physical properties,
- and so forth.

There are a variety of extrinsic descriptors which have no effect on the material properties but serve the purposes of identification, safety, and general information. These are:

- name (synonyms, registration number, specification codes, and so forth),
- generic class,
- application areas,
- handling procedures,
- general description,
- and so forth.

Whether the file system is an aggregate of segments (each can be considered as an independent but small file) arranged in a series with each segment consisting of all the applicable heterogeneous descriptors for a given material or is a collection of files arranged in parallel with each file containing a single homogeneous descriptor depends on the purposes of the data base and the preference of the designer. Both have their advantages and disadvantages. In the series file system, each segment of the file is exclusively assigned to a given material. Thus each material is self-contained. This type of file system is easier to implement and each material in the system is portable, but as a whole it has poor flexibility in updating and low throughput in operation. In the parallel file system, on the other hand, the information on a given material is distributed among various files, each dealing with a specific subject. This opens the possibility that information common to several materials can be defined in modules (segments in the file). If one module is not able to handle the situation, secondary modules can be defined to deal with the balances. Although this type of file system has high flexibility in data base update and high throughput in real-time operation, it is rather difficult to design and implement. The designer must have a full understanding of the subjects of the data base and be able to find the common denominators in various areas covered. Furthermore, the parallel-structured file system has less portability; it has to be transported as a whole, not partially. In principle, one of the above types can be transformed into the other and vice versa by means of properly implemented software. Such transformation is only possible and meaningful when the building blocks of the system are fully resolvable.

In the material file system, the most common problems are in material name, material class, and material applications. Problems in material naming arise in the codification of the materials. Some have a number of aliases, including synonyms, trade names, registration numbers, and other adopted codes. Unless the user is required to pick up the material codes listed in the user's manual for queries, the on-line string search for free-format queries may run into a problem—two or more aliases of the same material maybe prompted for further selection, thus confusing the user. The solution is to provide the user a single choice of that material with indication that there are other aliases. The user should be able to view the aliases. If the data base can be searched based on material classes or material applications, there can be too many materials in some classes or application areas, resulting in a long list to choose from. The solution is to combine the criteria of material class and application to narrow down the number of hits. If this still does not work, the material class or application under consideration should be further divided into subcategories. The building blocks contained in the materials file system are only pertinent to the materials, that is, intrinsic to them. Extrinsic information about the material pertinent to the measurement, such as sample preparation, sample orientation, and so forth, should be kept with the results of the experiments.

Building Blocks in a Variables File System

The building blocks of a variable file system have no direct effect on the experimental results, but serve the purposes of identification, data conversion, and general information. They are:

- variable name,
- units,
- units conversion factor,
- definition of variable,
- brief summary of available measuring techniques,
- and so forth.

Unlike the materials file system, the desirable file construction of a variables file system is in series. Building blocks in this system have far less overlap than those in the materials files do.

The problems in this file system are of a different nature from those encountered in materials files. Although the problem of aliases in the variable names is not serious, the meaning of the variable names needs careful consideration. In the most extreme cases, the use of variable name alone is not sufficient to express the real meaning of the variable. For example, if a data set is described by the two variables, voltage and distance, both are legitimate names. However, the meaning of "distance" is ambiguous. Whether the distance is measured from a positive point charge or from a negative point charge or perpendicularly from a positively charged rod must be determined before the user can make sense out of the data set. Some variables depend on additional information to clarify their sense. Since one cannot expand the list of variables without limit, the solution to the problem is to incorporate the appropriate context in the data set to reflect this effect.

The problem in the areas of units and unit conversion is that more than one code has to be assigned to some of the variables. A typical example is the variable "time," which is measured on the order of microseconds when used in connection with some electronic properties, but is expressed on the order of days when used for long-term property observations. Since data values in the data base are expressed in the standard units adopted, forcing the same unit to express the values of a variable of a wide span in magnitude is awkward, if not inappropriate. The solution to the problem is simply to assign different codes to the variable under consideration, with each code covering a different range of values of the variable. In other words, a variable is defined not only by its name, but also by the range it covers. For practical reasons, overlapping of ranges is also allowed.

Building Blocks in a Data File System

The accuracy of property data depends on the experiment. Every stage in the experimental process plays an important role in the reliability of the final results. Therefore, the measured results must be presented with a number of key experimental details. The experimental information required includes

- experimental method,
- special experimental technique,
- sample configuration,
- sample preparation,
- sample orientation,
- and so forth.

All such information is considered extrinsic to the material and is an integral part of a data set.

A data file used for storage is an aggregate of data sets. The term "data set" has different meanings as used by different data base designers. In some cases, a data set is defined in connection with the material. For each material, the available data on all properties comprise a data set. In some special projects, the data collected in a given project are considered a data set. Most commonly, however, a data set is designated for the data from one unique series of measurements on one material, one property, and one independent variable. No matter how a data set is defined, it must be constructed with building blocks containing the following information (when applicable):

- data set identifier,
- material,
- property,
- primary independent variable(s),
- affecting parameters,
- units of variables,
- material characterization (intrinsic and extrinsic),
- experimental details,
- data evaluation considerations,
- data analysis considerations,
- quality of data,
- data points,
- comments on data, and
- data source.

A data storage file consists of heterogeneous building blocks. Some are of the storage type and some are of the interface type.

To facilitate data search and subsequent data retrieval, a companion file (data search file) is constructed in which the homogeneous building blocks are of the functional type. For each data set entered into the storage file, a corresponding building block is created in the data search file. This block contains information about:

- material,
- property,
- primary independent variable,
- boundary values of the property,
- boundary values of the independent variable,
- data set identifier,
- data source identifier, and
- type of data.

Conclusions

Developing and maintaining a materials property data base is an everlasting and evolving art. The contents of the building blocks and file structures of the data base should never be taken lightly. There should be constant attempts to modify, enhance, and expand the data base to meet mounting demands from a wide spectrum of users. To construct a data base with all the capabilities mentioned, the building blocks of the data base must be fully resolvable.

Although the long-term objective for materials property data base development is to work out standards, standard procedures for data entry, standard terms for data description, standard formats for data exchange, and so forth, the real world cannot wait for such long-overdue decisions. While the various committees and task groups, international as well as domestic, are working around the clock to seek and set up standards, the data base development effort can move ahead as long as the data bases are constructed with fully resolved building blocks. Whenever the working standards are made available, transformation of such data bases into acceptable standard form is just a matter of appropriate interfaces.

References

[1] Westbrook, J. H. and Rumble, J. R., eds., "Computerized Material Data Systems," *Proceedings of the Materials Data Workshop,* Fairfield Glade, TN, 1982.
[2] Voronova, I. D., "Localization of Electrons in Compensated Gallium Arsenide," in *Electrical and Optical Properties of III-V Semiconductors,* N. G., Basov, ed., Proceedings (TRUDY) of the P. N. Lebedev Physics Institute, Vol. 89, English trans., 1978, pp. 1–62.
[3] Touloukian, Y. S., Powell, R. W., Ho, C. Y., and Klemens, P. G., eds., "Thermophysical Properties of Matter," *The TPRC Data Series,* Vol. 1 and 2, IFI/Plenum, NY 1970.
[4] Murdie, N., Hippo, E. J., and Kowbel, W., "Structural Characterization of Carbon-Carbon Composites," in *Metal Matrix, Carbon, and Ceramic Matrix Composites,* NASA Conference Publication 2482, NASA, Washington, DC, 1987, pp. 311–326.

Crystal H. Newton[1]

Consideration of a Preliminary Data Base for MIL-HDBK-17B

REFERENCE: Newton, C. H., "**Consideration of a Preliminary Data Base for MIL-HDBK-17B,**" *Computerization and Networking of Materials Data Bases, ASTM STP 1017*, J. S. Glazman and J. R. Rumble, Jr., Eds., American Society for Testing and Materials, Philadelphia, 1989, pp. 280–291.

ABSTRACT: Managing the data in a composite materials data base, such as the data base required for MIL-HDBK-17, is a complex task. MIL-HDBK-17, "Polymer Matrix Composites," is a fully coordinated military document currently supported by the Army, the Air Force, and the Federal Aviation Administration. The guidelines in Volume I of the handbook provide the requirements for a statistically based data base of polymer matrix composite materials properties.

The MIL-HDBK-17 data base must be capable of handling property values for fibers, matrix materials, prepregs, laminae, and simple laminates. A prepreg is a ready-to-mold or -cure material in sheet form. It may be fiber, cloth, or a mat impregnated with resin and stored for use. The properties of each of these types of materials can be further divided into chemical, physical, and mechanical categories. While the material types and property categories readily provide an outline for a data base, a number of properties must overlap the data base boundaries to adequately describe a given material.

Statistical analysis requirements for evaluating batch-to-batch variability and an interest in pooling data from various sources dictate the analysis and storage of all raw data in addition to the statistical parameters reported in MIL-HDBK-17. The capabilities for tracking these data use four types of files—raw data files, statistical parameter files, analysis files, and files that record the history of pooled data. In addition, a hard-copy record is retained for both analysis files and summary tables. The analysis files report raw data, the method of analysis as determined by batch-to-batch variability and the distribution of the data, and the resulting statistical parameters. The summary tables contain the statistical parameters in the MIL-HDBK-17 format tabulated by orientation, loading, temperature, and moisture conditions. These tabulated values can be readily transferred to a word processor for incorporation into the printed handbook.

Data-base management on personal computers was selected for several reasons. PC availability is desirable for the statistical analysis program since each company or laboratory that submits data to MIL-HDBK-17 will want access to this program. Management of the data base itself is conducted through the use of METSEL2.[2] METSEL2 is an interactive data-base program originally designed to store metals properties and currently being revised to include composite materials. Storage and manipulation of the statistical parameters are enhanced by using this program. The in-house data base is also considered a trial run for a data base that may be distributed to the members of the coordination group and other appropriate organizations. The establishment of this permanent, accessible data base depends on factors that have not yet been resolved, such as the export control of data.

KEY WORDS: materials properties, data base, composite materials, polymer-matrix composites

Several benefits result from the management of a materials properties data base on a personal computer. The number of users is not strictly limited by hardware capabilities or more loosely limited by the turn-around time for operation of the program. The hardware required

---

[1]Engineer, Materials Sciences Corporation, Spring House, PA 19477.
[2]Copyright ASM, International, 1987.

for the operation of the data base is relatively inexpensive and commonly available. If more than one level of data control is required, separate levels can be preserved by distributing different data disks for each level of control. In addition, the analysis package on which the data base depends on can be used at different workstations within a user organization.

Most of the requirements for a materials properties data base managed on personal computers are the same as those for a main-frame data-base system. An interim report of the Subcommittee on Evaluation and Analysis of the National Materials Property Data Network, Inc. has presented suggestions for a multistep evaluation and analysis procedure for materials properties data bases [1]. The steps in this procedure relevant to the general review of a data base are evaluating the maintainer and data generators, initially appraising the data record, and validating and analyzing the data. While most of the considerations within each of these steps are relevant to both on-line and PC data-base management systems, a few need to be modified to evaluate a PC-based system.

One example of a data base with a PC management system is the data base currently being developed in conjunction with MIL-HDBK-17. The evaluation procedures for materials properties data bases are being developed for metals data bases since the engineering community has had more experience with metals and data bases of metals properties. However, MIL-HDBK-17 is a military handbook on the properties of polymer matrix composite materials. The requirements for a data base associated with this handbook are different from those for a metals data base.

The effort to establish a handbook on the materials properties of polymer-matrix composites has continued for approximately three decades. MIL-HDBK-17, originally titled "Plastics for Aerospace Vehicles, Part I. Reinforced Plastics," was first published in 1959 [2]. This handbook was revised and issued as MIL-HDBK-17A in January, 1971 [3]. The materials properties presented were mean values of results for glass-fiber reinforced composite materials. In addition, standard deviation or minimum and maximum values were presented for most of the properties.

In 1976, Rockwell International, under contract to the Air Force, developed the DOD/NASA Advanced Composites Design Guide [4]. Volume 4 of the design guide presented materials properties primarily for carbon/epoxy and boron/epoxy composite materials. Typical values are presented for all properties, with the number of specimens and range indicated for strength properties. In addition, an A-basis value or B-basis value for strength is given for a limited number of cases.[3] The test method and the source of the data are also provided. However, the statistical analysis procedures are not clearly presented. The major drawback of this handbook is that, at the present time, there are no plans to provide updates of the information contained in the data volume.

A revision of MIL-HDBK-17 has been under consideration for approximately 10 years. Much discussion was necessary to settle such issues as whether the handbook would present design allowables or statistically based material properties, or if the handbook would present the results of lamina or laminate testing. As a result of the deliberations of the coordination group, the scope of the handbook has been increased to cover all polymer-matrix composite materials. In addition, the statistical analysis procedures have been improved and redefined. To cover the expanded range of materials, the handbook has been divided into three volumes. The first volume covers the guidelines, which define how material property data is to be generated and analyzed. The second volume will present the materials properties data. The third volume will include recommendations for quality assurance procedures, design, and analysis. Information on materials and processes is also presented in this volume.

[3]The A-basis and B-basis values are statistical parameters. An A-basis value is the property value above which at least 99% of the population of values is expected to fall, with a confidence of 95%. A B-basis value is the property value above which at least 90% of the population of values is expected to fall, with a confidence of 95%.

The revision of MIL-HDBK-17 is the responsibility of the MIL-HDBK-17 coordination group. From a small group at the beginning of the effort, the coordination group has grown to approximately 150 members. The coordination group is led by the MIL-HDBK-17 coordinator from the Army Materials Technology Laboratory. The organization of the coordination group is shown in Fig. 1. Five working groups currently exist within the coordination group. Much of the drafting of sections of the handbook is accomplished within the working groups. Once the drafted material is approved by a working group, it is submitted to the entire coordination group for approval and incorporation into the handbook. The effort of the coordination group is supported by the MIL-HDBK-17 secretariat. As the MIL-HDBK-17 secretariat, Materials Sciences Corporation is responsible for the editorial functions for the handbook as well as the statistical and technical analysis of data. Support for the MIL-HDBK-17 effort is provided by the Army, the Federal Aviation Administration, and the Air Force.

Handbook Requirements

The requirements for generating materials properties data for MIL-HDBK-17 are presented in Volume I of the handbook [5]. The areas of interest for the composite materials that are included in the handbook are shown in Fig. 2. Test matrices for generating these data and a test matrix for screening materials are provided. Separate chapters present the testing requirements for fibers, matrix materials, prepregs, laminae, laminates, and structural elements, although not all of these chapters will be included in the March 1988 issue of Volume I. The final chapter in Volume I provides the statistical analysis procedures and the framework for the presentation of materials properties in Volume II.

MIL-HDBK-17B will present statistically based materials properties values. The coordination group has decided that the handbook will not present any information as design allowables since that would limit each company's approach to design. Instead, B-basis materials properties will be presented for strength and strain-to-failure mechanical properties with enough information for a full statistical description of the data as analyzed. The statistical analysis program for MIL-HDBK-17 has been established and described elsewhere [6]. These statistical procedures for the analysis of the data are summarized in the flowchart in Fig. 3.

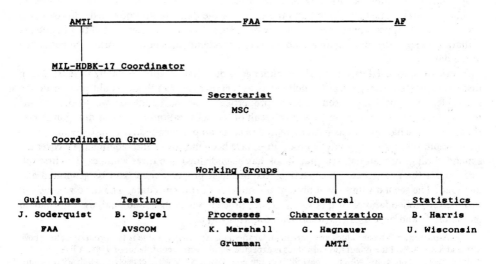

FIG. 1—*Organizational chart for the MIL-HDBK-17 coordination group.*

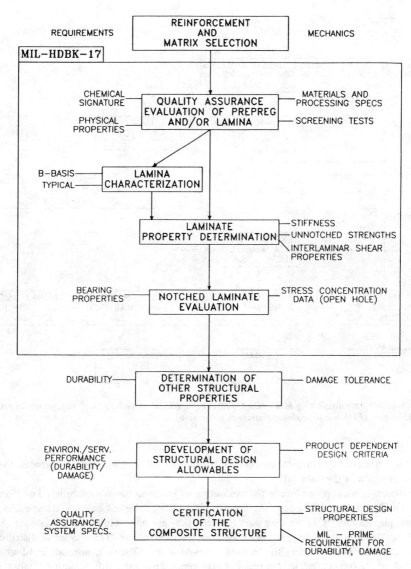

FIG. 2—*Focus of MIL-HDBK-17B* [5].

The procedure used to determine a B-basis value for a given material property depends on the characteristics of the data. For material properties to be published as approved data in MIL-HDBK-17B, data are required from a minimum of six specimens from each of five batches. Thus, the results from tests of a minimum of 30 specimens will be available for analysis. The first step in the computational procedure is examining the data for outliers, or observations that have been recorded in error. If an outlier is found, the data must be examined to determine a reasonable cause for the outlier. If the outlying value can be corrected, as in the case of a typographical error, it will be corrected. If an obvious cause can be traced, such as the

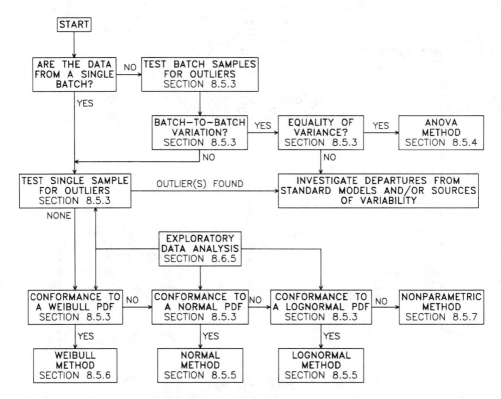

FIG. 3—*Flowchart illustrating MIL-HDBK-17B computational procedure for B-basis material property values [5] where (PDF is a probability distribution function).*

improper setting of experimental conditions, the datum will be eliminated. Otherwise, the outlier must remain in the data set.

The next statistical procedure is the evaluation of batch-to-batch variability. The k-sample Anderson-Darling procedure tests the hypothesis that the different batches are samples from the same population [7]. If the data pass this test, they are pooled into a single sample. At this point, the data are again checked for outliers as described above. Three types of distributions are evaluated for how well they fit the data. Two-parameter Weibull, normal, and lognormal distributions are considered, in that order of preference. In other words, if both a Weibull and a normal distribution adequately fit the data, the Weibull distribution is assumed. A B-basis value is calculated using the distribution that is selected. If none of the distributions adequately fit the data, then a nonparametric method is used to determine a B-basis value. If the data could not be pooled into a single sample and the variances of each batch are equal, an analysis-of-variance method is used to determine a B-basis value.

The full statistical analysis procedure is applied to the strength and strain-to-failure mechanical properties. The following information from each of these analyses is then included in the handbook:

(1) typical (mean) value,
(2) minimum value,
(3) maximum value,
(4) standard deviation,

(5) B-basis value,
(6) type of distribution or analysis,
(7) distribution constants,
(8) number of specimens,
(9) test method, and
(10) source code.

The distribution constants are the shape and scale parameters for the two-parameter Weibull distribution, the mean and standard deviation for a normal procedure, and the mean and standard deviation of the logarithmic values for the lognormal procedure. For the moduli, typical values, minimum and maximum values, standard deviation, number of specimens, test method, and source code are presented. Typical values, number of specimens, test method, and source code are presented for all other properties. The requirement that these values be presented affects both the handbook and the data base.

The properties that will be covered in summary tables in the handbook are listed in Tables 1 to 5 for each of the areas of fiber, matrix material, prepreg, lamina, and laminate characterization. Some of these properties apply only to continuous fiber-reinforced composite materials, such as twist, the number of turns about a yarn's longitudinal axis per unit length of a yarn. It should be noted that separate properties for each direction, that is, the longitudinal and transverse values of the tensile strength, are not given in these tables. Other types of material characterization, such as typical stress-strain curves, are included as graphs. If only the properties presented in the summary tables are considered and the statistical parameters are included as required, the number of parameters shown in Table 6 need to be presented in the handbook, and potentially in the data base as well. Obviously, the number of properties is a consideration in establishing the requirements for a data base.

Sources of data for the handbook and the data base will be primarily government contractors, government laboratories, and the open literature. Although six specimens from each of five batches are required for full approval according to the handbook, the coordination group has approved the inclusion of interim data in the handbook. Data in this category are useful and of interest to the composites' community, but do not meet all of the MIL-HDBK-17 requirements.

TABLE 1—*Reinforcement summary table properties.*

Condition	Property
Thermal	expansion
	conductivity
	specific heat
Tension	ultimate strength
	Young's modulus
	strain to failure
Physical	diameter
	density
	twist
	chemical resistance
	electrical resistivity
	sizing level

TABLE 2—*Matrix summary table properties.*

Conditions	Property
Physical	density
Compression	ultimate strength Young's modulus strain to failure
Tensile	ultimate strength Young's modulus strain to failure
Shear	ultimate strength shear modulus
Thermal	expansion conductivity specific heat glass transition temperature (wet/dry) melting temperature—crystalline thermoplastics crystallization rate—crystalline thermoplastics shrinkage during cure
Electrical	dielectric constant loss tangent
Processing	gel point gel time
Other	impact fracture toughness moisture expansion moisture diffusivity maximum equilibrium moisture content

TABLE 3—*Prepreg summary table properties.*

fiber areal weight	heat of reaction
resin content	tack
volatile content	drape
gel time	thickness, cured ply
flow	storage life at $-18°C$ ($0°F$)
viscosity	out time at ___ (temperature)

Requirements for an In-house Data Base

At this time, the MIL-HDBK-17 coordination group has not yet decided to develop and maintain a data base that will be available to the members of the coordination group and other interested organizations. The secretariat, however, needs a data base system to maintain the data records for the handbook. This in-house data base is being considered as a trial data base for the evaluation of the required capabilities for a distributed data base. Consequently, the most important concern in establishing the in-house data base for use with MIL-HDBK-17 was that the data be readily transferrable to another format if that becomes necessary.

The statistical analysis requirements for evaluating batch-to-batch variability and an interest in pooling data from various sources dictate the analysis and storage of all raw data. Data, in addition to the storage of the statistical parameters are reported in MIL-HDBK-17. Consequently, two data bases are used by the secretariat—one for the raw data and one for the statis-

TABLE 4—*Lamina summary table properties.*

Condition	Property
Physical	density thickness/ply
Compression	ultimate strength Young's modulus strain to failure Poisson's ratio
Tension	ultimate strength Young's modulus strain to failure Poisson's ratio
Shear	ultimate strength shear modulus strain to failure
Thermal	expansion conductivity specific heat glass transition temperature emissivity
Electrical	resistivity dielectric constant loss tangent
Other	processing data fiber volume resin content moisture expansion

tical parameters and materials properties. Raw data for mechanical properties in this context include batch numbers, test results, and specimen thickness values. Fiber-dominated properties are normalized with respect to fiber content or thickness before the statistical analysis is conducted. The normalized data are the "raw data" stored in a file and analyzed.

The original statistical software package from Battelle [6] has been modified to express the current requirements of the handbook, to reduce the memory requirements for a given amount of data, and to facilitate its use for large amounts of data. Capabilities for tracking these data involve five types of files—raw data files, statistical parameter files, analysis files, summary table files, and files that record the history of pooled data. Hard-copy records are retained for the original raw data, analysis files, summary tables, and files that record the history of pooled data. The analysis files report raw data, the method of analysis as determined by batch-to-batch variability and the distribution of the data, and the resulting statistical parameters. In addition, all possible statistical procedures are conducted in the program. A summary of the results of all procedures is included at the end of the analysis file for comparison. The summary tables contain the statistical parameters in the MIL-HDBK-17 format tabulated by orientation, loading, temperature, and moisture conditions. These tabulated values can be readily transferred to a word processor for incorporation into the printed handbook.

The statistical parameter files are transferred to the material properties data base. Raw or normalized data can also be transferred to the graphs section of the data base for exploratory data analysis, or for using graphic techniques for considering outliers and families of distributions. In addition to its other capabilities, the statistical properties data base can be used to track the raw data files. The revised software package handles all of the data on the mechanical

TABLE 5—*Laminate property summary table.*

Condition	Property
Physical	density thickness/ply
Tension	ultimate strength Young's modulus strain to failure Poisson's ratio
Compression	ultimate strength Young's modulus strain to failure Poisson's ratio
Shear	ultimate strength shear modulus strain to failure
Interlaminar shear	ultimate strength normal stress
Thermal	expansion conductivity specific heat glass transition temperature emissivity
Electrical	resistivity dielectric constant loss tangent
Other	processing data fiber volume resin content moisture expansion vacuum stability oxidation exposure, aging chemical resistance chemical resistance flammability fracture toughness compression after impact notch effects—tension notch effects—fatigue interlaminar tension—ultimate bearing—ultimate strength strain at failure damping characteristics

TABLE 6—*Number of parameters that must be presented in the handbook.*

Level of Characterization	Number of Properties
Fiber	69
Matrix	160
Prepreg	48
Lamina	249
Laminate	308

properties of a given material at once. This capability is a much-needed improvement over the handling of separate files and program runs for each material, property, temperature, and environmental condition.

Data-Base Structure

The data-base program used for the statistical parameters data base is ASM International's program MetSel2. As with many other data-base programs, the basic unit of the data base is the material record. A separate material record is devoted to each material in the data base. For example, information on T300/3501-6 graphite-epoxy composite would be contained in a single material record. Each material record can contain information on the material's designation and specifications, property values, graphs, and comments.

Once data are stored in a data base, the use of those data depends on the search capabilities of the data base program. All of the named fields in the material record except for quality and footnote fields may be searched. For example, users can search by composition, by designations or specifications, and by product forms or characteristics. Users can also search by property values at a given temperature.

In addition to numeric and text material, the material record can also maintain graphics material. The graphs maintained in the data base can only be viewed. However, by using the separate graphics program EnPlot,[4] users can edit graphs, add new graphs to the data base, or perform regression analysis on stored data points. Currently, the primary uses for the graphics capability are in the storage of typical stress-strain curves and in exploratory data analysis.

A MetSel2 data base can contain as many material records as there is data space available. For example, a 360K-byte diskette can hold between 100 and 200 records, depending on the amount of data in each record. In addition, the data-base program allows users to change data bases, which effectively removes all limits on the number of material records that can be compiled. The structure of the material record has been designed to provide as much flexibility as possible to the data-base builder. A schematic illustration of the material record appears in Fig. 4.

Requirements for a Distributed Data Base

Assuming that the coordination group decides that a data base should be available for distribution, the requirements for such a data base should be considered. Several such requirements are relevant based on the level of integrity demanded of the data and how user-friendly the data-base program can be. For MIL-HDBK-17 purposes, such a data base would include only the statistical parameters' data base and not the raw data. The raw data would be available only in very controlled circumstances. For instance, if one company wanted to pool data with data cited in the handbook and included in the data base, the handbook coordinator would contact the source of that data. If that organization were willing, an exchange would be worked out between the two companies.

The data-base program should differentiate MIL-HDBK-17 data from user-generated data, and restrict the incorporation of user-generated data into the MIL-HDBK-17 data base. One of the primary uses of the data base, when it has been distributed, will be to compare data generated within the company using the data base to data from MIL-HDBK-17. One of the strengths of the MIL-HDBK-17 data is that the testing requirements, statistical analysis, and technical review together provide the qualification of that data. The data must be clearly identified as MIL-HDBK-17 to avoid contamination that would be undesirable from the points of view of the coordination group and the user.

[4]Copyright ASM, International, 1986.

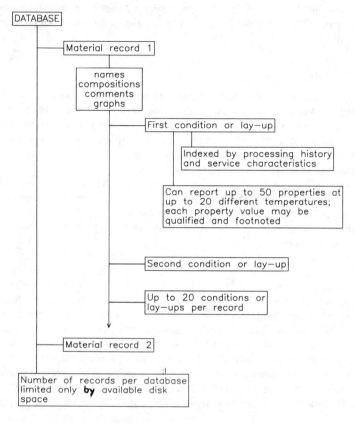

FIG. 4—*Schematic illustration of material record for MetSel2 as applied to MIL-HDBK-17 data base.*

It should be possible to compare and tabulate properties across material records and database boundaries. Design with composite materials is frequently an optimization process. Finding the best material for a given application may involve a trade-off between materials with ideal properties in one direction and lesser properties in another.

A data-base program for distribution should be able to search for materials with a minimum or maximum value of a given property over a range of temperatures. Again, this function is desirable in designing with composite materials.

Military handbooks must utilize units according to the International System of Units (S.I. units) and as dual units, that is, both S.I. and U.S. customary. The composites community prefers U.S. customary units, so dual units are necessary. It would be desirable to incorporate this function into the data base by making the data-base program able to convert properties from one system of units to the other.

Concerns that need to be resolved before the establishment of a distributed MIL-HDBK-17 data base primarily revolve around the data itself. The materials within the handbook subject to export control of data need to be clearly defined. The limitations on the release of the raw data have just been defined, as described above.

Summary

The management on the personal computer of a materials properties data base based on MIL-HDBK-17B has been considered. The requirements for such a system have been discussed using the composites materials properties data base being established in association with MIL-HDBK-17. These requirements are dependent on the military handbook analysis and data presentation guidelines. Although a distributed version of the data base has not yet been approved by the coordination group, an in-house data base is being established by the MIL-HDBK-17 Secretariat by Materials Sciences Corporation.

Acknowledgments

This work was supported by the Army, the Federal Aviation Administration, and the Air Force through Contract Number DAAL04-86-C-0061, with contract monitors Paul Rolston and Paul Doyle of the Army Materials Technology Laboratory. The technical contributions of Tim Gall of ASM International are gratefully acknowledged.

References

[1] Brister, P. M. in *Materials Property Data: Applications and Access*, J. G. Kaufman, Ed., American Society of Mechanical Engineers, 1986, pp. 141-147.
[2] Military Standardization Handbook, "Plastics for Aerospace Vehicles, Part 1. Reinforced Plastics," MIL-HDBK-17, 5 Nov. 1959.
[3] Military Standardization Handbook, "Plastics for Aerospace Vehicles, Part 1. Reinforced Plastics," MIL-HDBK-17A, Jan. 1971.
[4] DOD/NASA Advanced Composites Design Guide, First Edition, 1976 (distribution limited).
[5] Military Standardization Handbook, "Polymer Matrix Composites," MIL-HDBK-17B, 29 Feb. 1988.
[6] Rust, S. W., Todt, F. R., Harris, B., Neal, D., and Vangel, M., "Statistical Methods for Calculating Design Allowables in MIL-HDBK-17," presented at the Second Symposium on Test Methods and Design Allowables for Fiber Composites, Phoenix, AZ, 3-4 Nov. ASTM, 1986.
[7] Scholz, F. W. and Stephens, M. A., "K-Sample Anderson-Darling Tests of Fit for Continuous and Discrete Cases," *Journal of the American Statistical Association*, Sept. 1987, pp. 918-924.

Peter K. Schenck[1] and Jennifer R. Dennis[2]

PC-Access to Ceramic Phase Diagrams

REFERENCE: Schenck, P. K. and Dennis, J. R., "**PC-Access to Ceramic Phase Diagrams,**" *Computerization and Networking of Materials Data Bases, ASTM STP 1017,* J. S. Glazman and J. R. Rumble, Jr., Eds., American Society for Testing and Materials, Philadelphia, 1989, pp. 292–303.

ABSTRACT: A personal computer (PC)-based version of *Phase Diagrams for Ceramists* has been demonstrated by the joint National Bureau of Standards (NBS)/American Ceramic Society (ACerS) Phase Diagrams for the Ceramists Data Center. A selection of phase diagrams from the nearly 1100 diagrams of Volume 6 has been transferred from the Data Center's graphics workstations to a PC. Demonstration software has been developed for retrieving and plotting the phase diagrams on the PC's monitor from the PC-based Phase Diagram Data Base. In addition, the software allows the operator to retrieve data from the diagram by means of an interactive graphics cursor whose location is displayed digitally on the monitor in a choice of user units (for example, °C, °F, K). Areas of the phase diagram can be magnified and replotted to clarify features. The operator can also overlay a second diagram for comparison purposes, reverse the diagram, magnify or rescale the diagram, and apply an electronic lever rule or curve-tracking mode. Future refinements include conversions between weight and mole percent and retrieval of ternary or higher-order phase diagrams.

KEY WORDS: data base, phase diagrams, personal computer (PC), graphics, ceramics, digitizing software, chemical data bases

The National Institute for Standards and Technology (NIST) and the American Ceramic Society (ACerS) are cooperating to produce a critically evaluated data base of ceramic phase diagram information, which is published in the *Phase Diagrams for Ceramists* compilation series [*1*]. Currently there are two separate data bases: one contains the bibliographic information associated with ceramic phase diagrams; the other is the graphics data base, which contains the diagrams themselves. The bibliographic data base is described in the preceding paper of this volume [*2*], along with information concerning the NIST Phase Diagrams for the Ceramists Data Center (Data Center).

This paper describes prototype software for personal computer (PC) dissemination of the graphics data base. We envisioned that an appropriate and practical method of disseminating the graphics data base was through the PC market because of the proliferation of PCs in the workplace. Graphics data would be provided on floppy disks, accompanied by a program that retrieves the data, plots it on the monitor, and allows user manipulation of the diagrams. The software is designed to supplement the hard-copy volumes of the *Phase Diagrams for Ceramists* series, in which the phase diagrams and their accompanying bibliographic information and peer reviews are published.

---

[1]Physicist, Ceramics Division, National Institute for Standards and Technology, Gaithersburg, MD 20899.
[2]Ceramics Division A 229, National Institute for Standards and Technology, Gaithersburg, MD 20899.

Phase Diagram Data Base

Diagram Digitization

The graphics data base is currently being generated using software developed in the Data Center for technical desktop computers. The digitization program and procedure is described in detail elsewhere [3,4] and is only outlined here. Ceramic phase diagrams are identified in the literature by the staff in the Data Center and then sent out for peer review and evaluation. The diagrams are then digitized by a team of student technicians and added to the graphics data base.

The digitizing software is written in BASIC and allows the operator to digitize the diagram, edit the data file for clarity and accuracy, and obtain a hard copy of the diagram on a laser printer or multiple pen drafting plotter. The starting material in the digitization process is a photocopy of the phase diagram from the publication in which it appeared. After the diagram is digitized, the software performs automatic editing functions and scaling to remove aberrations introduced during the original drafting or photocopying process. The operator can also edit the data file for accuracy by keyboarding in known points, such as melting points and compound positions. This data is obtained from the original paper in which the diagram appeared and is only as accurate as the experimental techniques used to obtain the phase diagram. The program uses a digital smoothing algorithm on the digitized points of the curves. The smoothing is done because only a sparse sample of the points on the curve is digitized. The algorithm inserts five points between the actual digitized points by linear interpolation. A three-point smooth is then applied ten times to the augmented curve before plotting on the Hewlett Packard (HP) computer. The resulting curves are not only smooth but also more accurate than the curves of the original diagram since they incorporate any tabular data entered during the editing phase of the program. Labels are typed and positioned in the diagram using either X-Y pairs or an interactive graphics cursor. As an example of the capabilities of the software in capturing and reproducing phase diagram data, a photocopy of a phase diagram as it appears in the original publication is shown in Fig. 1 [5]. The digitized version is shown in Fig. 2 in the style of Vol. 6 of *Phase Diagrams for Ceramists*, which deletes certain features of the original diagram.

The phase diagram data files are stored in binary format on floppy disks and an average diagram requires 4K bytes of storage space. The data files contain information about the boundaries of the diagram, scaling factors, digitized points of each curve, line type (for example, bold, dashed, and so forth) and the location and content of all labels. The labels contain information about size, centering, orientation, super/subscripts, and alternate font sets, encoded as escape sequences.

Data Base Transfer

The digitizing software for the workstations was modified to provide tabular output data for a diagram via an RS-232 data link to a PC. The PC software, written in BASIC, inputs the tabular data and stores it for later use. The transfer process requires the PC to receive information concerning the file size and the number of curves and messages. The PC program then allocates the appropriate arrays and reads in the data via its serial port.

The coordinates and line type of the smoothed augmented curves are transferred to the PC, instead of the original or edited digitized points that make up the workstation data file. This is done because the smoothing algorithm is computationally intensive and would slow down diagram retrieval on the PC. Axes and label information is transferred as-is to the PC. A demonstration data base of 50 diagrams has been made with this technique and stored as ASCII files on a PC.

The size, aspect ratio, and resolution of both the diagram and its labels in the PC data base

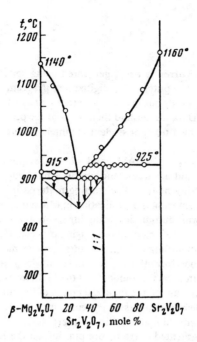

FIG. 1—*Photographic reproduction of original phase diagram of $Mg_2V_2O_7$-$Sr_2V_2O_7$ system from Ref 5.*

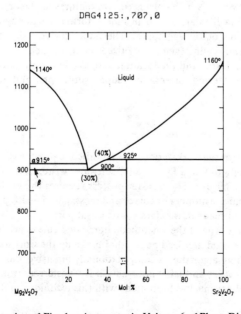

FIG. 2—*Digitized version of Fig. 1 as it appears in Volume 6 of* Phase Diagrams for Ceramists.

are different than they are in the hardcopy of Vol. 6. Minor editing allowed labels to be reordered and repositioned on the diagram. The edited diagram then replaces the original transferred diagram in the PC data base.

PC Program

Hardware Configuration

The required hardware configuration for the present PC software is essentially an IBM-compatible PC with at least 512K bytes of memory, one floppy disk drive, and a color graphics adaptor (CGA) or enhanced graphics adaptor (EGA) with appropriate monitor. An EGA graphics adaptor is required if all of the implemented program features are to be available. An abbreviated set of the program's features is available when using the CGA graphics adaptor. The program automatically selects the highest resolution mode available with the installed graphics card. The CGA card allows only two colors, white and black. The EGA card allows 16 user-selectable colors. Different colors can be used for labels, curves, and borders. Algorithms such as the LEVER RULE (discussed below) can distinguish between curve and label points. In addition, color can be used in special modes to highlight the diagram.

Data Storage Requirements

The entire set of diagrams in Vol. 6 will require between 3.6 and 5.4M bytes of storage for the PC data base, or 10 to 15 standard, 5.25-in. (133-mm) floppy disks. Preliminary tests involving packing the data files indicate that the storage requirements could be reduced by a factor of three. Most users of the data base would probably store a small subset of the diagrams on an internal fixed disk, if available, or on a single, working, 5.25-in. (133-mm) disk.

General Program Information

The PC program that retrieves and plots the diagrams on the PC's monitor is written in BASIC. The source code is compiled as an object file and then linked to a standard library to generate a standalone EXE file. The EXE program is currently 126K bytes. The primary function of this program is to draw the phase diagrams on the PC monitor. Curves and axis lines are readily plotted using the LINE statement. The greatest problem in doing this is the limited set of graphics commands available to PC programmers in high-level programming languages. Specifically, the lack of adequate commands for labeling PC monitor plots requires the development of algorithms for producing professional-looking labels on the diagrams. Drawing labels on the PC is tedious and slow and also results in a very crude-looking diagram. The internal character set of common graphics boards, usually used in the text modes of the PC, is very reasonable, especially on the EGA boards.

One way to employ the characters from the internal set in graphics applications is to coordinate drawn graphics, that is, curves and axes, with available text locations, that is, rows and columns, on the monitor. This technique limits the locations available and is only satisfactory for horizontal labels. Since the phase diagrams require vertical and slanted labels, the BASIC graphics statements, such as GET and PUT, and alternatively POINT and PSET, are used to retrieve the dot patterns of characters in labels and position them anywhere on the screen with any orientation. This process allows the generation of professional-looking vertical and slanted lettering as well as true superscripts and subscripts.

The program is interactive and uses the INKEY$ statement to trap keystrokes. (Nonsupported keystrokes are ignored.) Only the allowed keystrokes for the current mode are decoded and result in program action.

Program Operation and Features

The program starts by testing for the graphics board. If a graphics board is absent, the program returns to the operating system. Detection of either the EGA or CGA graphics adaptor sets appropriate internal flags for use by the graphics commands of the program. The program then displays an introduction and proceeds with several optional screens of instructions.

Figures 3 to 15 are screen copies demonstrating the program operation. The first screen (Fig. 3) lists the available binary phase diagrams identified by figure number (with the BPD extension) from Volume 6 of *Phase Diagrams for Ceramists*. The diagrams for Vol. 6 are organized in numerical order on the disks to correspond with the book. The message at the bottom of the screen directs the user to select a diagram by moving the pointer (shown next to Column 1, Row 9) with the keyboard's arrow keys.

In Fig. 4, FIG6741.BPD was selected, and the diagram is drawn on the screen. The compositional end members of the diagram, in this case $Mg_2V_2O_7$ and $Sr_2V_2O_7$, are the first and second labels of the diagram. All of the temperature-composition (T-X) diagrams in the PC data base will follow this labeling rule, providing an opportunity for searching the graphics data base for chemical systems. A limited number of special characters are available for display on the screen, that is, alpha, beta, and so forth.

The graphics cursor appears on the screen as a small cross as shown in Fig. 5 in the Liquid field, above the (40%) label. The cursor position is digitally displayed in the bottom-left corner of the screen. The graphics cursor can be moved with the arrow keys 1% (per stroke) of the range of the diagram in either the T or X direction. The relative cursor motion may be reduced to a pixel-by-pixel mode (by pressing key S) or at double the speed in the fast mode (by pressing the F key). The temperature units can be changed (with the T key) from °C (the standard units of T-X diagrams) to °F to K.

An important feature is a pop-down menu of available functions illustrated in Fig. 6. More detailed help is also available. Pressing one of the allowed access keys results in initiating that mode, that is, pressing N for NEW returns the operator to the diagram selection mode. Any unimplemented key clears the help menu. The help menu is abbreviated if a CGA graphics card is detected by the program.

The diagram can be reversed, as shown in Figs. 7 and 8. This feature eases the task of comparing two diagrams of the same system that were originally drafted with the end components reversed. The user must recognize that after reversing the diagram, some labels may override the curves, and the percent composition labels, for example, (72%), are not recalculated. The utility of the REVERSE feature will become obvious in the section that details the OVERLAY feature.

```
E:\
FIG6351 .BPD      FIG6352 .BPD      FIG6353 .BPD      FIG6356B.BPD
FIG6359 .BPD      FIG6361 .BPD      FIG6362 .BPD      FIG6363 .BPD
FIG6368 .BPD      FIG6369 .BPD      FIG6378 .BPD      FIG6381A.BPD
FIG6382A.BPD      FIG6382B.BPD      FIG6385 .BPD      FIG6386A.BPD
FIG6411 .BPD      FIG6412 .BPD      FIG6417 .BPD      FIG6421 .BPD
FIG6423 .BPD      FIG6428 .BPD      FIG6445 .BPD      FIG6464C.BPD
FIG6468 .BPD      FIG6479 .BPD      FIG6491 .BPD      FIG6497 .BPD
FIG6584B.BPD      FIG6505 .BPD      FIG6521 .BPD      FIG6641 .BPD
FIG6741 .BPD<     FIG6798 .BPD      FIG6806 .BPD      FIG6835D.BPD
FIG6909 .BPD      FIG6392 .BPD      TERN    .BPD
    51584 Bytes free

           SELECT DIAGRAM WITH CURSOR KEYS, THEN PRESS ENTER
```

FIG. 3—*Monitor screen dump showing available phase diagrams and selection cursor.*

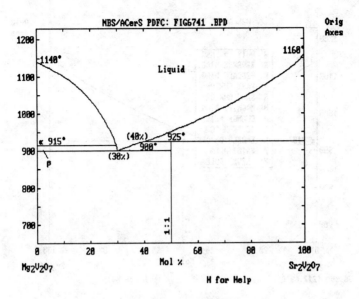

FIG. 4—*Phase diagram FIG6741.BPD as drawn on the PC monitor's screen.*

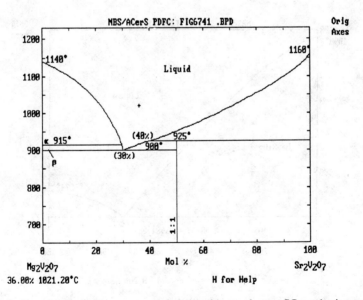

FIG. 5—*Interactive graphics cursor and digital position readout on PC monitor's screen.*

Any portion of the diagram may be magnified on-screen to increase the visual clarity of a region using either the MAGNIFY or RESCALE features. MAGNIFY enlarges the area immediately surrounding the cursor by a factor of two, as shown in Figs. 9 and 10. In Fig. 10 an overall magnification of four was obtained. Too many multiple magnifications can eventually result in the program aborting. The diagram may also be rescaled, the equivalent of variable

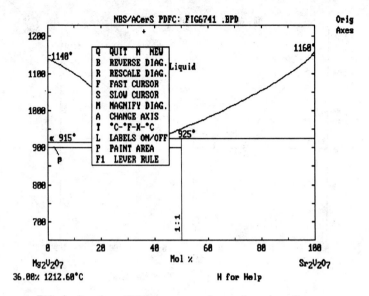

FIG. 6—*Pop-down HELP screen superimposed on phase diagram.*

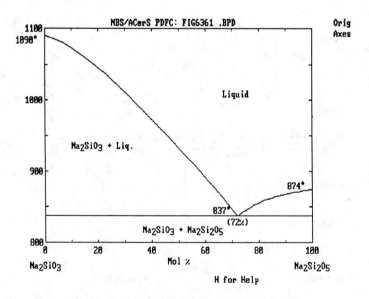

FIG. 7—*PC phase diagram of the Na_2SiO_3-$Na_2Si_2O_5$ system.*

magnification, by entering new X-Y ranges. For example, in Fig. 11 a new X range of 60 to 80% and a new Y range of 1200 to 1400°C have been selected, resulting in the exploded version seen in Fig. 12. The decade axes have been selected by pressing A to demarcate the diagram evenly. Rescaling instead of magnifying results in more control over the appearance of the enlarged diagram.

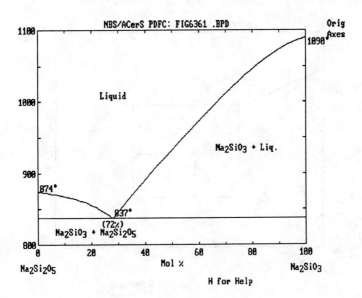

FIG. 8—*Reversed version of Fig. 7.*

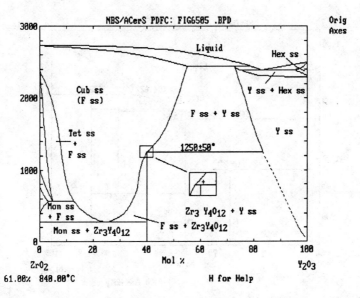

FIG. 9—*PC phase diagram of the ZrO_2-Y_2O_3 system with graphics cursor located on region of interest for magnification.*

There are a number of features in the PC program that are not implemented in the digitizing program, and vice versa. The purpose of the digitizing program is to capture phase diagram data, provide for computer storage, and produce camera-ready copy for publication purposes, whereas the purpose of the PC program is to disseminate the phase diagrams and provide interpretive aids. One of the features that gives an electronic version of the data base a significant

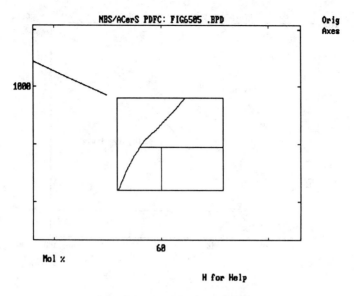

FIG. 10—*Magnified region of Fig. 9.*

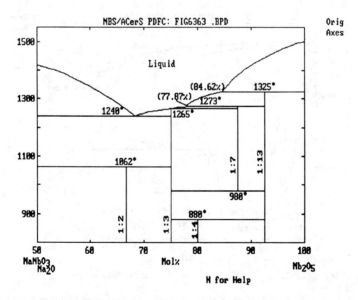

FIG. 11—*PC partial phase diagram of the Na_2O-Nb_2O_5 system.*

edge over the hard copy presentation is the ability to OVERLAY multiple diagrams for comparative purposes. Figure 13 shows the overlaid phase equilibria in the system calcium oxide/silicon dioxide (CaO-SiO_2) at two different pressures. On the screen, the second, or overlaid diagram appears in red if an EGA graphics card is present. If one of the diagrams had SiO_2 as the left-hand component, than the REVERSE feature can be used before OVERLAY. Though any two diagrams may be overlaid, the X-Y ranges must be compatible. The first diagram selected

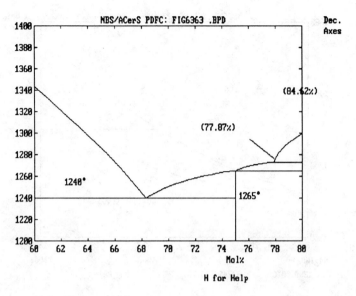

FIG. 12—*Rescaled phase diagram of Fig. 11 with decade axes.*

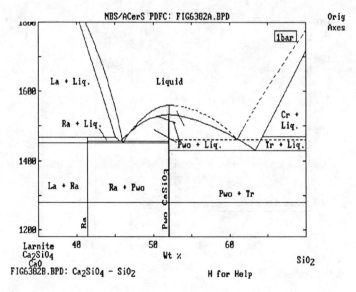

FIG. 13—*PC phase diagram of the CaO-SiO$_2$ system with an overlay of the phase diagram at a second pressure.*

sets the X-Y range, and the second diagram must have some portion that falls within that range.

Another program feature that aids in the interpretation of the phase diagrams is the application of an electronic lever rule. The lever rule, as applied to binary phase diagrams, is used to compute the relative amounts of each phase in a two-phase region. The graphics cursor is used

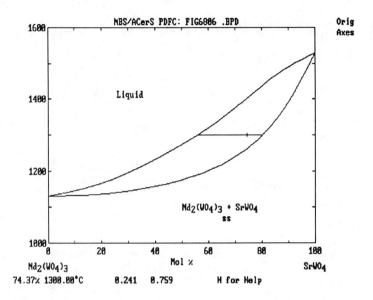

FIG. 14—*PC phase diagram of the $Nd_2(WO_4)_3$-$SrWO_4$ system with electronic lever rule.*

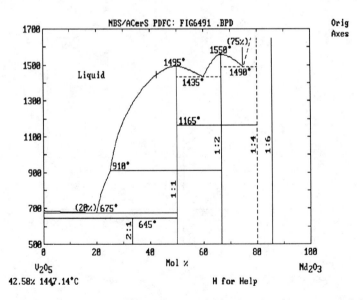

FIG. 15—*PC phase diagram of the V_2O_5-Nd_2O_3 system with a curve tracking cursor and digital curve coordinate readout.*

to identify the two-phase region of interest. A horizontal line is drawn connecting the two phases (the lever), as shown in Fig. 14, and the graphics cursor changes color to indicate the special mode. The ends of the horizontal line are identified by searching the adjacent left and right pixels from the cursor until pixels are located that are the color of the curves.

As noted above, with an EGA graphics card, the labels and other noncurve features are a different color and don't interfere with the LEVER. The cursor moves left to right in the LEVER RULE mode, and appears in Fig. 14 as the small vertical bar intersecting the horizontal LEVER. The digital readout of the cursor position (bottom left corner) also changes color to indicate the special mode. The relative percentage of the two phases (left, right) at any cursor position is displayed to the right of the cursor position. The LEVER RULE feature is not available with CGA cards.

Reading out the coordinates of a curve is facilitated by the CURVE FOLLOW or track mode (Fig. 15). Again, the graphics cursor is used to select a curve. The digital readout of the position and the cursor change color to indicate this special mode. The new cursor follows the curve to the left or right in response to the keyboard's left/right arrow keys. The coordinates displayed in the digital readout are all points along the curve. The CURVE FOLLOW mode is not implemented with CGA graphics cards since the presence of labels of the same color might interfere with tracking the curve in a manner similar to the LEVER RULE.

Additional features are available to users who would like to prepare slides or figures from user-modified diagrams. Users are allowed to "paint," that is, color fill, an area to highlight it. This kind of highlighting is best if the labels are turned off, the area is painted, and then the labels are restored. The PAINT AREA option is only available if an EGA graphics card is present.

Future Program Features

The current version of the prototype PC software supports only binary phase diagrams. Work is underway to expand the software to include ternary and other higher-order systems as well as schematic diagrams. Early feedback from users indicates that the ability to overlay user data on the diagrams is desirable. Utilities enabling the user to create a subset of the diagrams on their fixed disk or separate floppy disks are also being considered.

Conclusions

The prototype program for PC access to ceramic phase diagrams provides the user with an interactive environment to study the diagrams, retrieve data from them, and manipulate them for their own specific uses. The program augments the value of the compilation of critically evaluated phase diagrams in the volumes of *Phase Diagrams for Ceramists*.

Acknowledgments

We would like to thank A. C. Van Orden, D. W. Bonnell, and J. W. Hastie for their support and suggestions during the early development of the software. In addition, we also acknowledge the financial support of the NIST Standard Reference Data Program.

References

[1] Roth, R. S., Dennis, J. R., and McMurdie, H. F., Eds., *Phase Diagrams for Ceramists*, Vol. 6, American Ceramics Society, Westerville, OH, 1987.
[2] Ondik, H. and Messina, C., this volume pp. 304–314.
[3] Schenck, P. K. and Dennis, J. R., *Computer Handling and Dissemination of Data*, P. S. Glaeser, Ed., Elsevier, New York, 1987, pp. 184–188.
[4] Schenck, P. K. and Dennis, J. R., *User Applications of Alloy Phase Diagrams*, L. Kaufman, Ed., ASM International, 1987, pp. 255–261.
[5] Zhuravlev, V. D., Fotier, A. A., Zhukov, V. P., and Kristallov, L. V., *Russian Journal of Inorganic Chemistry*, Vol. 27, No. 4, p. 573.

Helen M. Ondik[1] and Carla G. Messina[1]

Creating a Materials Data Base Builder and Producing Publications for Ceramic Phase Diagrams

REFERENCE: Ondik, H. M. and Messina, C. G., **"Creating a Materials Data Base Builder and Producing Publications for Ceramic Phase Diagrams,"** *Computerization and Networking of Materials Data Bases, ASTM STP 1017*, J. S. Glazman and J. R. Rumble, Jr., Eds., American Society for Testing and Materials, Philadelphia, 1989, pp. 304-314.

ABSTRACT: The Ceramics Phase Diagrams Data Center faced several problems in the early stages of developing a computer data base to distribute to the ceramics user community. The Data Center had to create and store data files containing full scientific notation, provide for transferring data to undesignated computer(s) from the computer on which the files were created, and provide for installation of the data on undetermined Data Base Management System(s) for user access. The solution was incorporated in programs developed in the National Institute for Standards and Technology (NIST) Office of Standard Reference Data and expanded under the support of the American Ceramic Society. *All* information is expressed using *only* printable ASCII characters and is stored in dynamic arrays with all fixed length parameters encoded as program variables. Bibliographic and some chemical data for the next several volumes of *Phase Diagrams for Ceramists* are currently being stored in this form. The data are being manipulated to produce material in typeset form for the published volumes. They have also been transferred with no loss of information to a prototype data base for rapid retrieval.

KEY WORDS: data bases, data base builder, phase diagrams, typesetting

In late 1982 the National Institute for Standards and Technology (NIST) and the American Ceramic Society (ACerS) formalized a long-standing cooperative effort to provide high-quality phase equilibria data to the technical community. A major part of the effort included producing and publishing the compilation *Phase Diagrams for Ceramists*. Under the new agreement, both organizations expanded the support of the Ceramic Phase Diagram Data Center at NBS. NBS continued to be responsible for gathering resource data, for developing the program, and for overseeing the critical evaluation of the data. The ACerS assumed responsibility for the production aspects of the work and for all means of data dissemination.

Among several new goals set for the Data Center were three of direct interest to this symposium. They are (1) the development of a computer data base for direct access, (2) an increase in publication frequency, and (3) the development of computer graphics for the storage and retrieval of phase diagrams. The computer data base is to contain bibliographic and key word information, digitized graphics data, and commentaries evaluating each diagram. Some discussion of the development of the computer graphics task is covered in the paper by Schenck and Dennis in this volume.

The technical complexity of the textual data presented challenges to the Data Center and the actions taken to cope with these challenges are the subject of this paper.

---

[1] Research chemist, National Institute for Standards and Technology, and research associate, American Ceramic Society, National Bureau of Standards, Gaithersburg, MD 20899.

Challenges for Data Entry, Storage, and Dissemination

When the NBS-ACerS agreement was signed, Volume 5 of the compilation series was in press, and the Data Center had new references for about 6700 phase diagrams for subsequent volumes. It was important to begin computerization as soon as possible. The data entry and storage format, however, usually depend on the use for which data files are intended and the end uses of these files had been defined only in a general way. Decisions were yet to be made about the specific computers, the data base management system (DBMS), the typesetting procedures for book production, and the computer-dissemination methods to be used for the long term.

These decisions require time to consider all factors, but the backlog of information required more immediate action. It was necessary, therefore, to design and build data files that would be totally portable. The Data Center staff had to have data files with the following characteristics:

1. Data had to be formatted flexibly enough to permit insertion into any as-yet undefined DBMS and to permit the future development of a data base for distribution in any existing or future on-line or stand-alone computer system.
2. Data had to be encoded sufficiently so as to provide input to any typesetting routines presently available or yet to be developed.

Besides the required flexibility and portability properties of the data files, rapid keyboarding and easy editing methods were needed. A method also had to be devised to include full chemical notation and special character sets in a style that was machine independent. Printed copy, including all special characters for easy proofreading of the data, had to be readily available from computer printers. The computer facilities of the NBS Standard Reference Data Program (SRD) and their initial programming support were made available as a portion of that program's continuing contribution to the project.

Software Design Principles Used

To provide for flexibility and portability, these principles were followed in the writing of the software:

1. Data were to be divided into unique fields requiring no further subdivision.
2. Data fields were to be complete with no restrictions on the size or character of the field content.
3. All required coding information that did not apply to the whole field, but only to a part, was to be stored completely within the field. For example, codes to identify subscripts, superscripts, Greek and math characters, and bold and italic fonts had to be part of the data content of the field.
4. Only normally printable ASCII characters were to be used with the exception of the ASCII shift out (014) and shift in (015) codes. The shift out and shift in codes indicate an alternate character set. No use of escape sequences or characters unique to specific computers was to be allowed. This restriction permits use of the simplest computer terminal even though the alternate character set may not appear correctly on all devices.

The programs developed using these principles, the ABCUP PROGRAM SERIES, provide the data file characteristics required. The name ABCUP is derived from the properties of the data files and the software:

A—added value information created by the Data Center,
B—bibliographic information applicable to publications,

C—contents of the publications, both the original reference and the one being generated, stored in brief form, and

UP—update and editing procedures provided by the software.

The ABCUP Program Series

Data File Structure

The development of the ABCUP Programs was begun under SRD sponsorship, and has been expanded under ACerS sponsorship. The software is written in Fortran 77 and uses only the standard ASCII Character Set. Care was taken to make the programs as independent of the host minicomputer as possible.

To accommodate the highly variable data, the data files are structured to allow for a fixed (but indeterminate) number of records consisting of fields of unknown length. The fields, called data elements, are defined by the contents of the first record of a file. The definitions of the fields that identify each phase diagram at this time are listed in Fig. 1. Although 75 separate

```
Data element  1: Citation No.
Data element  2: Publication Type
Data element  3: Authors
Data element  4: Editors of Publication
Data element  5: Article No.
Data element  6: Article Title
Data element  7: Publication Series
Data element  8: CODEN
Data element  9: Vol., Ed. or No.
Data element 10: Issue
Data element 11: Page
Data element 12: Contract No.
Data element 13: Publication Date
Data element 14: Publisher; City, State
Data element 15: Sponsor and Address
Data element 16: Site Address
Data element 17: Patent Nos.
Data element 18: Application or Meeting Date
Data element 19: Dissertation Degree
Data element 20: Date of Last Data Record Update
Data element 21: Country of Patent issue
Data element 22: No. of Authors
Data element 23: Country (Authors)
Data element 24: Language
Data element 25: Eng. Transl. Journal Ref.
Data element 26: Abstract Source
Data element 27: Abstract Nos.
Data element 28: Related Citation Nos.
Data element 29: PDFC Working No.
Data element 30: PDFC Volume No.
Data element 31: PDFC Figure No.
Data element 32: Contributing Editors
Data element 33: CHEMICAL SYSTEM
Data element 34: No. of Chemicals defining System
Data element 35: Commentary Received
Data element 36: Diagram Done
Data element 37: Comments
Data element 38: Commentary Keyboarded
```

FIG. 1—*The first record of a data file defining the fields for ceramic phase diagram use.*

data elements have been defined for bibliographic and key word identification of a phase diagram, only 38 are currently in use. Data element 1, the citation number, is arbitrarily assigned, and identifies the record. It is used to track all information belonging to a given diagram that may reside in different data files—bibliography and key words, diagram graphics, and commentary texts.

The second data element, the publication type, identifies the reference as a journal paper, book, patent, conference report, and so forth. There are a total of 21 data elements employed to identify references. On the average, about nine of these are used for any particular bibliographic reference. The larger number takes into account the variety of fields needed to accommodate the different types of information identifying journal articles, books, conference reports, patents, and so forth. The complete flexibility requirement imposed on the data files is met by storing each item in a reference as a separate data field. For example, the reference for a paper in a technical journal consists of individual data fields for the title of the article, the title of the journal, the author(s), the volume number, the issue number, the page numbers, the year of publication, and the CODEN. The software displays the appropriate data fields during keyboarding to suit the different reference types.

Software commands permit the rearrangement and combination of data elements to satisfy the input requirements for a specific bibliographic citation system or DBMS. When a bibliographic list is created for typesetting, the computer automatically applies the appropriate format used by the ACerS to each reference using the information stored in the record. See Fig. 2 for examples of reference data elements and Fig. 3 for the same references printed in the format used in Volume 6 of *Phase Diagrams for Ceramists*. Note that in Fig. 2(b) combinations of the characters |, /, \, {, and } appear. These character combinations are the flags that provide for the bold and italic characters displayed in Fig. 3(b).

The records contain other data elements used by the Data Center to identify the chemistry, and to provide for data checking, record keeping, and information status tracking. One of the more important of the tracking data elements is number 20, which contains the date and time of the last update of an individual data record. Each time a data record is revised, the computer software changes this "date/time stamp" to the current date and time. The date and time will be used to determine which portion of the phase diagram data files must be entered into a data base management system. Only new information or records revised after the date of the last update need be considered.

Internal Computer File Structure

The internal structure of a data record within an ABCUP file is a dynamic array of data elements as opposed to a dimensioned array. Typical data records, as they appear on the screen, are shown in Fig. 4. The dynamic nature of the array of data elements is illustrated in Fig. 5, which shows the same records as they are stored within the computer.

The data records are stored in a format using 128-character ASCII lines. The first line of numbers in each record in Fig. 5 defines, in order, the number of data elements present in the record, the number of 128-character lines needed to supply the addresses of the data elements, the number of characters in the record, the number of 128-character lines in the record, the address of the beginning of the *next* data record, and the address of the beginning of the *previous* data record. The numbers in the next three lines of the records provide the addresses of the individual data elements within the records. The subsequent lines contain the actual ASCII characters for the data in the records. The files generated by the software are designed to be treated both as serial files for ready transfer to other computers or direct access files for rapid data access. For fast retrieval of individual ABCUP data records, these files are currently direct access files. Therefore, the total number of lines per record is set when a data file is first created, but this number can be expanded whenever necessary.

(a) Reference Data Elements

1. Citation No. : 03913
2. Publication Type : 1
3. Authors : Kimizuka, N.; Takayama, E.
6. Article Title : Survey of the phase formation in the $Yb_2O_3\text{-}Ga_2O_3\text{-}MO$ and $Yb_2O_3\text{-}Cr_2O_3\text{-}MO$ systems in air at high temperatures (M: Co, Ni, Cu, and Zn)
7. Publication Series : J. Solid State Chem.
8. CODEN : JSSCBI
9. Vol., Ed. or No. : 43
10. Issue : 3
11. Page : 278-284
13. Publication Date : 1982

(b) Reference Data Elements

1. Citation No. : 02999
2. Publication Type : 1
3. Authors : Pashinkin, A. S.
6. Article Title : Partial pressure diagrams of metal-tellurium-oxygen systems
7. Publication Series : Izv. Akad. Nauk SSSR, Neorg. Mater.
8. CODEN : IVNMAW
9. Vol., Ed. or No. : 11
10. Issue : 9
11. Page : 1650-1653
13. Publication Date : 1975
24. Language : Russian
25. Eng. Transl. Journal Ref. : |/Inorg. Mater. (USSR) (Engl. Transl.)|\, |{11|} [9] 1411-1414 (1975)

(c) Reference Data Elements

1. Citation No. : 04132
2. Publication Type : 6
3. Authors : Negas, T.; Roth, R. S.; McDaniel, C. L.; Parker, H. S.; Olson, C. D.
4. Editors of Publication : Lundin, C. E.
6. Article Title : Influence of potassium oxide on the cerium oxide-zirconium oxide system
7. Publication Series : Proc. Rare Earth Res. Conf., 12th
8. CODEN : 33WQAI
9. Vol., Ed. or No. : 2
11. Page : 605-614
13. Publication Date : 1976
14. Publisher; City, State : Univ. Denver: Denver, CO
16. Site Address : Denver, CO

FIG. 2—*Examples of reference information data elements.*

The data files generated with the ABCUP programs, currently referencing about 13 000 phase diagrams, are very flexible. Data elements can be combined and reformatted to meet many requirements. Complete advantage can be, and has been, taken of different computer facilities and optimum use has been made of other software, that is, editor, sort, and print routines. The input data are proofread using regular computer-printer output. Data are also checked by the computer against standard tables and against format standards defined for individual data fields.

(a) Formatted Reference

N. Kimizuka and E. Takayama, *J. Solid State Chem.*, 43 [3] 278-284 (1982).

(b) Formatted Reference

A. S. Pashinkin, *Izv. Akad. Nauk SSSR, Neorg. Mater.*, 11 [9] 1650-1653 (1975); *Inorg. Mater. (USSR) (Engl. Transl.)*, 11 [9] 1411-1414 (1975).

(c) Formatted Reference

T. Negas, R. S. Roth, C. L. McDaniel, H. S. Parker, and C. D. Olson, ''Influence of potassium oxide on the cerium oxide-zirconium oxide system,'' Proc. Rare Earth Res. Conf., 12th, Denver, CO, 2 p. 605-614 (1976).

FIG. 3—*Retrieved formatted references corresponding to the data elements in Fig. 2.*

Handling of Chemical Formulas and Special Characters

Note in Fig. 4 that the subscripts in the chemical formulas are correctly indicated as subscripts on the screen. The subscript display is a property of the software and is independent of the computer equipment used to access the data files. This capability is one example of the equipment-independence built into the software.

Data elements containing chemical formulas are easily keyboarded. Superscripts and subscripts are encoded using the ASCII slash characters to define them. For instance, the chemical systems in the data records shown in Fig. 4 were keyboarded using the characters shown in Fig. 6a and c. The direction in which the pairs of slashes point serves as a mnemonic tool for the operator. The operator can readily check the formulas as keyboarded and then confirm the correctness of the keyboarding by calling the record back on the screen. The formulas entered as shown in Fig. 6, items (a) and (c) then appear as shown in Fig. 6, items (b) and (d). The ^ character appearing at the beginning of a line indicates that the characters on that line are subscripts belonging to the line above.

Superscripts are handled similarly. For instance, cations and anions would be keyboarded as shown in Fig. 6, items (e) and (g) and can be observed on screen and in computer-printed form as shown in Fig. 6, items (f) and (h). The lower case v at the beginning of the top line serves as a pointer to indicate that the characters on that line are superscripts for the characters on the line below.

Superscripts can also be placed *above* subscripts if needed by the use of a makeshift back arrow consisting of the characters < and —. The keyboarding proceeds as shown in Fig. 6, item (g). The makeshift arrow instructs the computer to backspace before displaying the rest of the line. The subscript and superscript may be keyboarded in either order as illustrated. The ions are observed on screen and in computer-printed form as shown in Fig. 6, item (h). Any number of characters can be included between the pairs of slashes as illustrated in Fig. 6, items (i) and (j). The simple coding outlined here permits on-screen checking of the correctness of the formula by the keyboard operator and reduces the number of editing corrections necessary when hard copy is proofread.

The book editors are supplied with updated lists of the proposed chemical systems for new volumes regularly. The ability to provide chemical formulas on screen and computer printouts

(a) First Sample Record

1. Citation No. : 01167
2. Publication Type : 1
3. Authors : Get'man, E. I.; Marchenko, V. I.
6. Article Title : Refinement of the phase diagram of the molybdenum trioxide-bismuth oxide system in the bismuth molybdate ($Bi_2(MoO_4)_3$) composition region
7. Publication Series : Zh. Neorg. Khim.
8. CODEN : ZNOKAQ
9. Vol., Ed. or No. : 26
10. Issue : 4
11. Page : 1034-1037
13. Publication Date : 1981
22. No. of Authors : 2
23. Country (Authors) : UR
24. Language : Russian
25. Eng. Transl. Journal Ref. : |/Russ. J. Inorg. Chem. (Engl. Transl.)|\, |(26|) [4] 559-561 (1981)
26. Abstract Source : CAS
27. Abstract Nos. : 94:198296h
29. PDFC Working No. : EK-2018
30. PDFC Volume No. : 06
31. PDFC Figure No. : 6459
32. Contributing Editors : E.R.K.
33. CHEMICAL SYSTEM : Bi_2O_3-MoO_3
35. Commentary Received : YES
36. Diagram Done : YES
38. Commentary Keyboarded : YES
20. Date of Last Data Record Update : 2:24 PM TUE., 10 FEB., 1987

(b) Second Sample Record

1. Citation No. : 02633
2. Publication Type : 1
3. Authors : Fix, W.; Koch, K.; Bongers, U.
6. Article Title : Precipitation or iron oxide apatite from lime-phosphate melts with low silicic acid contents
7. Publication Series : Arch. Eisenhuettenwes.
8. CODEN : AREIAT
9. Vol., Ed. or No. : 46
10. Issue : 9
11. Page : 555-559
13. Publication Date : 1975
22. No. of Authors : 3
23. Country (Authors) : GE
24. Language : German
26. Abstract Source : CAS
27. Abstract Nos. : 83:198428a
28. Related Citation Nos. : 02633; 02634; 02635; 02636; 02637; 02587
29. PDFC Working No. : AM-0417A
30. PDFC Volume No. : 06
31. PDFC Figure No. : 6941A
32. Contributing Editors : A.M.
33. CHEMICAL SYSTEM : CaO-FeO-Fe_2O_3-SiO_2-P_2O_5
35. Commentary Received : YES
36. Diagram Done : YES
38. Commentary Keyboarded : YES
20. Date of Last Data Record Update : 1:58 PM MON., 16 MAR., 1987

Note that in (a), Data Element 25 contains ASCII flags used to signal bold and italic font changes for printing purposes. The inclusion of these flags follows the third software design principle.

FIG. 4—*Sample data records.*

ONDIK AND MESSINA ON CERAMIC PHASE DIAGRAMS 311

```
    45    3   432     4          4487       4461
     1    7     9   -42     0      42   191   208   215   218   220  -230   230  -235     0     0     0     0     0   235  -268
   268  270   273   281   350   354  -365   365   373   376   381   388  -411   411   415  -419     0     0   419  -429     0
     0    G   429  -433     0     0     0     0     0     0     0     0     0     0     0     0     0     0     0     0  -433
01167 1 Get'man, E. I.; Marchenko, V. I. Refinement of the phase diagram of the molybdenum trioxide-bismuth oxide system in the
bismuth molybdate (Bi\/2/\(MoO\/4/\)\/3/\) composition region Zh. Neorg. Khim. ZNOKAQ 26 4 1034-1037 1981  2:24 PM TUE., 10 FE
B., 1987 2 UR Russian |/Russ. J. Inorg. Chem. (Engl. Transl.)|\, |{26|} [4] 559-561 (1981) CAS 94:198296h EK-2018 06 6459 E.R.
K. Bi\/2/\O\/3/\-MoO\/3/\ YES YES Bi; O; Mo YES


    45    3   377     3         13795      13769
     1    7     9   -40     0     40   133   156   163   166   168  -176   176  -181     0     0     0     0     0   181  -214
   214  216   219  -226   226   230   241   282   291   294   300   305  -349   349   353  -357     0     0   357  -374     0
     0    0   374  -378     0     0     0     0     0     0     0     0     0     0     0     0     0     0     0     0  -378
02633 1 Fix, W.; Koch, K.; Bongers, U. Precipitation or iron oxide apatite from lime-phosphate melts with low silicic acid conte
nts Arch. Eisenhuettenwes. AREIAT 46 9 555-559 1975  1:58 PM MON., 16 MAR., 1987 3 GE German CAS 83:198428a 02633; 02634; 02
635; 02636; 02637; 02587 AM-0417A 06 6941A A.M. CaO-FeO-Fe\/2/\O\/3/\-SiO\/2/\-P\/2/\O\/5/\ YES YES Ca; O; Fe; Si; P YES
```

FIG. 5—*The internal data records corresponding to the records in Fig. 4.*

(a)   Bi\/2/\O\/3/\-MoO\/3/\

(b)   Bi$_2$O$_3$-MoO$_3$

(c)   CaO-FeO-Fe\/2/\O\/3/\-SiO\/2/\-P\/2/\O\/5/\

(d)   CaO-FeO-Fe$_2$O$_3$-SiO$_2$-P$_2$O$_5$

(e)   Na/\+\/, Ca/\++\/, S/\--\/, NO\/3/\/\-

(f)   Na$^+$, Ca^{++}, S^{--}, NO$_3^-$

(g)   NH\/4/\/\<-+\/, CO/\--\/\/<-<-3/\

(h)   NH$_4^+$, CO$_3^{--}$

(i)   SrMnO\/3-x/\, (Li\/0.5/\Ga\/2.5/\)O\/4/\)

(j)   SrMoO$_{3-x}$, (Li$_{0.5}$Ga$_{2.5}$)O$_4$

(a), (c), (e), (g), and (i) show the characters as keyboarded.

(b), (d), (f), (h), and (j) show the same characters as they appear on screen and simple computer printout.

The ^ character at the beginning of a line indicates that the characters on that line are subscripts belonging to the line above.

The lower case v at the beginning of a line serves as a pointer to indicate that the characters on that line are superscripts for the characters on the line below.

FIG. 6—*Examples of ABCUP chemical notation.*

in an easy-to-read form is of particular importance when dealing with these lists of chemical systems.

Commentary Texts

In *Phase Diagrams for Ceramists*, a commentary is published with each diagram that provides an evaluation of the data from which the diagram was constructed. These commentaries have to be keyboarded both for typesetting purposes and also for inclusion in the data base. The keyboarding of these texts was greatly influenced by the in-house minicomputer facilities available to the Data Center.

The resident editor utility was used to create the commentary texts. The same rules and conventions for the keyboarding of characters in the ABCUP data files were followed in keyboarding the commentaries. Following the fourth software design principle given above, the character set keyboarded was limited to ASCII numbers less than 127, and the ASCII shift out (014) and shift in (015) were used to indicate changes into and out of the alternate character set. The lists in Fig. 7 show examples of special symbols as they are encoded in the data files and also the markup codes to which they are translated during typesetting.

Data elements from the ABCUP data files are extracted in a processing step, which precedes the entry of the texts. In that processing step, the chemical system designation and a properly formatted reference are retrieved and inserted into the commentary file. Since a text processor utility, resident on the host computer, is available, formatting commands are inserted, and the text keyboarded. The formatting commands, known as dot lines since they begin with a period, are then used by the text processor to obtain laser-print proof copy for the book editors. This copy provides the same variety of special characters and type fonts, for example, roman, italic, bold, and so forth, as typeset copy. The same dot lines are also interpreted by the ABCUP typesetting program, which converts them to the appropriate typesetting commands.

Figure 8 contains an example of commentary text displayed on the equipment in the Data Center. The shift out and shift in codes cannot be seen in this fashion on all computer devices. The computer file of this commentary contains all the information needed for computer-controlled typesetting. Until a set of diagrams and commentaries is chosen for publication in a volume, the diagrams and commentaries are identified by arbitrarily assigned working num-

Symbol Name	Symbol	Current Shiftout Code	Current Typesetting Markup Code
lowercase alpha	α	^S_Oa^S_I	|lalpha|
lowercase beta	β	^S_Ob^S_I	|lcbeta|
uppercase sigma	Σ	^S_O?^S_I	|ucsigma|
degree	°	^S_O0^S_I	|degree|
infinity	∞	^S_O<^S_I	|infin|
plus or minus	±	^S_O%^S_I	|pom|
rightward arrow	→	^S_O\\^S_I	|arwrt|

FIG. 7—*Examples of special symbols in the current computer code and corresponding typesetting markup code.*

```
.page 0
.indent 2
|{Bi\/2/\O\/3/\-MoO\/3/\|}
.skip 2
.indent 2
Fig. |(6459|)-System MoO\/3/\-Bi\/2/\O\/3/\.
.indent 2
[1167] E. I. Get'man and V. I. Marchenko, |/Zh. Neorg. Khim.|\, |(26|)
       [4] 1034-1037 (1981); |/Russ. J. Inorg. Chem. (Engl. Transl.)|\,,
       |(26|) [4] 559-561 (1981).
.skip 2
.indent 2
Fifteen compositions in the range 20-32% Bi\/2/\O\/3/\ were prepared
from chemically pure Bi\/2/\O\/3/\ and MoO\/3/\.# Samples were
contained in open quartz (fused silica) crucibles.#
Phase boundaries were established primarily
on the basis of DTA experiments run at 1$0$C/min.# Specimens
were equilibrated for up to 300 h prior to the DTA runs.# Phases were
identified by XRD.# Although an attempt was made to obtain high
temperature XRD patterns for high-Bi\/2/\(MoO\/4/\)\/3/\, this was
unsuccessful.# The solidus curve from 656$0$S
to 600$0$S$C was shown incorrectly in the original
diagram.# A correct configuration is shown here by a dotted line,
but the exact location of this part of the solidus is
uncertain.# The solvus curves are based on changes in the transition
temperature as determined by DTA on various specimens and on phase
analyses of samples which had been heated to the temperatures of
interest.# X-ray lattice parameter changes were too small to be used
for this purpose.# High-Bi\/2/\(MoO\/4/\)\/3/\ crystallizes from
supercooled melts and can be cooled to room temperature without
transforming to the low form.# The congruent melting of
Bi\/2/\(MoO\/4/\)\/3/\ is not in agreement with earlier work by
Chen and Smith (Fig. 5196), but is in agreement with the observations
of Kohlmuller and Badaud (Fig. 4393).# The high bismuth oxide end of
the diagram is shown in Figs. 330 and 4393, which are not in agreement.
.margin left 62
E.R.K.
```

FIG. 8—*Example of a commentary text.*

bers. A special data file is generated from the ABCUP data files and information is added by which the computer recognizes those figures which contain multiple diagrams and the different types and sizes of diagrams. During the computer procedure that typesets galley copy, the figure numbers are assigned, the temporary working numbers are replaced, suitable space for each diagram is reserved and labeled, and the text material is fully typeset.

In addition, at the final typesetting stage, the computer reprocesses the commentaries, again extracting from the ABCUP data records those data elements containing the bibliographic reference information. Because of this reprocessing, it is possible to edit and correct the ABCUP data files, if necessary, in parallel with the keyboarding of the commentary texts.

The currently used book production procedure involves transferring the galley-copy files from the computer to five-inch PC floppy disks. These disks are then transferred to a composition system on which the galley copy is paginated and a direct driver tape for typesetting equipment is prepared.

Index Preparation

The order in which phase diagrams are published in the volumes of *Phase Diagrams for Ceramists* follows complex rules dependent upon valence and other chemistry. Checking the system designation (Data Element 33) for the validity of the chemical element, checking for the order of the components within the designation and reordering them within the data element when necessary, and sorting lists of chemical system designations are completely under computer control.

Data Element 33 is also used to generate the chemical systems book index. The preparation of this index involves a step which permutates the individual components of the chemical system designation and adds each of the permutated versions to the list of chemical systems. A two-component system then appears twice; a six-component system appears six times. The permutation of chemical components permits the reader to find a desired system under any one component without having to know the rules that establish the order used in the data base or in the published volume.

For the author index, the names are taken from Data Element 3. Although all of the authors for a reference are listed in one data element, the computer procedure lists each author separately for entry in the author index. After the last page of a volume has been typeset and approved, so that there is no longer a possibility of change in the figure numbers, the author and chemical systems indexes are prepared. The complete author and chemical systems indexes both were prepared and typeset on the same day that the final page of Volume 6 was finished.

A Prototype Bibliographic Data Base

An ABCUP data file corresponding to Volume 6 of *Phase Diagrams for Ceramists* and the data files for subsequent volumes have been transferred from the SRD minicomputer data files to a DBMS, available in-house at NBS, operating on a personal computer (PC). Not all of the data elements were transferred to this test data base. Some data elements used for tracking and record keeping were omitted since they are only used within the Data Center. This data base is a prototype of the larger, more complete data base, which is one of the goals of the Data Center operation. The formation of this data base has tested the portability property of the data files and shows that the transfer can be made readily with no loss of information.

Summary

The data base builder and the output routines contained in the ABCUP Program Series have been successfully used to produce Volume 6 of *Phase Diagrams for Ceramists* and form a rapid-retrieval system. To accomplish these results, the phase diagram ABCUP data records and commentary texts had to be successfully transformed by the software for transfer to a laser printer, a typesetting composition system, and a PC DBMS. Because of the data architecture, similar transformations can be achieved for data transfer to any hardcopy production equipment or any DBMS by adding other software translation subroutines.

Cooperative Data Programs in the United States

David B. Anderson[1] *and Glenn J. Laverty*[2]

Corrosion Data for Materials Performance Characterization

REFERENCE: Anderson, D. B. and Laverty, G. J., **"Corrosion Data for Materials Performance Characterization,"** *Computerization and Networking of Materials Data Bases, ASTM STP 1017*, J. S. Glazman and J. R. Rumble, Jr., Eds., American Society for Testing and Materials, Philadelphia, 1989, pp. 317–321.

ABSTRACT: A corrosion data program sponsored by the National Association of Corrosion Engineers and the National Institute of Standards and Technology includes computer data base development to characterize the corrosion performance of engineering materials over a wide variety of environments and exposure conditions. The important features of corrosion data are reviewed from both the data source and the user standpoints. Guidelines are provided for data input format development, multiple source data compilation, data validation and evaluation, and interpretive data output schemes.

KEY WORDS: corrosion, data base, data evaluation, data format, material performance, material selection, metals, nonmetals

Durability requirements are key factors in materials selection, particularly for corrosive environments. Identification of viable materials and corrosion control methods requires a review of corrosion data relevant to the candidate materials, the specific application, component fabrication, exposure conditions, and the environment of interest. Corrosion characterization requires a careful analysis of the susceptibility of each candidate material to a variety of corrosion or degradations mechanisms, such as uniform wastage, pitting, crevice attack, stress corrosion cracking, corrosion fatigue, swelling, and so forth, within the framework of the particular material/exposure condition/environment matrix.

Characteristics of Corrosion Data

Corrosion data come from many sources, including standardized laboratory tests (often characterized as "accelerated" tests), nonstandard tests, field tests, simulated environment tests, the examination of operating equipment, failure analyses, static exposures in natural environments, and theoretical studies. Data are generated by researchers and materials engineers in industrial, independent, and government laboratories. Data appear in technical publications, textbooks, handbooks, maintenance records, laboratory notebooks, research reports, and promotional literature. Few, if any, of these records are maintained in a systematic electronic format. Many data sources emphasize specific forms of corrosion or degradation, and thus the data output does not fully characterize the overall performance capabilities or limitations of the materials evaluated. Quality and completeness of data vary widely, stressing the need for careful assessment and evaluation of data sources if multiple source data compilations are to secure

---

[1]Director, NACE-NIST Corrosion Data Center, National Institute of Standards and Technology, Gaithersburg, MD 20899.
[2]Associate director, National Association of Corrosion Engineers, P.O. Box 218340, Houston, TX 77218.

user confidence. Characteristics of several of the more important sources are summarized in Table 1.

The natural variability of existing corrosion data must also be recognized. Data are generated with inhomogeneous materials all too often exposed in inhomogeneous environments with inadequate statistical sampling.

Data Input

Task Group discussions within ASTM Committees E 49 on the Computerization of Material Property Data and G 01 on the Corrosion of Metals have identified general categories for data collection for inclusion in a comprehensive corrosion data base. Categories are listed in Table 2. Subheadings outline specific fields that need to be defined for each category. These groupings should serve as useful guides for any comprehensive corrosion data base development activity. Fields related to material identification should be consistent with other materials property data bases and information sources.

"Essential fields" must be clearly identified both for data bases structured around single test methods and for more complex, multiple-source data bases designed to accommodate input relating to the full spectrum of engineering materials and types of corrosion. As an example, an essential field requirement could reject data for "aluminum" or "stainless steel" without specific alloy or temper identities. The essential fields are defined as the minimum for which data are required for reliable characterization of results obtained from evaluation of a single corrosion test sample or inspection. Once defined, the essential field list can be used as the basis for the acceptance or rejection of candidate data for inclusion in a specific data base. Rejected data can often be upgraded to an acceptable level through inquiries to data sources soliciting added detail and thus assuring maximum data quality in the data base. Data failing to meet the accep-

TABLE 1—*Characteristics of corrosion data sources.*

Data Source	Source Characteristics
Standardized laboratory tests (ASTM, NACE, and so forth)	well-defined environmental conditions and test procedures often developed as "accelerated tests" with uniquely aggressive environments. Data are difficult to relate to natural or industrial environments. Standards documents provided limited guidance on data interpretation; frequent emphasis on a specific corrosion mechanism (pitting, and the like).
Nonstandard tests	same as standardized tests, provided procedures and environmental conditions are clearly defined.
Field tests in industrial process streams	environmental factors and exposure conditions generally not well described. Provide reliable comparisons of materials tested simultaneously. Difficult to compare data from other tests in similar environments.
Static exposures in natural environments	generally involves long-term exposures with variable environmental conditions. Macro and micro environmental variables generally not well characterized; no control over environmental variables
Theoretical studies	generally emphasize specific corrosion mechanisms; often difficult to extrapolate to numerical material performance data from other sources.

TABLE 2—*Corrosion data base inputs.*

Type of test
  standardized, laboratory, field
  relation to specific process or application
Test emphasis
  metals—general corrosion, specific type of corrosion (pitting, parting, etc.)
  nonmetals—volume change, property change, dissolution
Environment
  generic description
  concentration and state of principal components
  contaminants
  aqueous or gaseous
Exposure conditions
  duration, temperature, pH, hydrodynamic conditions, aeration, agitation
Material identification
  class, common name, standard designation, condition, manufacturing process, product form
Specimen identification
  specimen number, size, surface condition, composition, properties
Material performance
  weight change, property change, localized attack (identified by corrosion mechanism)
Source/reference
  published and unpublished data

---

tance criteria established for a given data base may still be considered acceptable for data bases designed for less stringent applications.

Users wishing to classify data to satisfy specific needs can add essential fields during data searching, or the data base can be structured to classify data by predetermined input requirements. However, a classification scheme should not serve as a substitute for careful data evaluation, thus placing the user in the position of having to assess overall data quality independently.

The essential field lists can also serve as guides for the standardization of corrosion test methods to assure adequate data recording and future data base acceptance. Statistical evaluations of significant compilations of data from a standardized test can, in turn, identify needs for modifying essential fields and specified test procedures. The use of corrosion data bases to define guidelines for the development of new or modified standardized corrosion tests provides a unique opportunity for corrosion scientists and data base managers.

Data Searching

Ideally, a corrosion data base should accommodate the search strategies outlined in Table 3. These meet defined user interests as follows:

(1) statistical comparisons of multiple samples and materials included in a single test,
(2) data generated using a standardized test method,
(3) identifying material compatibilities in specified environments,
(4) identifying the environmental compatibilities of specific materials,
(5) retrieving data relating to a specific industrial process, material application, or corrosion mechanism, and
(6) identifying data from a designated reference or source.

Other search strategies can be defined to meet specific user needs and to accommodate any limitations of the data base management system used.

TABLE 3—*Data search strategies.*

Duplicate samples in same test
—permit statistical comparisons of duplicate samples tested simultaneously
Different materials included in same test
—provide direct comparisons of relative performance of different materials
Standard test (by specification)
Environment—generic description or specific component(s)
seawater can be described as "seawater" or an aqueous chloride solution, and so forth
Material—class, specific designation, common name, composition
aluminum alloys (class), UNS A96061 (designation), 6061 Al (common name), Al-Mg-silicide alloy
Related application or process
heat exchanger tubes, sour gas production tubing, pulp bleaching, and so forth
Type of corrosion or degradation mechanism
pitting, corrosion fatigue, stress corrosion cracking, etc.
Reference
include published and unpublished data

Data Validation and Evaluation

Validation is confirming that data was developed using documented test procedures and that material identifications conform to established published or internal standards. Necessary clarifications can be sought through source interviews during the validation process along with requests for additional data to satisfy essential field requirements.

Evaluation is expert judgmental appraisal and statistical analysis for consistency within material/environmental domains. Data bases may incorporate unevaluated data as a means of making data compilations available for independent evaluations. The extent of evaluation should be clearly documented for user guidance.

For broad-based corrosion data bases, data sets relating to specific materials, applications, environments, or corrosion mechanisms should be extracted for individual evaluations. Evaluators should be given guidelines reflecting data base objectives and the needs of the anticipated user base. For a fully evaluated data base, these should encompass the following:

(1) the assessment of data quality, and consistency with known service experience,

(2) completeness in terms of a reasonable summary of candidate materials and key environmental factors that will not mislead user, and

(3) the identification of significant gaps and alternate data sources to fill gaps.

The diversity of types and sources of corrosion data, users, user needs, and user interpretations emphasizes the importance of a comprehensive evaluation program that can accommodate multiple source inputs. This can be viewed as a data base upgrading process to improve user confidence and should eliminate any user bias requiring searching by preferred data sources.

Data Output

Corrosion data user needs and interests vary widely, generally focusing on specific environments, materials, or corrosion failure mechanisms. These are used as an aid in (1) service life projections, (2) isolation of environmental or test variables, (3) identification of candidate materials based on user defined durability requirements, or (4) reviewing experience and laboratory data relative to a specific industrial process, application, or form of corrosion.

Customized outputs directed toward each need can be particularly useful along with interpretive graphics and tutorial or help screens. Outputs can also be in the form of user defined graphics, summaries, or reports to help meet specific user applications. For maximum effectiveness, outputs should aid a user in identifying operative corrosion or degradation mechanisms and their relative criticality for the material/environment/application of interest.

NACE-NIST Corrosion Data Program

A joint program sponsored by the National Association of Corrosion Engineers (NACE) and the National Institute of Standards and Technology (NIST) has been initiated to provide a focal point for the computerization of corrosion behavior data on engineering materials. Data needs were defined during a workshop for corrosion specialists in industry, academia, and government [1]. The development of a corrosion data base as a major component of this program was a particular recommendation of the workshop.

To ensure continuing input to the program, format standards are being established through ASTM Committees E-49 on the Computerization of Material Property Data and G-1 on the Corrosion of Metals and through NACE Committee T-3 on Corrosion Science and Technology. Many of the comments and concerns expressed in this paper reflect the deliberations of these committees.

Multiple source data are being compiled for evaluation and subsequent distribution in a personal computer (PC) format. Separate software programs are anticipated, each with data relevant to a specific topic, for example, corrosion in soils, sulfuric acid, or reactor environments, corrosion-fatigue, the characterization of a specific material or class of materials, and so forth. The topics selected will reflect both user interests and data availability. Each program will stand alone with custom outputs to guide the user.

Summary

Design requirements have been defined for structuring a multiple source corrosion data base. A variety of user defined needs and interests are accommodated to assure broad applicability for identifying candidate materials and quantifying their performance capabilities for given environments and exposure conditions. The guidelines developed should serve as the basis for corrosion data bases designed to serve a variety of needs while assuring compatibility with other materials data base resources.

Reference

[1] Verink, E. D., Kolts, J., Rumble, J., and Ugiansky, G. M., *Materials Performance,* Vol. 26, No. 4, April 1987, pp. 55–60.

William W. Scott, Jr.,[1] Hugh Baker,[1] and Linda Kacprzak[1]

Development of Data Bases for the ASM/NBS Data Program for Alloy Phase Diagrams

REFERENCE: Scott, W. W., Jr., Baker, H., and Kacprzak, L., "**Development of Data Bases for the ASM/NBS Data Program for Alloy Phase Diagrams,**" *Computerization and Networking of Materials Data Bases, ASTM STP 1017*, J. S. Glazman and J. R. Rumble, Jr., Eds., American Society for Testing and Materials, Philadelphia, 1989, pp. 322-328.

ABSTRACT: The joint efforts of ASM INTERNATIONAL (ASM) and the National Bureau of Standards to form the ASM/NBS Data Program for Alloy Phase Diagrams are traced back to a 1975 report prepared by an ad hoc committee of the Numerical Data Advisory Board of the National Academy of Sciences/National Research Council. The Committee's recommendations for a program to collect, evaluate, and disseminate all existing alloy phase diagram data are presented. This paper discusses the progress related to the recommendations of the committee, particularly regarding the computerization of alloy phase diagram data. The management aspects of data base development are discussed as well as the various alternative delivery systems that have been explored. The current status of each of the data bases is presented, including the search capability, degree of completion, and plans for maintaining currency.

KEY WORDS: alloys, bibliographies, computer programs, crystal structure, data storage, phase diagrams

In 1975, a data base including graphics was a futuristic dream. While data bases of bibliographic information were well established and offered for sale by data base vendors, graphic information was only available through print media. Computer storage space was a limiting factor in even contemplating a graphic data base. Also, the complex software that was needed was simply not available.

In that year, however, a report was prepared by an ad hoc committee of the Numerical Data Advisory Board of the National Academy of Sciences/National Research Council. This committee identified alloy phase diagram data as priority information to be developed and disseminated in the technical community. The blue ribbon panel made the following specific recommendations:

• A worldwide, coordinated effort is required. The program would be far too large to be accomplished in one country.
• Data must be critically evaluated for reliability.
• An International Council should be established to represent all areas of the world where phase diagram activity exists. Committees should be appointed to select evaluators and establish priorities of systems.
• A computerized bibliographic data base to aid in evaluation should be developed.

---

[1]Technical director, manager, Alloy Phase Diagram Program, and supervisor, Alloy Phase, Diagram Products, respectively, ASM International, Metals Park, OH 44073.

- A computerized numerical data base needs to be established so that all data and diagrams may be digitized and made available interactively.

The recommendations of the committee formed the basis for a joint program between ASM INTERNATIONAL (then the American Society for Metals) and the National Bureau of Standards (NBS). The ambitious venture was called the ASM/NBS Data Program for Alloy Phase Diagrams. ASM's responsibility was to provide overall management of the program, to compile the data, to publish and deliver the data, and to develop funding for the critical evaluation of phase diagram data. NBS's role was to assure the technical integrity of the data and diagrams that would be published, and to have primary responsibility for guiding the development of a computer graphics data base. ASM Research Associates were also to be placed at NBS to assist in the development and building of the computer data bases.

Today, not quite 10 years after ASM and NBS joined in this important activity, significant milestones have been passed related to the initial goals of the program. In this paper we discuss the important developments in computerized data bases, explain what phase diagram data are, and provide a degree of understanding regarding the challenge of producing numerical and graphic data bases encompassing these data.

What is a Phase Diagram?

Phase diagrams are simply graphic representations of the thermodynamic relationships of two or more chemical elements at different temperatures or pressures. They have been described as "roadmaps" or "blueprints" for the metallurgist and material scientist, as important to him as *Gray's Anatomy* is to the physician, codified law to attorneys, or generally accepted accounting principles to accountants. Phase diagrams can be found in the literature of more than 100 years ago, from the time of J. Willard Gibbs.

To provide a reasonable understanding, it is best to refer to an actual phase diagram and explain some of the information. Figure 1 is a phase diagram of the lead-tin (Pb-Sn) system [1], and Fig. 2 is a much more complex phase diagram of the aluminum-copper (Al-Cu) system [2]. Each area bounded with lines or curves within the diagram represents a temperature and composition region in which one or more phases exist. For instance, region A is a mixture of liquid and solid phases (a slush) that has a specific crystal structure. Region B is entirely (Pb), thus indicating a single, solid phase in which the crystal structure remains the same over these temperature and composition ranges. Region C is a mixture of two crystal structures, (Pb) and (β Sn). Note that on the graph, there are a number of points—terminal points of curves—that are extremely important. The knowledge of melting points, such as D, transformation points from one phase to another, and so forth, are very important. Also, the terminal points where several curves meet have specific names (for example, the eutectic, E) because of their special properties.

Diagrams such as these are extremely important to many facets of engineering and science, such as the heat treatment of metals, where metals are heated in the solid state to generate small amounts of other phases, thus improving the properties within a given metal or alloy. Phase diagrams are also critical in alloy design, in surface treating and coating operations where high temperatures might cause diffusion between metals, in welding, and in slag-metal reactions for refining metals. They can also be important in analyzing failures, since through use of the diagrams one can predict thermal history by the nature of the phases present in a given metal.

Because of the importance of phase diagram data, a diagram must be reliable to maximize its utility. Inaccurate phase diagram information can cause literally millions of dollar's worth of effort to be wasted, can cause premature failures, or can result in overdesign. It is for these reasons that the committee mentioned earlier determined that the development of reliable alloy phase diagram data should be a priority international effort.

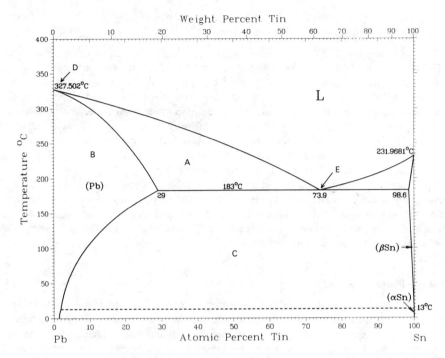

FIG. 1—*Example of a simple binary phase diagram.*

The Challenge

For the purposes of this paper, all of the start-up tasks—the background to the fundraising effort, the effort of finding appropriate experts who could sift the literature and come out with a reliable phase diagram, and the development of an approach to the international community—will be set aside to focus on the elements of the program that are of interest to this group. Four million dollars was raised from the industrial and government community in the United States, experts were found, and critical evaluations began to be submitted for publication. In 1980, a periodical, *The Bulletin of Alloy Phase Diagrams,* made its debut, and in 1986, the first major compilation of binary phase diagrams evaluated and reviewed under the ASM/NBS Data Program was produced. ASM and NBS have cooperatively developed the data bases and have worked closely to provide the data to the engineering and scientific community.

The challenge, from the standpoint of data bases, was three-fold: (1) to produce a bibliographic data base so that the literature could be searched easily for metal combinations producing literature citations of interest, (2) to produce numerical data bases that would include the important numerical data related to the diagram such as crystal structure, terminal points, melting points, transformation points, and so forth, and (3) to provide a graphics data base for binary (two-metal) phase diagrams, of which there were known to be about 2000. Development of the data bases was approached in parallel fashion with the second and third items combined into a single project.

Bibliographic Data Base Development

The bibliographic data base was developed by taking selected records from METADEX, a data base that ASM and The Institute of Metals in London have operated since 1965 and that

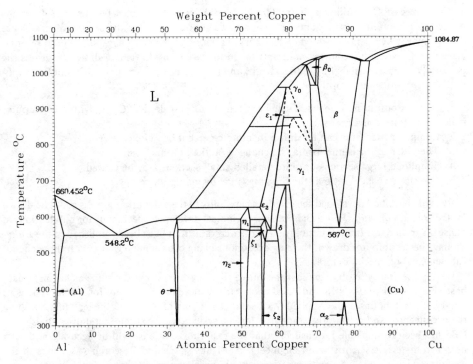

FIG. 2—*Example of a complex binary phase diagram.*

incorporates most of the known literature on metals, abstracts from the files of the Alloy Data Center of the National Bureau of Standards, and other files that had been collected by Elliott for earlier compilations of phase diagram data in the 1960s [3]. The approach had the advantage of an established updating mechanism, since the mechanics of updating METADEX were already in place at ASM and The Institute of Metals.

Also, in some situations where needed journals were not being accessed by METADEX, they were easily incorporated into the METADEX collection system and abstracts provided to the METADEX files from that point on. The primary task was developing a search strategy for efficiently searching the METADEX citations relating to phase diagrams and crystal structure. This was done with the help of technical consultants, and has resulted in a compact, bibliographic data base of more than 35,000 records, which ASM has been using to supply literature to the experts evaluating phase diagram data. Recently, ASM announced that the bibliographic data base, PhaseLit®, is now available to the public and searches can be made of the data base on an off-line basis by calling or writing to ASM and describing the systems to be searched. Because of the small size of the user market, on-line availability does not currently seem warranted.

Numerical and Graphics Data Base Development

The initial objective of the computer graphics and numerical data elements of the program was to produce accurate diagrams for publications. Computer-produced phase diagrams promised to be highly reliable and also economical, especially when compared to the cost of draftsmen's renderings. It was estimated, for instance, that drawing one set of all binary diagrams would have cost two million dollars. The high estimate arose in part because the diagrams needed to be plotted both in atomic percent and weight percent. (Atomic percent diagrams are used primarily by scientists who do R&D work, and weight percent diagrams are used by engineers who produce the resulting materials.) Add to the cost of draftsman-drawn diagrams the potential for error, mislabeling, and so forth, and they have even less appeal. Computer graphics, however, offer many options:

1. Input or output may be made using any temperature scale (K, °C, °F) with the computer making accurate conversions.
2. Computer diagrams can be shown by atomic percent or weight percent.
3. The order of the elements may be changed on the graph.
4. Simplified large sections or quite detailed small sections may be viewed.
5. The size or aspect ratio of the drawings may be manipulated.

By 1981, NBS had developed the graphics capability to the extent that graphs could be drawn for the *Bulletin of Alloy Phase Diagrams,* a periodical developed specifically for publications of phase diagram data. In fact, virtually all of the diagrams that have been published in the *Bulletin* have been computer drawn, the only exceptions being occasional ternary (with three chemical elements) or special diagrams.

Having the ability to digitize diagrams into a file, however, is not the same as having a data base. For several years, NBS and ASM cooperatively developed software for a comprehensive binary-alloys data base. The Applied Mathematics Division of NBS took the lead in developing the architecture and setting the standards for the system, whereas the Metallurgy Division of NBS has been responsible for ensuring that the resulting data base will be accurate and fully useful to materials scientists and engineers. ASM has sponsored Research Associates in the Metallurgy Division to work with NBS scientists in developing the software for the system, and additional ASM Research Associates also residing there digitize the diagrams and enter the data.

Descriptions of many elements of the data base are available in the literature, having been provided through another forum [4]. From a data management standpoint, however, the resulting data base can be thought of as a relational, numerical data base that includes graphics files.

The data base is organized according to data pertaining to the alloy system and data pertaining to an individual phase of the system.

For each alloy system, the properties are

- Phase-diagram graphics and captions
- Diagram type (for example, simple eutectic)
- Reaction table

    (a) three-phase equilibria, congruent, critical, tricritical points
    (b) pure-metal transformation
    (c) phases, compositions, temperature, reaction type
    (d) phase descriptions
    (e) publication and update documentation

For each phase, the properties are:

- Crystal structure

    (a) space group, prototype, Pearson symbol
    (b) Strukturbericht designation, comment

- Equilibrium, metastable or high-pressure phase
- Stability range

(a) homogeneity range and type (for example, terminal solution, line compound)
(b) temperature range and decomposition type (for example, peritectic, congruent melting, solid-state)

Figure 3 is a simplified schematic of the software and data files produced at NBS. The custom graphics program written for this project uses the GKS standard and was designed for use with Tektronix 4115 terminals and a VAX 11-750 computer. The structure that stores input data uses Linear Encoding for Multi Media (LEMM) to provide data independence and self-documenting files. These files are used to produce hard copy for the *Bulletin of Alloy Phase Diagrams* and other APD Program publications, and they also provide the means for editing and updating the graphics.

Once data are digitized and initial quality-control checks are complete, the data are "challenged." "Challenge" is a program written to make sure the diagram and data conform to a number of fundamental criteria, such as standardized melting and transformation temperatures. This program also organizes the data to prepare it for loading into the numerical data base. After the "challenge" program is run and the final quality control checks are made, subroutines are used to create the picture files and "load" the data into the numerical data base.

Numerical and crystal structure data in the data base are accessed via a user-friendly program (EZRIM) and by using a commercial relational data base management software package (RIM, Boeing Computer Services). EZRIM allows the user to search and display all the graphics data and provides access to the picture files. The numerical data can also be queried to select metal systems having various features. As an example, one could ask for all "systems with Cesium Chloride (CsCl)-type phases melting congruently above 1400°C."

The numerical data base consists of the following relations, or tables:

- BINARY describes the phase diagram, that is, the type of diagram, number of solid phases, number of invariant reactions, and volume and issue of the *Bulletin* in which the complete evaluation is published,
- PHASES contains detailed crystal structure data and some phase-equilibrium information,
- INVARS gives the type of each invariant reaction, along with the compositions of the phases and the temperature at which the reaction occurs,
- CONG gives the composition and temperature of congruent points and allotropic transformations,

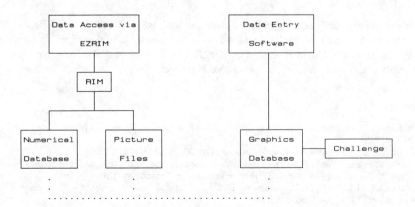

FIG. 3—*Schematic diagram illustrating the relationship between software and data files.*

- CRITICAL is a table of the critical points of miscibility gaps,
- HORDER gives the composition and temperature for magnetic and second-order transitions, and may include a minimum and maximum, and
- ELEMENTS contains the atomic mass, the melting and boiling points, and the group and row in the periodic table for each pure element.

The graphics in the data base are provided in the form of picture files for approximately 1575 binary systems at the present time, including the main phase diagram plotted in both atomic and weight percent. In some cases, solubility diagrams, enlargements of complex areas, or diagrams showing experimental data are also included.

Data Dissemination

While phase diagram data are basic to solving many metallurgical problems, the market for them is limited and in our opinion does not justify on-line availability through traditional data base vendors, if, in fact, they were prepared to handle a complex graphics data base of this type. Currently, plans are to develop two data base offerings, one for use on a PC, and one for use on a VAX. The PC version will include the numerical data base and use about 4M bytes. All criteria searches would need only this data base. It will be available on several floppy disks. Since picture files take up significant storage space in the computer, they will initially only be available on the VAX version, which is intended to incorporate most of the elements shown in Fig. 3.

The loading of currently available data into the numerical and graphics data bases has just recently been completed, so that test marketing can begin, as well as off-line servicing of customers who do not wish to purchase software and hardware.

Conclusion

As you can see, ASM INTERNATIONAL and the National Bureau of Standards took a bold step almost 10 years ago when they undertook to develop and build a computerized bibliographic data base of phase diagram literature and a computerized data base of critically evaluated scientific data that would include graphic information. We are pleased to report these tasks have now been successfully accomplished, and we are confident that they will serve the materials community well.

Acknowledgments

The authors gratefully acknowledge the efforts of the NBS staff and the ASM Research Associates who have labored to develop and load the data bases we have described. We particularly wish to recognize Dr. David F. Redmiles, Dr. James S. Sims, and Dr. Joanne L. Murray for their creative and long-standing devotion to this project.

References

[1] DiMartini, C., in *Metals Handbook*, 8th ed., Vol. 8, American Society for Metals, Metals Park, OH, 1973, p. 330.
[2] Murray, J. L., *International Metals Reviews*, Vol. 30, No. 5, pp. 211-233.
[3] Elliott, R. P. and Shunk, F. A., *Reviews of the Constitution of Binary Alloys*, IITRI Report No. B6082-5 (Vol. 1, Part A, B, C and Vol. 2), Illinois Institute of Technology, Chicago, 1972.
[4] Bhansali, K. J., Murray, J. L., Redmiles, D. F., and Sims, J. S., *Proceedings—29th National SAMPE Symposium*, 1984, pp. 1450-1464.

Jerald E. Jones[1] and H. H. Vanderveldt[1]

Welding Information Systems

REFERENCE: Jones, J. E. and Vanderveldt, H. H., "**Welding Information Systems,**" *Computerization and Networking of Materials Data Bases, ASTM STP 1017*, J. S. Glazman and J. R. Rumble, Jr., Eds., American Society for Testing and Materials, Philadelphia, 1989, pp. 329–339.

ABSTRACT: This paper presents recent developments in Welding Information Systems at the American Welding Institute (AWI). First, it covers the traditional data and knowledge representation and use of welding technology. Then it presents the use of advanced computer technology to rapidly retrieve and use of new welding technology. Specifically, the paper discusses applications developed by AWI in the use of 4GL data-base systems for welding data; the use of knowledge representation techniques in the development of expert systems for welding application; and the use of neural network simulators for associative memory storage and retrieval of welding-related image data. The paper presents an overview of the WIN® system, a welding engineering workstation developed by AWI.

KEY WORDS: welding, data base, expert system, neural network, computerization, information, WELDSELECTOR, WELDSYMPLE, WIN, artificial intelligence

The Need For a Welding Information System

The American Welding Society now classifies over 100 separate welding and joining processes. The total number of alloys that can be welded using those processes exceeds 5000. Each one of these alloys, having different compositions and mechanical properties, will exhibit a different behavior when heated to the melting point and allowed to solidify. Consequently, the specific welding parameters used to join a material will be unique to that material or at the very least, unique to a small, select group of alloys exhibiting essentially the same composition and mechanical properties. A welding engineer needing to specify a welding process and procedure for a fabrication or repair application must be able to specify the particular welding procedure and parameter values for his application. Depending on the application, a mistake in that choice of welding procedure can have catastrophic results, leading to loss of life, significant damage to property, or removal of the machinery from service for an extended period of time.

An experienced engineer who works with the same group of materials over a period of time will soon become familiar with the welding procedures required to produce adequate welds and comply with applicable codes and standards designed for those materials. However, it is difficult for an engineer to maintain a complete mental record of the welding procedures applicable for even one broad category of alloy such as chromium-molybdenum steels. It is virtually impossible for one engineer to recall the information necessary to specify welding procedures for the entire range of metals and all possible applications. A well-versed welding engineer, given a sufficiently complete library of reference information and a sufficiently long period of time, can, by diligent effort, find a welding procedure for almost any material; however, this process can result in a less-than-optimum choice. As the complexity of weld-fabricated structures and com-

---

[1]Manager of technology transfer and president, respectively, American Welding Institute, Route 4, Box 490, Louisville, TN 37777.

ponents, as well as the number of welding processes and engineering alloys, continues to grow, the likelihood that optimum choices will be made in the specification of welding procedures continues to decrease. If the United States wants to maintain and increase its competitiveness in the world marketplace, efficient methods for providing welding information must be found. This approach will result in the optimum use of available resources and will not require excessive numbers of personnel. Recent work has indicated that nearly one-half of the gross national product of this country depends on welding. Even small changes in the ability to optimize the use of welding information will have a dramatic impact on the commercial viability of this country. Clearly then, a computerized welding information system that can be used to efficiently search for and retrieve optimum welding procedures for a wide variety of materials and welding processes and can also provide other useful information about welding and welding materials is an important undertaking for the United States.

Representation of Welding Knowledge

If a computerized welding information system is to function, methods must be employed that allow that information to be stored and retrieved. Computers should also be able to apply this information to problem solutions. Welding information, as a whole, is the collection of a large body of knowledge about numerous aspects of welding and welding processes. That knowledge exists in a variety of forms. Each form has historically developed because of its usefulness when applied to the knowledge it is used to represent. Therefore, it is important that a computerized information system have the capability to use a variety of forms of knowledge representation.

Tabulated Data

The simplest form of welding information is tabulated data. This information has traditionally been archived as relational tables published in reference books. However, these tables rarely exist "out of context" and are often only useful when combined in an overall information structure. For example, a table of recommended preheating temperatures related to carbon equivalent values is only applicable to a specific set of steels and only for one carbon equivalent equation.

Consequently, the representation of tabulated welding information in a computerized system would be inadequate using a simple relational table management system. Instead, a more complex system consisting of a series of relational tables organized in a hierarchical structure must be used.

Rules for Applications of Information

Much of the information about welding consists not of simple data but of that data combined with the rules that define the application of the data. For example, a welding code may include tabulated data regarding the types of steels that can be grouped together in applying a single welding procedure. But there are rules stipulated in that code that also specify the variation allowed in the procedures for the manufacture of those steels. That allowed variation consists of a series of rules that, when combined, form the codified application of that information.

A production-system knowledge representation schema provides an adequate method for representing such data. By applying a series of "If . . . , Then . . . " production system rules, a welding code, for example, can be represented in a computerized system.

The use of a production-system knowledge representation schema requires that the rules be applied in a logical order and that the logic be exhibited by a complete path connecting the antecedents (premis) and consequents of the rules applied to a given situation. That path must extend from the initial data given to the system to the point at which all of the goals (requested information requirements) have been met. Thus, the system that applies such a production-system representation schema must have appropriate paradigms (algorithms) embedded to provide for all of the activities needed to achieve the goals of the system.

Graphic Analogic Knowledge Representation

The design of a welded joint not only requires the information regarding the procedure, techniques, and materials, but also must include data that describes the joint configuration itself. The joint groove must be machined to useful tolerances, especially for automated welding applications. Small changes in joint design can result in substantial cost differences in producing a weld, which are magnified as the length of the weld seam increases.

For example, the angle of the edges of the groove is very critical. If the groove is too wide, machining costs increase and the additional metal and time required to fill the joint will also add to the overall costs. In addition, the residual stress and distortion will increase with joint angle, causing potentially serious engineering problems in the structure being fabricated. However, if the groove is too narrow, the welding apparatus may not fit properly within the joint, which can cause welding defects such as incomplete penetration or inadequate fusion. These defects can cause a weld to catastrophically fail in service and are often difficult to detect without expensive and complex nondestructive testing. Also, defects can be expensive to repair, both during initial fabrication as well as in service.

Welding design information is generally used in the form of engineering drawings and graphic welding symbols. The representation of this form of knowledge is not well suited to alpha-numeric data-base or to production-rule information storage. Instead, graphic and photographic data are required for computerized welding design information systems. Figure 1 is a typical weld joint and welding symbol graphic representation. The welding information system requirements must include provisions for storing, manipulating, and retrieving/displaying these types of data.

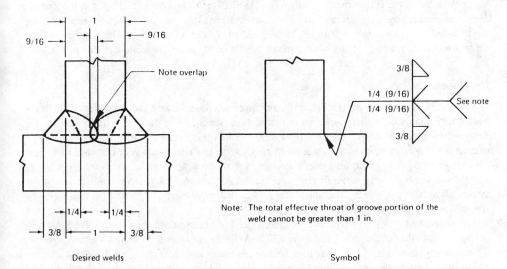

FIG. 1—*Diagram showing a typical weld joint design and the welding symbol that would be used to describe the weld on a mechanical drawing.*

Frame-Based Information Representation

Objects can be stored in a computerized representation schema by using several techniques. However, one method that can be used both to represent objects as well as to structurally organize that representation is the technique of frames. In a frame-based representation schema, objects are stored by maintaining information files about the attributes of the object. The frames technique varies from the data-base technique in that in the frame system, an inheritance hierarchy is used to describe the objects. Thus, for example, there could be an object known as "welding electrode" that has attributes such as diameter. Then, there is an object known as "SMAW electrode" that has attributes such as coating, position, and current type (AC, DC, or both). "SMAW electrode" is related to "welding electrode" just as a child is related to a parent. Thus, the attributes of "SMAW electrode" include the inherited attribute of diameter.

Organization of the Win System

To provide welding information services that are useful for application problem solving, an integrated system needed to be developed. Such a system had to include the ability to work with a wide variety of knowledge representation techniques so that all of the information types required can be used. The system also needed to be flexible and expandable, so that future enhancements can be included, as well as take best advantage of the latest available computer technology, which is accessible by a large portion of the potential users.

Based on these criteria, the American Welding Institute, working with the Colorado School of Mines Center for Welding and Joining Research, has designed and implemented the Welding Information Network (WIN®) System. The system operates in a desktop computer environment, which makes it accessible to the large majority of potential users. The system is also being made available on selected mainframe computers for distribution over time-share networks.

The WIN system is organized as a series of data bases coupled with production rule-oriented knowledge bases and graphic data systems. Figure 2 is a schematic diagram of the WIN system's organization showing the relationships between the various represented knowledge sources (AWIKBS) and data bases (AWIDBS). Figure 3 gives an example of several rules found in the WELDSELECTOR expert system. WELDSELECTOR has over 2000 rules in the knowledge base used for welding electrode selection. In each case, the technique used for representing the knowledge was chosen to optimize the eventual use of the information. The knowledge representation schemes applied are all discussed briefly in the previous section of this paper.

User Interaction

The user interface built for the WIN® System presents the user with either conventional data-base query menus or with expert system interaction screens. The data-base query menus are designed to interface with complex software that builds the data-base query and presents the user with a report or table of information as requested. The user is always presented with simple, natural language menu and query screens so that no complex codes or query language is needed and so no training is required to operate the systems. An interactive menu driven query system allows users to select data, sort data, and completely design the format for hard-copy or screen-based report generation.

The expert system user interface technique is used to present data applied to solve a specific problem in welding engineering and design. The WIN® system uses an analysis expert system. Thus, the expert system queries the user for the necessary information. Then it generates queries of the data base system to gather knowledge about the problem and potential solutions. The

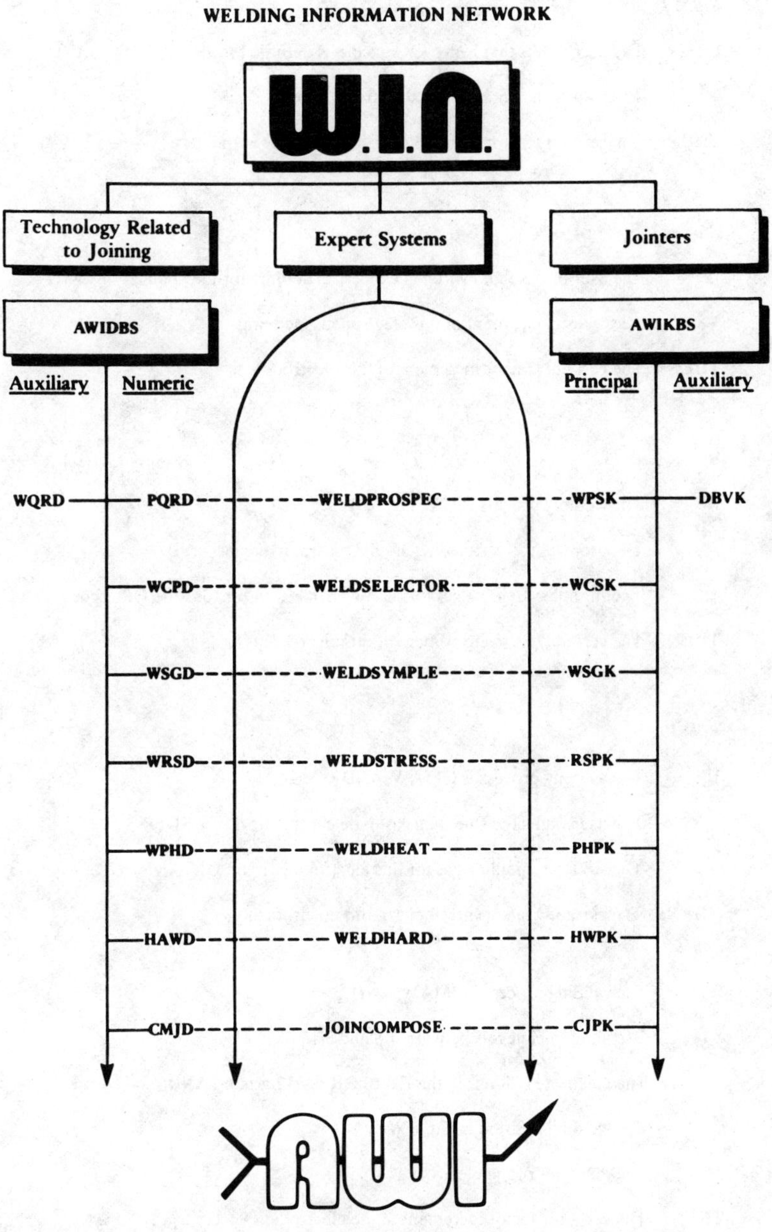

WELDING INFORMATION NETWORK

W.I.N.

- Technology Related to Joining
- Expert Systems
- Jointers

AWIDBS — AWIKBS

| Auxiliary | Numeric | | Principal | Auxiliary |

WQRD — PQRD — — — — WELDPROSPEC — — — — WPSK — DBVK

— WCPD — — — — WELDSELECTOR — — — — WCSK —

— WSGD — — — — WELDSYMPLE — — — — WSGK —

— WRSD — — — — WELDSTRESS — — — — RSPK —

— WPHD — — — — WELDHEAT — — — — PHPK —

— HAWD — — — — WELDHARD — — — — HWPK —

— CMJD — — — — JOINCOMPOSE — — — — CJPK —

AWI

FIG. 2—*Schematic representation of the WIN® system developed by the American Welding Institute.*

Rule 21

IF: The user thinks that hydrogen concern is definite for the application, **AND**

The hydrogen content of the electrode is low;

THEN: The certainty factor increment of this electrode is 10.

Rule 33

IF: The user thinks that hydrogen concern is definite for the application, **AND**

The hydrogen content of this electrode is medium;

THEN: The certainty factor increment of this electrode is 5.

Rule 42

IF: The welding process is GMAW, **AND**

The user wants the shielding gas to be carbon dioxide, **AND**

The suitable shielding gas for this electrode is carbon dioxide **OR** argon;

THEN: The certainty factor increment of this electrode is 8.

Rule 10

IF: The welding process is GMAW, **AND**

The user wants the shielding gas to be carbon dioxide, **AND**

The suitable shielding gas for this electrode is argon **OR** oxygen;

THEN: The certainty factor increment of this electrode is 6.

Rule 5

IF: The welding process is SMAW, **AND**

The plate thickness is between 1/8 and 3/16 inch, **AND**

The electrode belongs to the Exx10 OR Exx12 groups, **AND**

The welding position is flat, **AND**

The diameter of the electrode is 1/8 inch;

THEN: The certainty factor increment of the electrode is 9.

FIG. 3—*Example rules from the WELDSELECTOR Expert System.*

system then analyzes the problem and subsequently presents the user with that information that is best suited for problem solution.

Data Base Development

The developers of data bases for the WIN® system used a systematic approach to data needs. First, they formed a national committee to provide input on data-base development activities. This committee consists of individuals who are knowledgeable about welding or information systems or both and represent a broad cross-section of the technical community ranging from industry to government to academia. Then they implemented a software and hardware selection process. Finally, the data bases were developed.

Committee Recommendations

The guidance committee developed a strategy for the implementation of a welding data base system. One of the primary aspects of that strategy was the convening of the Workshop on the Computerization of Welding Information. The workshop was held in Knoxville, TN, in August of 1986. The workshop recommendations are documented in the workshop report, which is available from the American Welding Institute or the National Bureau of Standards. That report identified four top categories of national data needs:

(1) general welding procedures,
(2) materials property data,
(3) procedure qualification records, and
(4) welding variables.

In addition to enumerating priorities, the workshop provided a guide for the full development and implementation of U.S. computerized welding information system. The overwhelming majority of attendees at the workshop enthusiastically supported the development of this national resource. Thus, a mandate was provided to continue development work on the WIN system.

Data Base Subject Areas

Within the WIN system there are two categories of data bases under development. The first covers those data bases that stand alone and are designed for direct user access. The second includes data bases accessed by other software that then transfer findings or results to this other software. Using this approach, these data bases provide necessary input data to expert systems. These expert systems in turn provide the requested decision support to the user. The subject areas covered by the WIN® data bases generally conform to the guidelines established by the workshop. The data-base systems that are stand-alone include:

(1) PQRDB - a data base of welding procedure qualification records (PQR),
(2) WPSDB - a data base of welding procedure specification records (WPS),
(3) CMJDB - a data system of composite material joining procedures, and
(4) WELDERTRACK - a data system that includes multiple data bases and is designed to track weld and welder capability and quality.

The data bases designed to interface with other WIN software include

(1) WELDMATDB - a data base of weldable base materials, including all of the ASTM classified steels,
(2) WELDELDB - a data base of welding electrode information,
(3) WELDSSDB - a data base of stainless steel base materials,

(4) WELDSSELDB - a data base of welding electrodes for use in stainless steel joining,
(5) WELDSYMGB - a data base of graphics objects used to generate welding symbols,
(6) WELDJNTGB - a data base of graphics objects used to generate welding joint designs,
(7) WELDHEATDB - a data base of welding preheating and post-weld heating information, and
(8) WELDHARDDB - a data base of hard-facing materials and wear information.

Knowledge Base Development

Knowledge bases are information storage systems designed to provide expertise to expert systems. In the WIN® system, two types of knowledge representation are applied to welding information to develop knowledge bases. First, information that defines the procedural rules for applying data to solve problems is represented as production rules. The rules are written as a series of "if. . . , then. . . " statements (see Fig. 3). This collection of statements provides a thorough representation of this information in some small, clearly defined region of the total knowledge domain for welding.

These production rules and the data they use must be organized to provide an efficient system. By treating both the rules and the data as objects, which have quantifiable attributes, a frame-based organizational structure can be applied. Frames allow objects to be represented by the attributes of the object. In addition, the objects represented within a frame structure carry the inherited traits (attributes) of objects located above (parents) them in the frame network. Thus, knowledge bases can be constructed that allow a subject domain to be organized in a hierarchical substructure. This technique of combining production system with frame system knowledge representation has been used extensively in the WIN® system.

Knowledge bases that have been developed or are under development for WIN® include

(1) *Electrode Selection* (WCSK) - a knowledge base of information used in selecting electrodes,

(2) *Symbol Development* (WSGK) - a knowledge base of graphically oriented information used to develop welding symbols,

(3) *Stainless Steel Welding* (WSSK) - a multiple segmented knowledge base of information abut the selection of electrodes, ferrite number prediction, preheating and post-weld heating, joint design, and procedure development for stainless steel welding,

(4) *Preheat and Postweld Heating* (PHPK) - a knowledge base of information about the prediction of preheating temperatures and post-weld heat treatment,

(5) *Welding Procedure Development* (WPSK) - a knowledge base of information regarding the selection of PQRs and subsequent development of welding procedures based on the AWS D1.1 structural welding code,

(6) *Hardfacing Selection* (WHPK) - a knowledge base of information about the selection of hard-facing material and wear mechanisms, and

(7) *Welding Equipment* - a knowledge base of information about the selection of welding equipment.

Expert Systems

All of the expert systems in the WIN® software network must make extensive use of data. Consequently, the expert systems available in WIN® are all coupled with large data bases, and are known as hybrid expert systems. In each case, a knowledge base is coupled with one or more data bases to provide an informational expert system. The expert system then provides engineering decision support, emulating the behavior of a human expert with a large library of data books.

Within the WIN system, there are two expert systems being used outside of the American Welding Institute:

(1) WELDSELECTOR – an expert system for optimum selection of welding electrodes. It includes a data base of all of the weldable ASTM classified steels and all of the AWS classified welding electrodes for shielded metal welding (SMAW), gus metal welding (GMAW), and flux-cord air welding (FCAW), and

(2) WELDSYMPLE – an expert system for the development of welding symbols. Based on the ANSI/AWS A2.4-86 document on welding symbols, it produces a graphic representation of a welding symbol in a computer-assisted design (CAD) computer environment.

The WIN system also contains several prototyped expert systems, including:

(1) WELDSTAINLESS – an expert system for the development of stainless steel welding procedures. It features ferrite number prediction using both the Schaeffler and the DeLong approaches. This expert system provides comprehensive decision support for the entire welding engineering of similar or dissimilar stainless steel welds,

(2) WELDHARD – an expert system for the selection of hard-facing materials. It can discern wear types and select materials for optimum performance,

(3) FERRITEPREDICTOR – an expert system for selecting welding electrodes based on ferrite number according to either Schaeffler or DeLong analysis,

(4) WELDHEAT – an expert system predicting optimum preheating and post-weld heat treatment of steels,

(5) WELDPROSPEC – an expert system for the development of welding procedure specifications based on the AWS D1.1 structural welding code and on PQRs from a data-base system.

Also currently under development are expert systems for other joining related decisions such as:

(1) WELDEQUIP – an expert system for the optimum selection of welding equipment and

(2) JOINCOMPOS – an expert system for the optimum selection of composite material joining procedures.

Automation Applications

As welding automation becomes more sophisticated, the requirements for information in automated systems will grow dramatically. This need can only be met by computerized welding information systems. A variety of computerized information will be required by automated welding systems. The requirements include information such as welding procedures, process control data, digital signal template data for sensor control, and welding design data for joint tracking control.

Recent work by AWI and the Colorado School of Mines on the WIN® system has included the investigation of methods of storing, retrieving, and manipulating visual image data for welding uses. Rapid processing of visual image data can be a monumental task using traditional computational methods. Using a parallel processing environment with neural network simulation (NNS), the time consumed by task can be considerably reduced. Because an NNS system does not require a point-by-point or pixel-by-pixel comparison, the time required to identify images is drastically reduced and appropriate decision-making can be executed without requiring an exact match. Thus, by using a rule matrix and an image matrix and then teaching the system in a parallel processor environment, the system can learn to identify images and begin to "understand" those images by binding conceptual ideas to the elements of the image.

Neural networks are well suited to robot vision tasks, especially in welding. Neural networks have several advantages over traditional programs. They deal with noise well, can handle more input with fractional time increases, and can be taught new configurations without reprogramming. In a typical welding situation, spatter and other inconsistencies such as poor lighting conditions combine to give a relatively high noise level. This requires either a great deal of filtering in a standard vision program, or an algorithm that deals with noise. Since scaling up is often required to encompass the data necessary to make appropriate welding decisions, a method that increases computation time fractionally rather than linearly is preferable. Figure 4 shows a typical interpreted graphic display of a weld joint using the neural network to identify the angle quadrant of the joint.

Tests have been conducted regarding the ability of NNS systems to recognize noisy images. Images were created that contained randomly distributed "black" spots (spatter). The amount of noise was determined as the quantity of pixels contained in the image of the weld joint itself, not in the total image. This measure of noise is independent of magnification or other camera characteristics. It also allows reporting of data that are collected at noise levels greater than 100%.

The results of the tests indicate that images of weld joints can be processed at a rate exceeding 400 images per minute. At that processing rate, 100% accuracy was achieved at noise levels of up to 130%. The system was being operated at approximately 10% of capacity; thus, it appears likely that completely accurate image processing capacity using this NNS system could exceed 2000 images per minute at noise levels over 100%.

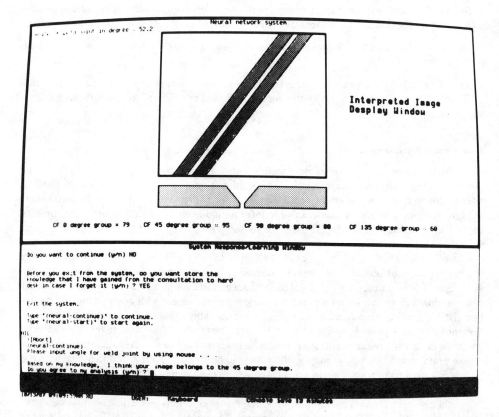

FIG. 4—*Interpreted image display window showing a typical weld joint display.*

Conclusions

1. The American Welding Institute has begun a major effort to supply welding information to the U.S. welding industry through the Welding Information Network (WIN®) system.

2. By applying advanced computer technology, that information can easily be used to meet a wide variety of data needs.

3. Advanced expert systems combined with large data bases or graphics (symbol) bases can provide engineers with high-quality decision support information.

4. New computer capabilities have produced such techniques as neural network simulators, which will be applied to welding information systems in the future.

Said Jahanmir,[1] *Stephen M. Hsu,*[1] *and Ronald G. Munro*[1]

ACTIS: Towards a Comprehensive Tribology Data Base

REFERENCE: Jahanmir, S., Hsu, S. M., and Munro, R. G., **"ACTIS: Towards a Comprehensive Tribology Data Base,"** *Computerization and Networking of Materials Data Bases, ASTM STP 1017*, J. S. Glazman and J. R. Rumble, Jr., Eds., American Society for Testing and Materials, Philadelphia, 1989, pp. 340-348.

ABSTRACT: Tribology is being increasingly recognized as a critical discipline that can play a key role in raising the level of U.S. competitiveness. The transfer of tribology research information into general engineering practice can be made quickly and efficiently by the computerizing of tribology data. The tribology community has been participating in an international effort to centralize and computerize tribology data that goes beyond the establishment of bibliographic data bases. The community is developing a computerized tribology information system as a self-sustaining activity, with government providing the initial funding for research and prototype construction of a PC-based system that contains six data-base components: numeric, design, newsletter, research-in-progress, bibliography, and product and services directory. The numeric data base, which contains 'best judgment' values compiled by experts, and the design data base, which contains validated design programs, are being developed in the first two years of a six-year program. These two data bases present a formidable technical challenge, since not only must they be self-contained, but they must also interact.

KEY WORDS: data base, tribology, friction, lubrication, wear, design, technology transfer

This paper outlines the planning and coordination of an international effort to standardize and disseminate computerized tribological data. It also covers the implementation of a self-sustaining effort that will eventually be supported by industry. In addition, it presents a technical plan for the design of the computer system architecture and development of standardized data-base formats and a framework for data-base communication and dissemination. This activity requires a major advance in the system architecture of technical data bases because tribology is complex. Computerization will provide a critical technology transfer medium whereby ready access by engineers to proper tribological information will have significant impact on the national economy.

Tribology encompasses cross-disciplinary research and practice in materials, lubricants, and component and system design; as such, the research results are published in a variety of specialized journals. This, coupled with the diversity of data, makes it difficult for researchers and engineers who work in different fields to locate pertinent data. As a result, advancement in tribology that could be incorporated into engineering practice has been slow. The centralization of tribological information in a computerized tribology information system (ACTIS) will facilitate access to validated numeric data, design calculations, bibliographies, research-in-progress, available products and services, and electronic mail. These objectives require a coordinated international effort to standardize tribology data so that the system can be easily used and disseminated.

---

[1]Group leader, division chief, and physicist, respectively, Ceramics Division National Bureau of Standards, Gaithersburg, MD 20899.

ACTIS is designed to become self supporting, that is, users will pay for the services of the system. Government sponsors are providing the initial funding for the planning of this system and for the construction of a prototype. Technical areas under the current plan include the development of system architecture, data entry modules, software interfaces, data base formats, linkages between data bases, and communication between the system and the data base users.

Background

The development of ACTIS is being planned in accordance with the recommendations of the international tribology community. In the last three years, the Industrial Research Institute-Conference on Tribology [1], The Canadian Associate Committee on Tribology [2] and the American Society of Mechanical Engineers' Research Committee on Tribology (ASME-RCT) [3] have recommended that the development of a computerized tribology information system receive the highest priority. These organizations have recognized the rapid expansion and growth of tribology and the critical need to organize, standardize, and coordinate tribology data. With the advent of user-friendly data base management software [4] and the proliferation of personal computers [5], a computerized tribology information system is now possible [6].

The Tribology Program of the U.S. Department of Energy's Energy Conservation and Utilization Technologies Division (ECUT) and the Tribology Group of the National Bureau of Standards (NBS) conducted a workshop in August 1985 at NBS in conjunction with the Research Committee on Tribology of the American Society of Mechanical Engineers (ASME-RCT). An international group consisting of over 60 tribologists and information specialists representing industry, universities, and U.S. national laboratories and government agencies participated in planning a computerized tribology information system [7]. The workshop attendees discussed and evaluated the utility and practical problems associated with the implementation of such a system. They recommended that the needs of the tribological community could be met with an information system containing six data base components: numeric, design, newsletter, bibliographic, research-in-progress, and product and services directory. The primary users of such a system were identified as tribology researchers of materials, lubricants, and lubricating systems, as well as design and applications engineers. In addition, the data systems were also envisioned to be useful to the broader engineering community as well as to new product and component developers, failure analysts, and purchasing agents. After the workshops, the DOE-ECUT Tribology Program and NBS Tribology Group led the implementation of the project. To date, it has become an interagency (DOE, DOC, DOD, NSF), intersociety (ASME, STLE) program, with many yet to join.

Description of the Data Bases

The numeric data base will consist of critically evaluated numeric data on the basic properties and performance of materials, lubricants, components, and systems. A wide diversity of tribological data, which include subjects such as well-defined material properties as well as the performance data of complex systems, will be included. Data in each of the areas will be critically evaluated by tribology experts, who will review, distill, and compile a listing of evaluated "best-judgment" parameters and properties in a standardized format. In many cases, the scope and the bounds of the subject area will need to be defined, including the standardization of variable names and units.

The design data base will consist of a selector guide for materials, lubricants, and components, design analysis programs for components and systems, and design calculations. These computer codes will incorporate an expert system front-end for the nontribologist as well as large extended modeling codes for system design. Many of the codes are available now and will be evaluated and validated before incorporation into ACTIS.

The newsletter data base will be the communication link for ACTIS. This component will consist of electronic mail as well as a hard copy newsletter. The newsletter data base will serve as an exchange of old and new technical information, including the latest research results and breakthroughs in tribology, meeting notices, calls for papers, requests-for-proposal, new products, book reviews, and summaries of pertinent technical topics.

The bibliographic data base will be designed so that tribologists, materials scientists, design engineers, librarians and other information specialists, and students can search the literature through a single point of entry. This data base will also serve the needs of the broad industrial community so that technology transfer can be more easily accomplished.

The research-in-progress data base will contain abstracts of current (unpublished) tribology research being conducted by government, industry, universities, and research institutes. The initial source material for this data base will be compiled by conducting a survey that contains key words specific to tribology. The Tribology Program of the National Science Foundation will perform this survey.

The product and services directory data base would provide a single source of commercially available tribology products and related services most often used by application, maintenance, and design engineers and purchasing agents. The data base will contain information on tribocomponents, tribosystems, materials, and lubricants, as well as available services such as consultation and maintenance. The input data will be obtained in machine-readable form from vendors who will pay a data storage fee. This data base will not be funded by government sources.

Technical Challenges for the Implementation of ACTIS

The development of ACTIS requires programmatic and technical plans that incorporate input from the international tribology community. Many of the issues involved have never been addressed before. Their resolution will require new and innovative approaches [8].

Each of the six data-base components requires a different data base management system. These have never before been incorporated into a single comprehensive computerized technical information system that is publicly marketable. The initial technical challenge then is structuring the computer system architecture to facilitate access to information in each of the data-base components. In addition, some of these data-base components must be linked together to allow the user to move information between different data-base components when needed.

Figure 1 shows the conceptual plan for the computer system architecture that will be used for ACTIS. A user can access any one data-base component through a main menu or through an interactive interface that connects the data-base components. This connection must be transparent to the user. For example, a designer using the design data base may require numeric data for a particular design analysis program. These data will be called-up in the proper format and units, and the user can select the best judgment value or analyze the available data to suit the purpose.

Each data-base component also requires a unique structure. For example, the numeric data base will consist of such diverse technical subjects as materials properties, lubricant and additive properties, and tribological performance data on components and systems. The data within these subject areas must be organized in standardized formats with text so that a nonexpert can use the data. Figures 2 and 3 show possible formats for properties of materials, lubricants, and additives that will be used in the numeric data base. Each item must be recorded in proper units and contain explanatory text and references. The computer codes that will contain these data can be queried for any of the data items, and the numeric data can be statistically analyzed.

The design data base will contain a component, material, and lubricant selector guide, calcu-

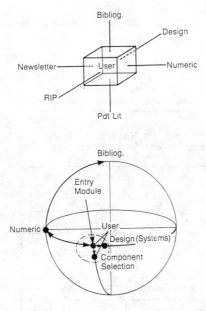

FIG. 1—*Schematic diagram of the computer system architecture.*

lation software for design parameters, and design analysis programs. The software for many of these programs has not yet been developed. Since each program is unique in complexity, the system architecture will include a front-end so that all the available programs appear with a common menu format when the system is entered. The programs must be evaluated and verified for accuracy and ease of use.

The newsletter, research-in-progress, bibliographic, and product and services directory data-base components will also require a complex computer structure. The newsletter will eventually be an on-line news distribution system. The individual news items must be linked to additional text that guides the user into the system. The research-in-progress data base must be updated annually. The bibliographic data base also needs to be updated and expanded as new materials or designs are developed. The product and services directory will serve a critical need for the engineering community by providing data on new products or services. This data base will serve as a technology transfer link, so the structure of this data base must be flexible and easy to use.

Methodology

The construction of ACTIS is an outgrowth of the workshop held at NBS in 1985 [7]. At that meeting, the workshop attendees committed to the development of a plan that incorporates the six data-base components. Since that time, the organizational structure of ACTIS and the technical contents of the prototype system have been developed.

ACTIS has been designed to be built in three phases, with Phase I funded principally through government agencies. Figure 4 shows how the system has evolved. A PC-based prototype system is being developed in Phase I. Self-contained PC disks will be available for distribution through technical societies. Funds from the sale of these disks will be used to support additional efforts related to the development of ACTIS. Phase II will be managed jointly by a Government Steering Committee and ACTIS, Inc. During Phase II, a minicomputer gateway system will be devel-

Material Identification
Physical/Chemical Properties

	Solid 1	Solid 2
Material Designation (AISI, SAE, ASTM) Chemistry (major element/minor elements, additives)		

Surface Treatment
    Ion Implantation (ion type/ion energy/density)
    Atmosphere/Heat Treatment (gas/temperature/time)
    Other		

Grain Size (diameter, mm) (ASTM GS No.)		
Texture (surface orientation [hkl])		
Density (g/cm^3)		
Melting Point (°C)		
Linear Thermal Expansion Coefficient (°C^{-1})		
Modulus Young's (MPa) of Bulk (MPa) Elasticity Shear (MPa)		
Yield (flow) Stress (MPa, ksi)		
Elongation (percent)		
Poisson Ratio		
Hardness (Type of test, N/m^2, MPa) { RT / T > RT }		
Fracture Toughness (psi $\sqrt{in}$, MPa $\sqrt{m}$), K_{IC}		
Thermal conductivity (Wm^{-1} C^{-1})		
Fatigue Limit		

FIG. 2—*Example of data base format of physical and chemical properties of metals and ceramics.*

oped; users will access this gateway through PCs linked via telephone lines. In Phase III the activities will be completely turned over to ACTIS, Inc. It will manage the system and carry out a business and marketing strategy.

ACTIS is currently being managed by the Government Steering Committee, which consists of representatives from the DOE, NSF, Army, Air Force, and NBS. This committee is composed of tribology managers whose organizations have contributed funds or other resources for the initial stages of this project. The Government Steering Committee is advised by a Technical Advisory Committee and an Industrial Liaison Committee. The membership of these committees strongly emphasizes the level of commitment by U.S. Government agencies, the industrial community, and internationally recognized tribology researchers and engineers. Additionally,

Lubricants

Trade Name _____
Base Lubricant _____
Additives _____
Viscosity (state T & Test Method) _____
    40°C kinematic _____
    100°C kinematic _____
Viscosity/Temperature Prop. (Viscosity Index) _____
Viscosity Pressure Coeff. (Measured or Predicted - give Prediction Method)
API Gravity _____
    Specific Gravity 60/60°F _____
Boiling Point (BP) or BP Range by Gas Chromatography _____
Molecular Weight (average) _____
Flash Point _____
Fire Point (Flame or Cleveland Open C.O.C.) _____
Cloud Point _____
Pour Point _____
Foam Test _____
Color (ASTM) _____
Specific Heat _____
Bulk Modulus _____
Surface Tension _____
Carbon Residue (state Test used) _____
Ash _____
Sulfated Ash _____
Saponification _____
Neutralization (Acid) No. _____
Base No. _____
Lubricant Content _____
    Paraffin, Naphtine, Aromatic, Polars _____
    Sulfur _____
    Nitrogen _____
    Oxygen _____
    Phosphorus _____
    Calcium _____
    Magnesium _____
    Zinc _____
    Barium _____
    Polymeric VI Improver _____
Lubricant Properties _____
    Thermal Stability (state Test Method) _____
    Oxidation Stability (state Test Method) _____
    Lubricity Stability (state Test Method) _____

FIG. 3—*Example of data base format of lubricant property data.*

the technical program is being coordinated by an executive director and a technical coordinator, supported by and located at NBS.

Figure 5 shows the technical program plan for a four-year period. The effort in year one has been devoted to the development of the computer system architecture and the numeric and design data-base components. These two components are initially being incorporated in a pro-

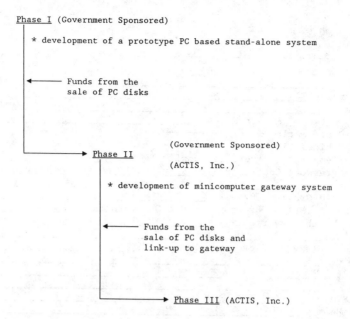

FIG. 4—*A schematic diagram showing the development of ACTIS.*

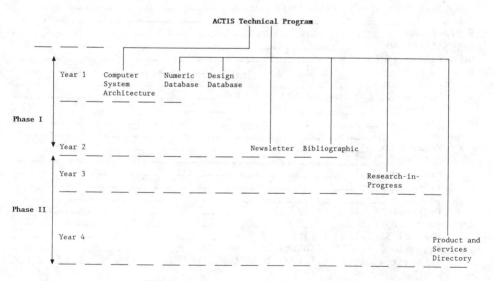

FIG. 5—*Technical program plan for ACTIS.*

totype system that will run on an IBM PC-compatible system. In later years, the data bases will also be stored on a minicomputer and accessed via telecommunications. In the second year, the newsletter and bibliographic data bases will be initiated. The newsletter component will be distributed in hard-copy form as well as by electronic mail. In the fourth year, the product and services directory data base will be assembled. The support for this data base will not involve government funding.

Present Status of ACTIS

The technical plan for development of the prototype system in Phase I is shown in Fig. 6. The activities during this phase include the development of the computer system architecture, designed to be self-contained so the software can be used directly with a stand-alone, PC-based system and to be easy to demonstrate and market. The interface is written in C programming language for an IBM-compatible PC with 640K bytes of random-access memory, two floppy-disk drives, and a hard-disk drive, graphics card, and math coprocessor. The user interface will use windows and be menu driven with forms-oriented data entry.

The prototype system will contain representative numeric and design data bases. The numeric data base subject areas recommended by the Technical Advisory Committee are as follows:

(1) metals and ceramics properties,
(2) lubricants and additives properties,
(3) lubricated friction and wear,
(4) abrasive wear,
(5) erosive wear,
(6) metal cutting and machining, and
(7) rolling element bearings.

The data in each of these areas are currently being collected and evaluated by experts in each area. The Technical Advisory Committee will review these contributions for uniformity and completeness before the data are incorporated into the system.

The design data base subject areas were recommended by the ASME/Joint Societies Design Task Force, a committee composed of members from universities, technical societies (ASME, STLE, and others), and industrial research organizations. The Technical Advisory Committee will evaluate and validate the design programs for accuracy and ease of use. This data base component will contain the following design programs and design calculations:

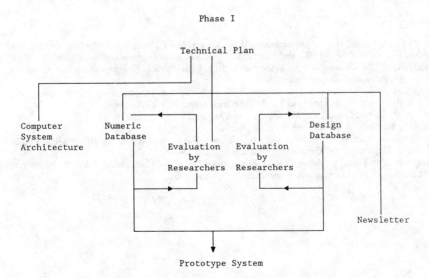

FIG. 6—*Phase I technical plan.*

(1) component selector guide,
(2) hydrodynamic journal bearings,
(3) ball bearings,
(4) spur gears,
(5) mechanical face seals,
(6) elastic contact stress analysis,
(7) elastohydrodynamic film thickness calculations, and
(8) failure analysis of tribological components.

The data entry module for both data-base components will be similar; the design data base will contain a standard set of input values, variable names, designation of output destination, output formats, and error handling. In addition, the design data base will contain a specialized, abbreviated, numeric data base specific to each design program.

This first year of ACTIS has involved programmatic planning, the development of the computer system architecture, and details of the prototype system. Data compilation in seven subject areas in the numeric data base was scheduled to be completed in March 1988. The group plans to evaluate and validate one design program and to make it available for distribution. The prototype system scheduled for January 1989 will contain at least one set of numeric data and one design program.

Acknowledgment

The authors gratefully acknowledge the support of the Tribology Program of the DOE-ECUT, the Tribology Program of the NSF, the U.S. Army and Air Force, and the Tribology Program of the NBS. Mr. Dave Mello, Dr. Bobby McConnell, Dr. Elbert Steven M. Marsh, Dr. Maurice Lepera, and Dr. Steve Hsu, who are members of the Government Steering Committee, have provided the management support as well as personal and organizational support for this project. The members of both the Technical Advisory Committee and the Industrial Liaison Committee have volunteered their time and efforts for this activity. The technical and financial support of ASME and STLE are gratefully acknowledged. This activity relies on the volunteer efforts of more than 60 individuals who have supported the ACTIS concept since its inception.

References

[1] Dake, L. S., Russell, J. A., and Debradt, D. C., *Journal of Tribology,* V. 108, Oct. 1986, p. 497.
[2] Sibley, L. B., Peterson, M. B., and Levinson, T., *Mechanical Engineering,* Sept. 1986, p. 68.
[3] Peterson, M. B., Chairman, The American Society of Mechanical Engineers Research Committee on Tribology, private communication 1987.
[4] Fong, J. T., Ed., *Engineering Data Bases: Software for On-Line Applications,* ASME Special Publication H00310, American Society of Mechanical Engineers, New York, 1984.
[5] Toong, H. D. and Gupta, A., *Scientific American,* V. 247, Dec. 1982, p. 82.
[6] Fong, J. T., *Computers in Mechanical Engineering,* V. 5, 1986, p. 42.
[7] Rumble, J. and Sibley, L., Eds., *Towards a Tribology Information System,* NBS Special Publication 737, U.S. Department of Commerce, Washington, D.C., Dec. 1987.
[8] Danyluk, S. and Hsu, S. M., *Managing Engineering Data: The Competitive Edge,* R. E. Fulton, Ed., American Society of Mechanical Engineers, New York, 1987.

Author Index

A–B
Alashqur, A., 109
Anderson, D. B., 317
Baker, H., 322
Barkmeyer, E. J., 126
Barrett, A. J., 99
Bathias, C., 92
Burte, H. M., 197

C–D
Coyle, T. E., 200
Dennis, J. R., 292

F–G
Falco, P. M., Jr., 185
Fan, S., 75
Furlani, C. M., 126
Grattidge, W., 151

H–J
Harmsworth, C. L., 197
Ho, C.-Y., 272
Hsu, S. M., 340
Iwata, S., 175
Jahanmir, S., 340
Johnson, P. M., 185
Jones, J. E., 329

K
Kacprzak, L., 322
Kanao, M., 80
Kaufman, J. G., 7, 55
Kröckel, H., 63

L
Laverty, G. J., 317
Lees, J. K., 265
Li, H. H., 272
Libes, D., 126

Little, C. D., 200
Lu, Y., 75

M
Martini-Vvedensky, J. E., 211
Marx, B., 92
McCarthy, J. L., 135
Messina, C. G., 304
Michaud, R. J., 265
Mitchell, M. J., 126
Moniz, B. J., 239
Monma, Y., 80
Munro, R. G., 340

N
Newton, C. H., 280
Nishijima, S., 80

O–P
Ondik, H. M., 304
Petrisko, L. S., 229
Pilgrim, R. A., 185

R
Ranger, K., 253
Reynard, K. W., 43
Roberts, B. K., 265
Rumble, J. S., Jr., editor, 1, 216

S
Schenck, P. K., 292
Scott, W. W., Jr., 322
Steven, G., 63
Su, S. Y. W., 109

V–W
Vanderveldt, H. H., 329
Westbrook, J. H., 23
Wool, T. C., 239

Subject Index

A

ABCUP Program Series, 305
ACTIS tribology information system, 340
Aerospace applications, 197, 200
Alloys (*See* Metals and alloys)
American Ceramic Society, 60, 292, 304
American Chemical Society, 59
American Society of Mechanical Engineers, 60
American Society of Metals, 60, 322
American Welding Institute, 60, 329
American Welding Society, 60
Anisotropic materials, 265
Artificial intelligence interfaces, 175, 329, 332
ASM International, 60, 322
ASTM committees
    E-28: 11
    E-49: 7, 231, 239, 318–319
    G-01: 318–319
ASTM standards
    D 648: 232
    D 1600: 231
Audience profiles, 216
Automated manufacturing, 109, 126
Automated Manufacturing Research Facility (AMRF), 126

B

Basic Data Administration System (BDAS), 131–132
Bibliographic data bases, 324–325, 342–343
Bilingual data bases, 89

C

CAD/CAM, 109, 197, 200
Ceramics data base, illus., 90
Ceramics phase diagrams, 292, 304
Characterization of materials
    categories required, 10, 33
    composites, 270
    corrosion data, 317
    definition, 25
    measurement units, 137–139
    metals and alloys, 23
    tables (data), 135
Chemical Abstracts Service, 59
Chemical data bases, 292
Chinese activities, 75
Classification of materials, 268–269 (*See also* Characterization of materials)
Commission of the European Communities (CEC) activities, 43, 63, 70–72
Committee on Data of the International Council of Scientific Unions (CODATA)
    activities, 99
    China, activities in, 79
    cooperation with Versailles Project for Advanced Materials and Standards (VAMAS), 68
    history, 43, 99
    survey, 92
Composite materials, 92, 253, 265, 280
Computer-aided design (CAD)/computer-aided manufacturing (CAM), 109, 197, 200
Computer-integrated manufacturing (CIM), 109, 126
Computer programs (*See also* Data bases)
    ABCUP Program Series, 305
    DATATRIEVE, 240–241
    METSEL2, 280
    SPIRES, 58
Corrosion data base, 239, 317
Cost-benefits studies, 99
Coupon corrosion testing, 239
Creep of materials, 80
Crystal structure, 322

D

Data administration, 126
Data banks (*See* Data bases)
Data base software
    ABCUP Program Series, 305
    DATATRIEVE, 240–241
    METSEL2, 280
    SPIRES, 58
Data bases
    ABCUP Program Series, 305
    ACTIS tribology, 340
    alloys (*See* Metals and alloys)
    anisotropic materials, 265
    architecture (structure), 140–143, 326–328, 342–343
    bibliographic, 324–325, 342–343

Data bases (*cont.*)
  bilingual, 89
  builders, 304
  building blocks, 272
  ceramics, 90, 292, 304
  chemical, 82, 292
  Chinese, 75
  composite materials, 92, 253, 265, 280
  computer-aided design (CAD)/computer-aided manufacturing (CAM), 109, 197, 200
  corrosion, 239, 317
  cost-benefits studies, 99
  creep of materials, 80
  data description, 135
  data structure complexity, 136–137
  DATATRIEVE, 240–241
  directories, 63, 101, 135
  distributed, 126, 135, 175, 289 (*See also* Networks)
  Engineering Materials Property Data Bases, 200
  fatigue of materials, 80
  file systems, 272, 306–312
  fourth-generation language (4GL), 329
  incompatibilities, 63, 224, 272
  indexing, 314
  input format, 12–15, 317 (*See also* Interface design)
  Integrated Chemical Database System, 82
  integration, 63, 272
  interest in, survey, 92
  interfaces (*See* Interface design)
  Japanese Information Center for Science and Technology, 83–86
  management, 216
  Materials Engineering Center (MEC) of the Dow Chemical Company, 229
  Materials Information for Science and Technology (MIST) program, 23, 135
  MATUS, 211
  METADEX, 324–325
  metals (*See* Metals and alloys)
  METSEL2, 280
  MIL-HDBK-17B, 280
  modular construction, 272
  National Materials Property Data Network, Inc. (MPD), 7, 32, 55, 281
  National Research Institute for Metals, 82–86
  nomenclature, 15, 43, 109, 135
  nonmetals, 92, 229, 239, 317
  object-oriented data representation, 147–149
  Open System Interconnection (OSI) model, 80
  personal computer based, 280, 292, 340
  Phase Diagrams for Ceramists, 292, 304
  PhaseLit, 325
  problems in development, 216
  publication, 200, 304, 342
  RUST, 239
  Science and Technology Agency, 82
  Semantic Association Model (SAM*), 109
  Society of Materials Science, Japan, 81
  software, 58, 240–241, 280, 305
  SPIRES system, 58
  STN International, 59
  structural design, 253, 272, 289, 306–312
  terminology, 15, 43, 109, 135
  unified life-cycle engineering, 197
  welding, 329
  WIN system, 332–339
Data capture, 151
Data description (*See* Characterization of materials)
Data dictionaries, 109, 135, 178
Data exchange formats, 293
Data evaluation, 320
Data files, 272
Data representation, 151, 317
DATATRIEVE, 240–241
Demonstrator program, materials data banks, 63
Descriptors (*See* Characterization of materials)
Designations (*See also* Characterization of materials)
  composite materials, 266–267
  definition, 24
  metals and alloys, 23
  VAMAS Task Group findings and recommendations, 43
DIANE information service network, 64
Dictionaries, 109, 135, 178
Digitizing software, 293
Directories, 63, 101, 135
Directories of Data Sources for Science and Technology, 101
Distributed Data Administration System (DDAS), 132–133
Distributed data bases, 126, 135, 175, 289 (*See also* Networks)
Dow MEC data base, 229

E

Electronic mail systems, 200, 342
Engineering Materials Property Data Bases, 200
EURONET-DIANE, 63
European Communities activities

INDEX 353

Code of Practice, 70-72
Materials Data Banks Demonstrator Program, 63
VAMAS Task Group findings and recommendations, 43, 44
European Host Network (EHN), 66
Expert systems, 175, 332

F

Fairfield Glade workshop, 100
Fatigue of materials, 80
File systems, 272, 306-312
Formats, data, 12-15, 317
Fourth-generation language (4GL) data bases, 329
Frame-based information organization, 332
French activities, 92
Friction, 340

G

Graph metadata, 151
Graphics, 287, 289, 292, 325-328, 331

H

Handbooks, 135, 198-199, 280
Hierarchical agglomerative clustering, 185

I

Identification of materials, 23, 24 (*See also* Characterization of materials)
Incompatibilities of data bases, 63, 224
Indexing, 314
Information market, 63
Information systems (*See* Data bases)
Input format (*See* Interface design)
Institute of Metals, 60
Integrated Chemical Database System, 82
Integrated Manufacturing Data Administration System (IMDAS), 130-134
Interface design
  artificial intelligence, 175, 329, 332
  expert systems, 175, 332
  fourth-generation language (4GL), 329
  input formats, 12-15, 317
  issues and standards, 15-16
  types, illus., 184
  WIN system, 332-335
International cooperation (*See* Commission of the European Communities (CEC) activities; Committee on Data of the International Council of Scientific Unions [CODATA])
International Numbering System for Metals (INSM), 31

J-K

Japanese Information Center for Science and Technology, 83-86
Knowledge base, 175

L

Laboratory information management systems, 200
Lamina, 265
Lay-up, 265
Liability, 99
Life-cycle engineering, 197
Local area networks, 200
Lubrication, 340

M

Manufacturing systems, 109, 126
Master Data Administration System (MDAS), 133
Material selection, 317
Materials Engineering Center (MEC) of the Dow Chemical Company, 229
Materials Information for Science and Technology (MIST), 23, 135
Materials Property Data Network, Inc. (MPD), 7, 32, 55, 281
MATUS materials data bank, 211
Measurement units, 137-139
MEC data base, 229
Metadata
  data capture, 163-167
  definition, 152
  graphs, 151
  schema, 151
  system, 56-57
  tables, 151
  uniform treatment, 109
METADEX data base, 324-325
Metals and alloys
  alloy phase diagrams, 322
  characterization, 23
  corrosion, 239, 317
  France, interest in data base, 92
  Japanese data base, 82-86
  numbering systems, 23
METSEL2 program, 280
MIL-HDBK-17B data base, 280
Military specification handbooks, 135, 198-199, 280
MIST program, 23, 135

N

National Association of Corrosion Engineers (NACE), 60, 317
National Bureau of Standards (NBS) ASM/NBS Data Program for Alloy Phase Diagrams, 322
  Automated Manufacturing Research Facility (AMRF), 126
  corrosion data base, 317
  Integrated Manufacturing Data Administration System (IMDAS), 126
  Phase Diagrams for Ceramists data base, 292, 304
National Materials Property Data Network, Inc. (MPD), 7, 32, 55, 281
National Research Institute for Metals, 82-86
Natural clustering, 185
Networks
  aerospace applications, 200
  EURONET-DIANE, 63
  European Host Network (EHN), 66
  local area, 200
  neural, 337-338
  Open System Interconnection (OSI) model, 80
Neural networks, 337-338
Nomenclature, 15, 43, 135
Nonmetals, corrosion of, 239, 317

O

Object-oriented data representation, 147-149
Open System Interconnection (OSI) network model, 80

P

Personal computer data bases, 280, 292, 340
Phase diagrams, 292, 304, 322
PhaseLit data base, 325
Plastics, 92, 229
Polymers, 92, 229
Programs (*See also* Data bases)
  ABCUP Program Series, 305
  DATATRIEVE, 240-241
  METSEL2, 280
  SPIRES, 58
Published materials capture, 151

R

RUST data base, 239

S

Schluchsee workshop, 100
Science and Technology Agency, 82
Semantic Association Model (SAM*), 109
Semantic capacity, 177-180
Simulators, 181-182
Society of Materials Science, Japan, data base, 81
Socioeconomic problems, 216
Software (*See also* Data bases)
  ABCUP Program Series, 305
  DATATRIEVE, 240-241
  METSEL2, 280
  SPIRES, 58
Specification (*See* Characterization of materials)
SPIRES data base system, 58
Standards for data bases
  ASTM Committee E-49, 7
  ASTM D 648: 232
  ASTM D 1600: 231
  recommended, table, 9
  VAMAS Task Group findings and recommendations, 43
Steels (*See* Metals and alloys)
STN International, 59
Strength of materials, 80

T

Table metadata, 151
Table structures, 151
Technology transfer, 340
Terminology, 15, 43, 135
Test data reporting, 11
Thesauri, 143-146
Tribology, 340
Typesetting, 304

U

Unified life-cycle engineering, 197
Unified Numbering System (UNS), 23, 243
Usage patterns, 216
User friendliness, 104-105 (*See also* Interface design)
User interfaces (*See* Interface design)
User profiles, 92, 216

V

Validation file, 23
Variables files, 272

Versailles Project for Advanced Materials and Standards (VAMAS), 43, 44, 68

Welding information systems, 329
WELDSELECTOR expert system, 332–335
WELDSYMPLE expert system, 337
WIN system, 332–339

W

Wear, 340